Russian Energy Security and Foreign Policy

This book provides an original and thoroughly academic analysis of the link between Russian energy and foreign policies in Eurasia, as well as offering an interpretation of Russia's coherence on the international stage, seeking to understand Russia and explain its behaviour.

The authors analyse both energy and foreign policies together, in order to better grasp their correlation and gain deeper understanding of broader geopolitical issues in Eurasia at a time when things could go either way – towards producers or towards consumers. Questioning the concept of 'energy deterrence' which aims to fuel uncertainty in Russia's relations with its partners, as well as projecting its overall power on the international scene, this provocative volume seeks to stimulate debate on this very important issue.

Assessing the weight that energy has in Russia's foreign policy and in its pursuit of power on the international stage, this book will be of interest to students and scholars of international relations, energy politics, geopolitics and Russian and Central Asian Studies.

Adrian Dellecker is Manager of Research and Policy at WWF International in Switzerland and formerly in charge of the energy portfolio at Ifri's Russia/NIS Center.

Thomas Gomart is Vice President for Strategic Development at Ifri, Director of Ifri's Russia/NIS Center and editor of the three-language digital collection *Russie. Nei. Visions*.

Routledge/GARNET series: Europe in the World

Edited by David Armstrong

University of Exeter, UK

and

Karoline Postel-Vinay

Centre for International Studies and Research (CERI), France.

The Routledge GARNET series, *Europe in the World*, provides a forum for innovative research and current debates emanating from the research community within the GARNET Network of Excellence. GARNET is a Europe-wide network of 43 research institutions and scholars working collectively on questions around the theme of 'Global Governance, Regionalisation and Regulation: The Role of the EU', and funded by the European Commission under the 6th Framework Programme for Research.

1. EU Foreign Policy in a Globalized World
Normative power and social preferences
Edited by Zaki Laïdi

2. The Search for a European Identity
Values, policies and legitimacy of the European Union
Edited by Furio Cerutti and Sonia Lucarelli

3. The European Union and the Social Dimension of Globalization
How the EU influences the world
Edited by Jan Orbie and Lisa Tortell

4. Governance and the Depoliticisation of Development
Edited by Wil Hout and Richard Robison

5. The European Union and International Organizations
Knud Erik Jørgensen

6. The EU and Global Governance
Edited by Mario Telò

7. External Perceptions of the European Union as a Global Actor
Edited by Sonia Lucarelli and Lorenzo Fioramonti

8. A Global Security Triangle
European, African and Asian interaction
Edited by Valeria Bello and Belachew Gebrewold

9. European Union Democracy Aid
Supporting civil society in post-apartheid South Africa
Lorenzo Fioramonti

10. Civil Society and International Governance
The role of non-state actors in global and regional regulatory frameworks
Edited by David Armstrong, Valeria Bello, Julie Gilson and Debora Spini

11. Debating Political Identity and Legitimacy in the European Union
Edited by Sonia Lucarelli, Furio Cerutti and Vivien A. Schmidt

12. Comparative Environmental Regionalism
Edited by Lorraine Elliott and Shaun Breslin

13. Russian Energy Security and Foreign Policy
Edited by Adrian Dellecker and Thomas Gomart

Russian Energy Security and Foreign Policy

Edited by Adrian Dellecker and Thomas Gomart

Routledge
Taylor & Francis Group
LONDON AND NEW YORK

First published 2011
by Routledge
4 Park Square, Milton Park, Abingdon, Oxon OX14 4RN

Simultaneously published in the USA and Canada
by Routledge
605 Third Avenue, New York, NY 10017

Routledge is an imprint of the Taylor & Francis Group, an informa business

British Library Cataloguing in Publication Data
A catalogue record for this book is available from the British Library

Library of Congress Cataloging-in-Publication Data
Russian energy security and foreign policy / edited by Adrian Dellecker and Thomas Gomart.
p. cm. -- (Routledge/GARNET series, Europe in the world)
Includes bibliographical references and index.
1. Energy policy--Russia (Federation) 2. Russia (Federation)--Foreign relations--Former Soviet republics. 3. Former Soviet republics--Foreign relations--Russia (Federation) 4. Russia (Federation)--Foreign relations. I. Dellecker, Adrian. II. Gomart, Thomas.
HD9502.A4R825 2011
333.790947--dc22
2010047165

ISBN: 978-0-415-54733-8 (hbk)
ISBN: 978-0-203-81673-8 (ebk)

Typeset in Times New Roman
by GreenGate Publishing Services, Tonbridge, Kent

Contents

Figures and Tables

Figures

Tables

Contributors

Dr Pavel K. Baev is a Research Professor at the International Peace Research Institute (PRIO), Oslo, Norway, where he specializes in energy interests in Russia's foreign and security policy.

Martha Brill Olcott is a Senior Associate in the Russian and Eurasian Program at the Carnegie Endowment, where she specializes in the problems of transitions and the security challenges in Central Asia and the Caucasus.

Dr Maureen Crandall is a Professor of Economics in the Industrial College of the Armed Forces at the National Defense University in Washington, DC. She has written extensively on energy and political issues in the Caspian region.

Adrian Dellecker is Manager of Research and Policy at WWF International in Switzerland and formerly in charge of the energy portfolio at Ifri's Russia/NIS Center.

Folkert Garbe is Project Director of the Minsk Forum and specialist in Russian–Belarusian relations.

Dr Thomas Gomart is Vice President for Strategic Development at Ifri, Director of Ifri's Russia/NIS Center, and editor of the three-language digital collection *Russie.Nei.Visions*.

Dr Leonid Grigoriev is a professor at the High School of Economics in Moscow and the President of the Institute for Energy and Finance, a think-tank based in Moscow.

Dr Jérôme Guillet is a French investment banker who specializes in the structured finance activities in the oil and gas, power and renewable energy sectors.

Felix Hett is a Program Manager at the Moscow Office of the Friedrich Ebert Foundation.

Dr Rainer Lindner is Executive Director of the Committee on Eastern European Economic Relations, Professor at the University of Konstanz and Chairman of the Minsk Forum.

Vladimir Milov is a Russian politician and the president of the Institute of Energy Policy, a Moscow-based think-tank. He is a former Deputy Energy Minister of the Russian Federation.

Dr Andrew Monaghan is a Research Advisor in the Research Division of the NATO Defense College (NDC) in Rome. He is also the NDC's senior researcher on energy security matters.

John M. Roberts is an Energy Security Specialist with Platt's, the world's leading independent source of energy information.

Dr William Tompson is Head of the Regional and Rural Development Unit, Regional Development Policy Division, OECD Directorate for Public Governance and Territorial Development.

Dr Frank Umbach is a Senior Associate and Head of the Programme 'International Energy Security' at the Centre for European Security Strategies (CESS, GmbH), Munich, Berlin, and Associate Director at the European Centre for Energy and Resource Security (EUCERS) at the King's College, London.

Abbreviations

AIOC	Azerbaijan International Operating Company
BTC	Baku–Tbilisi–Ceyhan pipeline
BTE	Baku–Tbilisi–Erzurum pipeline
CAC	Central Asia–Centre pipeline
CACR	Central Asia and the Caspian Region
CDC	Caspian Development Corporation
CIS	Commonwealth of Independent States
CNG	Compressed Natural Gas
CNPC	China National Petroleum Corporation
CPC	Caspian Pipeline Consortium
CSTO	Collective Security Treaty Organization
DRA	Drag Reduction Agents
EAP	Energy Action Plan
EBRD	European Bank for Reconstruction and Development
ENI	Ente Nazionale Idrocarburi
EU	European Union
EurAsEC	Eurasian Economic Community
FDI	Foreign Direct Investment
FSU	Former Soviet Union
GDP	Gross Domestic Product
GECF	Gas Exporting Countries Forum
GUUAM	Treaty embracing Georgia, Ukraine, Uzbekistan, Azerbaijan and Moldova
IEA	International Energy Agency
IGI	Interconnection–Greece–Italy
INOGATE	Interstate Oil and Gas Transport to Europe
IPO	Initial Public Offering
KMG	KazMunaiGas
LNG	Liquefied Natural Gas
MAD	Mutual Assured Destruction
NATO	North Atlantic Treaty Organization
NIS	New Independent States
NOC	National Oil Companies

OECD	Organization for Economic Cooperation and Development
OIC	Organization of Islamic Conference
OPEC	Organization of the Petroleum Exporting Countries
PSA	Production Sharing Agreements
SCO	Shanghai Cooperation Organization
SCP	South Caucasus Pipeline
SOCAR	State Oil Company of the Azerbaijan Republic
SPV	Special Purpose Vehicle
TACIS	Technical Aid to the Commonwealth of Independent States
TAP	Turkmenistan–Afghanistan–Pakistan pipeline; Trans-Adriatic Pipeline
UGSS	Unified Gas Supply System
UN	United Nations Organization
US	United States of America
USEIA	United States Energy Information Administration
USGS	United States Geological Survey
USSR	Union of Soviet Socialist Republics
WTO	World Trade Organization

Preface

Since the creation of the Russia/NIS Center at Ifri (French Institute for International Relations) in July 2004, the exploration of the interactions between Russia's energy and foreign policies has been a cornerstone of our research. It was already clear at that time that the Russian authorities had decided to leverage energy resources and energy wealth to consolidate power at home and to project it abroad. We initially focused on Russian political, security and business elites, and noted the emergence of a widely supported 'resource nationalism'. After the first gas crisis between Russia and Ukraine in January 2006, we decided to set up the Eurasian Energy Taskforce to create a pole of expertise and to pool the views of researchers, policy makers, officials and energy executives. In this framework, we organized two roundtables entitled 'The Restoration of Russian Power: Toward an Energy-Based Deterrence?' and 'What Transit Strategies?'. These titles did much to encapsulate the mood of the time, and had the added benefit of being provocative. We expected lively exchanges after having gathered people coming from Norway, Lithuania, Germany, France, the UK, the US, and obviously Russia. We were not disappointed, to say the least, by the tone and quality of these sometimes difficult and passionate debates.

The first roundtable took place in December 2006, and the second one in November 2007. A group of leading experts, the authors of the chapters that follow, were chosen to submit papers on pre-selected transversal or regional topics. These draft papers were then distributed to all participants several days before the seminar, so that their content would fuel the debate. Trying to go beyond dry analyses of regional production numbers or individual bilateral relations, the roundtables – and finally this book – aimed to anchor the debate within a Eurasian context, and to focus on Russian policies. The intention to provide deeper and clearer understanding of this context remains with us today. With each meeting the original papers were amended, and they were again revised and updated once the book's general structure was elaborated by us (the bulk of revisions were made in winter 2009, and certain papers again in summer 2010).

We are grateful to all of the contributing authors for their enthusiastic participation in the roundtables, and for having shaped their presentations into chapters. For understandable reasons, the updating was sometimes a painful supplementary task, which deserves special thanks. We benefited enormously from comments and observations made during sessions by Ambassador Yolanta Balciunene, Maïté Jaureguy-Naudin, Senior Research Fellow at Ifri; Jacques Lesourne, President of the Energy Program at Ifri; Philippe Lorec from the French Ministry of Industry; Thierry de Montbrial, Ifri's President, and Ambassador Pierre Morel. All of them should be thanked for their time and contribution. We also wish to extend our thanks to Hannah Shakespeare, Karoline Postel-Vinay and David Armstrong. We would like to thank them for their support and patience. Nothing would have been possible without the consistent involvement of the Russia/NIS Center at Ifri, namely Tatiana Kastueva-Jean, Catherine Méniane; special thanks are due to Dominic Fean and Julien Nocetti.

In writing about Russia there is always the risk of missing a moving target. However, we think that intelligent analysis requires a consistent effort, straining to combine topical issues with long-term insight. We hope that this book responds to that challenge.

Adrian Dellecker and Thomas Gomart
August 2010
Paris, France

Introduction

Solving the Eurasian energy equation

Adrian Dellecker and Thomas Gomart

In Eurasia as anywhere, energy and geopolitics are closely intertwined. The Eurasian energy equation is of key importance to help anticipate, rather than predict, the future development of Europe and the former Soviet Union (FSU), but also of China, India, Turkey and, even if it is an ocean away, the US. The fundamental problem of this equation is that it is full of unknown factors and ever-changing. Given its massive size, its vast resources, its history, international ambitions, and its rentier organization, Russia is undoubtedly the central multiplying factor of this Eurasian equation. However, even this basic fact has been subject to many different interpretations in the last decade. This is the book's starting point: there is a dramatic need to better understand Russia's organization, capacities and motivations.

More specifically, the book's primary concern is to investigate the link between Russia's foreign and energy policies. Indeed, the latter is commonly characterized as the former's driving force, even its *raison d'être*, while the former is often reduced to the latter. Published in 2003, Russia's official energy strategy to 2020 urges the state to be deeply involved in the energy sector so as to protect the country from both internal and external threats. Added to this, the official strategy notes that 'energy security is the most important element in Russia's national security'.[1] This book will try to identify the complex relations between these two policies. On the one hand, energy impacts various manifestations of power such as political, military, economic and technological power. On the other hand, given its great power background, Russia can neither be compared to Norway – a 'quiet' energy producer – nor to Venezuela – a vocal and aspiring power. Russia's international influence is not limited to its energy supplies the way it is for Saudi Arabia, but spans from its nuclear status to participation in the G8 and G20, not to mention its permanent seat on the United Nations Security Council. This book therefore also aims to assess Russia's actual potential on the international scene, for example its capacity to alter its environment, by identifying both the weaknesses and strengths of its foreign and energy policies.

To do so, this book will try to avoid one particular analytical bias: European energy security. This is not to say that this topic is unimportant, but it is more to say that Europe, as the main solvent market for Russian natural gas, is only a part of the equation. The EU's objective of liberalizing its own domestic gas market may in fact have clouded our perception of the different actors, Russia first and foremost, particularly since the EU is inherently uncomfortable dealing with a country which has a policy blatantly based on the maximization of power.

This bias towards European energy security may explain why the policy-oriented debate has so far mainly focused on Europe's increasing dependence on Russian energy, on the possible forms of economic–political blackmail made by the Kremlin, the Russian monopolistic control of pipelines between certain energy producers and the European market, and on the injection of corruption into European politics.[2] What is more, the academic debate is often Eurocentric, much of it focusing on the issue of a possible 'gas deficit' in Russia in the coming years. Facing increasing external and domestic demand and stagnation of its production due to the depletion of the major West Siberian fields and lack of massive investments in new fields, Russia's very capacity to meet contracted demand at home and abroad is thus often questioned.[3]

Yet to best seize the Eurasian energy equation, it is essential to understand the Russian external energy system within the FSU, and not only through the concerns for European energy security, however justified they are. Clearly, there is the issue of a legacy coming directly from the USSR, for example central planning, which helps explain some factors of the equation. The present work is therefore conducted in both space and in time. As explained above, many chapters of this book deal directly or indirectly with Russia's relations with both producer countries (Azerbaijan, Kazakhstan, Uzbekistan and Turkmenistan) and transit countries (Azerbaijan, Belarus, Georgia and Ukraine). Russian diplomacy has very often favoured bilateral approaches owing to the primacy given to state sovereignty in the international system and to the sanctity of state borders. This is not to say that it has neglected multilateralism, but it has been used less as a way to manage global issues than as a way to promote Russia's own national interests. At the same time, Russia often argues in favor of regional economic and political integration within the FSU, especially in the energy field, but the outcomes of these integrationist policies are highly questionable. Our determination to elaborate this project has been clearly framed by our consistent analysis of the Russian security and foreign policies.[4]

The context of this analysis, as well as time factors, is also crucial to understanding the Russian external energy system. Even if it addresses the consequences of the war in Georgia in August 2008 and the gas crisis between Russia and Ukraine in January 2009, this book is mainly focused on the period from 2000 to 2008 for two specific reasons. Firstly, international energy prices rose dramatically during in that period. Oil prices began to rise above US$ 30 per barrel in early 2003, peaking at US$ 147 per barrel in July 2008. Since natural gas is not traded on an open market responding to spot changes in supply and demand, natural gas prices in Europe are indexed on oil prices and updated periodically (generally four times a year) within long-term take or pay contracts whereby the supplier and the buyer agree on fixed volumes and pricing formulas. Consequently, the European net-back price for Russia/NIS gas, which includes transit costs, jumped from approximately US$ 140 per thousand cubic metres (tcm) in 2004 to over US$ 360 in 2008. In February 2009, with oil prices at US$ 40 per barrel, Russia entered a severe recession and faced its first budget deficit in a decade, effectively marking the end of an era, and carrying with it domestic and foreign policy implications.

The 2000–2008 period also corresponds to Vladimir Putin's years as President of the Russian Federation, and are thus telling of what may be construed as 'the Putin era'. From the very beginning of his tenure, Putin intended to restore the Russian economy using the energy industry as a lever. In 1997, the then-future president submitted a thesis to the University of Saint Petersburg on the role of natural resources in economic development. According to his piece, published in 2000 (but made unavailable since), the development of Russia required the exploitation of energy reserves by the state to leverage the state. The calculation was simple: in order to catch up with the more advanced economies of the G7, Russia needed to achieve an annual growth rate of four to six per cent per year; this in turn it could only achieve through imposing industrial–financial consortiums capable of competing with Western majors.[5] This programme, which Putin implemented upon acceding to the presidency, seems to have borne fruit, as Russian annual growth from 2000 to 2008 was around six per cent, with companies such as Gazprom, Rosneft, and Lukoil in effect acting out the Kremlin's domestic and international ambitions.

In retrospect, it is also interesting to observe that the level of Russia's international ambitions seems to have increased proportionately to energy prices and the re-nationalization of its energy sector. When elected president in March 2000, Putin faced a considerable foreign policy challenge. As aptly explained by Bobo Lo, Putin had 'to find a kind of "third way" between the Westernizing foreign/domestic policy linkage of the later Gorbachev and early Yeltsin years, and the anachronistic, over-geopoliticized foreign policy of the second half of the 1990s'.[6] He successfully did so in the early years of his presidency, before embarking on a path that stressed Russia's specificity and, ultimately, coming to present Russia as an emerging country belonging to the so-called BRIC (Brazil, Russia, India and China) group – in effect sidelining pure 'specificity' in favour of selective multilateralism.[7]

In fact, Russian 'specificity' in terms of its economic, social and political development, has gradually become a doctrinal framework of Russian strategic thinking. 'Specificity' has even come to frame the Russian approach to international relations. Russia draws a large portion of its specificity from an interwoven political, economic, judicial, media and security network, as well as from a strategic culture of great power politics. Beneath all this lie complex civil–military relations inherited from both the Czarist and Soviet past characterized by the concentration of power in the leader's hands. These civil–military relations directly impact the decision-making process within the ruling elite, as well as the links between the presidential leadership and the security and business communities.[8] This mindset is crucial to understand Russia's foreign and energy policies.

Given his own personal background, Putin as Russian leader was naturally oriented towards security-related issues which have always occupied a central place in Moscow's world outlook. As Pavel Baev convincingly explains, Putin based his policy of re-establishing and re-constituting Russia's 'greatness' on the two pillars of military security and energy security. This third theme may have been the main driver of Putin's policy 'that in some ways contradicts his devotion to political "pragmatism", which in fact is often reduced to a mix of cynicism, commercialism, and vulgar materialism'.[9] Under Putin's Russia (2000–2008), there has clearly been a dynamic interplay between restoring military power, instrumentalizing energy resources and rebuilding 'greatness'.

The energy industry fuelled Russia's resurgence, but Russian strategic thinking is not limited to energy issues. This observation raises four main questions. First, how do Russia's foreign and energy policies relate to one another? The subordination of the latter to the former may well depend on the situation in each state, but this relation may also contribute to the Russian specificity. Second: should Russia's policies be understood as the product of a coherent strategy or as short-term tactics? Do Russian policies reflect an active agenda or a reactive reflex? Russian policies sometimes appear to be uncoordinated and, to some, incoherent. This may be explained by the alignment of the Kremlin's objectives and those of the giant energy companies the state sponsored – the interests of which do not always coincide. Third: are popular notions such as 'energy superpower', 'resource nationalism' or 'state capitalism' appropriate to explain Russia's behaviour?[10] While these notions may be overly simplistic, they may nevertheless help outline a Russian model of development. Fourth: how are Russia's foreign and energy policies articulated at the global, regional and bilateral levels? The connection between geo-economics and geopolitics may be observed differently at these three levels. This last point is key, as it can help us assess whether Russia's energy policy contributes to further integration or fragmentation of the FSU – an area still of vital interest to Russian foreign policy.

Focus of the book

The book thus begins by examining broad issues in the post-Soviet energy landscape under the section *Energy's link to domestic and foreign policies*. This opens with a broad discussion of concepts of energy security as they apply in the post-Soviet space (Frank Umbach), the impact of policy on production (William Tompson), the financial realities of pipeline construction (Jérôme Guillet), and ends with a concrete and timely example of rival policies for bringing Caspian gas to market (Maureen Crandall). The book's second section, *Russian foreign and energy relations with NIS net exporters*, takes a closer look at Russia's relations with the NIS producers it depends upon for its westward contractual obligations, namely, Turkmenistan (Vladimir Milov), Kazakhstan (Martha Brill Olcott), Uzbekistan (Andrew Monaghan) and Azerbaijan (Pavel Baev). The book's third and last section, *Russian export policy and relations with transit countries*, focuses on Russia's most recent relations with transit countries as a whole (Leonid Grigoriev) and Georgia (John Roberts) and Belarus (Rainer Lindner, Felix Hett and Folkert Garbe) in particular.

Frank Umbach's opening chapter delivers a broad picture of energy security in Eurasia, which is useful to identify the main players and their respective interests. Umbach analyses energy security from the point of view of producers and consumers, as well as transit countries. Various definitions of energy security have produced many misunderstandings and conflicts. The interactions between all of them, added to the evolution of global demand, have fuelled a highly complex situation: the EU has strived to liberalize its internal energy market whereas Russia has based its resurgence on the control of its energy potential. Umbach takes his argument from the notion of 'resource nationalism' to explain these tensions. This notion has been popularized to describe the political behaviour of states, which as energy producers manage energy flows in accordance with their own security interests. In this respect, the EU is in an uncomfortable position given its political identity, and its future energy needs, especially in natural gas. Beyond Norway and Algeria, Russia is, and will remain, the EU's principal natural gas supplier.

Umbach's argument has four main strands. First, the Kremlin through Gazprom has been exploiting energy dependencies within the EU (mainly between old members and new members) as a foreign policy tool. Second, Russia alone will not be able to fulfill the EU's increasing natural gas demand, and needs to position itself in a global energy contest. Third, the Central Asia and the Caspian Region (CACR consists of Georgia, Armenia, Azerbaijan in the Caucasus, and Kazakhstan, Uzbekistan, Turkmenistan, Kyrgyzstan and Tajikistan in Central Asia) is of crucial importance to anticipate the future of energy security in Eurasia. CACR should be a significant supplementary supplier of natural gas for the EU. Fourth, even if Russia cannot be removed from considerations of EU energy supplies, it can no more afford to neglect the CACR in its energy security strategy, not only regarding the EU but also China.

Indeed, during the last decade China has emerged as a very active energy player in Central Asia. Securing energy supplies has been the main driver of Chinese foreign policy in the Middle East, Africa and Latin America. Central Asian states, particularly Kazakhstan and Turkmenistan, have been targeted by Beijing as credible partners. In December 2009, the Turkmenistan–China pipeline crossing Kazakhstan and Uzbekistan was inaugurated. It represents a major shift, dramatically changing the traditional East–West approach in terms of energy supplies. Umbach also analyses the well-known debate on the 'Southern Corridor' with the competing projects Nabucco and South Stream. It seems clear that the Kremlin tries to block the development of the Nabucco pipeline by aggressively promoting South Stream. Umbach gives his insights into the consequences for different countries of the CACR. According to him, the war in Georgia (August 2008) was a way for the Kremlin to discredit Georgia's role as a transit country. Using its old links in the CACR, the Kremlin strives to enhance its position vis-à-vis the EU, which failed to develop a consistent strategy, both in the political and energy fields, towards this region. However, according to Umbach, Russia is focusing on short-term economic and diplomatic gains rather than on long-term strategic interests.

This final point has been debated by other contributors, and was tabled many times during our workshops. In fact, there is another possible reading of Russian behaviour, one which underlines the Kremlin's capability to forecast Russia's future. It is worth noting that many official strategy documents run to 2020, and not only in the energy field. This means that the juncture between Russian energy policy and Russian energy policy should be explored across a generation, from 2000 to 2020. From this point of view, renationalization of the energy field can be much more easily understood: it is one of the main engines of Russia's domestic and international resurgence. Domestically, it has been the easiest way to regain wealth thanks to the rise of energy prices since 1998. Internationally, energy's enduring connection to political decision-making has made it the natural way to project power. It also allows power to be used differently according the country targeted. Fundamentally, Russian policy favours bilateral approaches. It is worth wondering if the very concept of the CACR makes sense for the Kremlin given the deep heterogeneity of this area. CACR makes sense for the EU regarding Russia, but this grouping may be strongly confusing when it is time to set up real energy deals. Given the fact that Russia will remain the main EU gas supplier and CACR can become an important supplementary one, the relationship between both of them should be consistently explored by the EU. This book participates to this challenge.

William Tompson offers a wider picture dealing with both oil and gas. Tompson does not deal with CACR as a coherent area, but prefers to focus on Caspian Sea producers – Turkmenistan, Azerbaijan and Kazakhstan – and delivers some thoughtful insights on Uzbekistan. Tompson questions the long-term elasticity of production in these countries, and its consequences for Russia. Added to this, he analyses the respective types of ownership and their impact on foreign investors. Having underlined the uncertainties related to the future political orientation of these four countries, Tompson argues three main points. First, it is absolutely sure that oil and gas consumers, and in particular the EU, will wish to maintain diverse supply channels. Second, Tompson recalls the potential for instability or threats to supply from these countries, and consequently the possibility for Russia to reinforce its position as the most reliable gas supplier. Third, Russia's reliability depends in equal part on the production elasticity of these countries. Obviously, this broad picture should be adapted to the evolution of international demand.

Within this grouping, Kazakhstan has certainly become the leading energy producer, and the most promising (some regions are still underexplored). In the 1990s, Kazakh authorities privatized most of the existing oil and gas enterprises, but never went as far as privatizing the pipeline network. In 2002, Kazakhoil and Transneftegaz were merged to set up a new national energy company able to face international majors: Kazmunaigaz. Tompson notes that the atmosphere for foreign investors started to change in 2003. Kazakh authorities became tougher and determined to capture a larger share of energy rents by renegotiating PSAs (production-sharing agreements) agreed during the 1990s. However, this situation seems manageable to Tompson who proclaims that it 'should not be over-dramatized'. Due to the opacity of the Turkmen regime, it is much more difficult to assess the volume of reserves precisely, especially since some regions are still under-explored. The future of Turkmenistan is the greatest unknown factor in predicting possible developments in this complicated area. Needless to say, the political situation in Ashgabat reminds tricky to decipher, as are the real contents of energy deals concluded with Beijing. Uzbekistan is also an important piece of the energy puzzle in Central Asia. However, having no direct access to the sea, its situation is qualitatively different from those of Turkmenistan and Kazakhstan. Added to this, Uzbek fields developed in the Soviet period are now in decline, and new fields require exploration. By comparison with these three countries, Azerbaijan has much more valued its tradition in the oil industry as well as its strategic location. SOCAR was created in 1992 precisely to favour agreements with foreign companies. Azeri authorities succeeded in establishing the most investor-friendly climate to attract foreign capital. Azerbaijan played a leading role in the development of the East–West energy corridor through the Baku–Tbilisi–Ceyhan (BTC) pipeline and the SCP (South Caucasus Pipeline).

As Tompson argues, these four countries share common troubles that increase political and business risks: weak institutional framework, pervasive corruption and opaque policy-making within the ruling elites. These four countries inherited both infrastructure and institutions from the Soviet period. They had to manage post-communism on their own, facing attempts by international majors to enter either their market or their production systems. The dramatic decrease of Russian influence during the 1990s masked the close links, not only within elites but also in terms of industrial organization, between these four countries and Russia. Through a nuanced approach, Tompson identifies an overall trend to dirigisme shared by authoritarian regimes. The interesting point is to observe each of them establish its own balance between dirigisme and rule of law. Consequently, foreign investors should be able to adapt their behaviour and ambitions to the country in question and its political and economic context. Because of a fast-moving background and predominance of political decisions, foreign investors should be ready to accept a lower rate of return in an environment of increased risk. This a basic factor which should be taken into consideration when there is a desire to enter these countries. In parallel, Tompson questions the performance of large state-owned companies throughout the NIS, remaining rather sceptical about their long-term competitiveness.

Tompson's chapter is useful to understand the very different situations in Turkmenistan, Azerbaijan, Kazakhstan and Uzbekistan, as well as their common development problems. Many of them can be explained by the organization of political power. From this point of view, it is worth mentioning the death of Saparmurat Niyazov in December 2006 to underline that political transition remains a highly sensitive issue in these four countries. There is no doubt that a change of leadership can dramatically alter the atmosphere within each of these countries, and directly impact the investors' position. Regardless of the leader, the bargaining power of these states vis-à-vis international majors has been reinforced during the 2000s due to increasing demand and better organization. From this point of view, an interesting point still deserves to be explored extensively: the decision-making process not only within the authorities, but also within national companies as well as the manner in which they interface with their international business partners.

Given his experience as a researcher and as a banker in the energy field, Jérôme Guillet explains the process of getting a gas pipeline built. As can be expected, it is not easy. Unexpectedly, political factors are of far less importance than business rationale. Guillet regrets that the debate on Russian energy policy is very often reduced to an abstract geopolitical reading without understanding two key drivers: economic and financial conditions of pipeline construction, as well as profound interdependencies with pipelines which already exist. Guillet reminds us of a basic fact: pipelines cost a lot to build, but are rather cheap to operate. A lot depends upon the amortization of the initial investment. The value of a project is not only related to the cost of the pipeline, but also to the value of the gas transported by it over 20 years or more. Investment rationale is also based on pricing, which Guillet analyses in depth.

His argument is presented in three main ways. Having observed a geopolitical consensus about Europe's increasing dependency on Russian gas, Guillet strongly argues in favour of a much more balanced reading. First, starting from the very inappropriate concept of the 'energy weapon', Guillet explains why nobody can win a so-called 'gas war'. According to him, there is one single example of a highly asymmetrical relationship in the gas field, between Russia and Armenia. Second, Guillet points out a flaw in European expert analysis: Russian decisions are not analysed according to Russia's interests and needs, but according to the EU projects for gas market liberalization, as if Russia can only share these objectives. The third argument is made in terms of methodology: the more one talks about a given project, the less likely it is to happen. Given the complexity of setting up a gas pipeline from scratch, Guillet explains that political decision-makers, in any country, 'cannot get a pipeline built on their own'. No project makes sense if a pipeline already exists and is not already used to capacity. Politicians are involved as just one component in a very complex decision-making process involving multiple actors. Guillet suggests that diplomacy and PR campaigns can only initiate and drive a project when there is an actual need for new transportation capacity that could not be filled by existing alternatives, and more than one economically sustainable project is under consideration.

Guillet's chapter is helpful to our understanding of the different components of a gas pipeline. He lists 11 prerequisites for a pipeline to be built and underlines how the combination of all of them requires strong managerial and financial capabilities: 'all of the relevant components and parties need to be present at precisely the same time for the project to exist'. Guillet recalls that there is a basic difference between paying for construction and actually financing a pipeline, and also gives some key figures to keep in mind regarding capacity and pricing. Given this complexity, any project relies on strong leadership, which can be either an energy company or a consortium controlling most components of the chain.

Geopolitics of energy is quite often very primitive, using maps to describe a supposed grand strategy which does not exist in the field. Guillet's chapter helps to explain how the energy business is constrained by the need for financing, and consequently he discredits simplistic readings of the Russian approach. However, the role played by governments is undoubtedly crucial in this matter. The granting of authorization for pipelines to cross their territory, direct or indirect funding of projects, support for favoured operators when dealing with other governments, and the protection of critical infrastructure are all examples of how governments can play a role. Business risk is about how each segment of the pipeline's value chain can be realized and made profitable in the long run. Given the fact that some of these components depend upon matters directly related to state sovereignty, political factors should also be integrated in our understanding of pipeline politics.

Maureen Crandall's chapter maps both existing pipelines and possible new ones within their respective political backgrounds. Crandall establishes January 2006 as the first turning point regarding Russian energy behaviour, and obviously January 2009 as the second one. The gas cutoffs to Ukraine, and consequently to some downstream EU customers, underscored Moscow's political leverage over EU gas supplies. These twin energy crises explain why both Russia and the EU 'are seeking alternative export–import routes'. On the EU side, there is a strong political wish to diversify sources of gas supplies. On the Russian side, there is a strong political wish to avoid transit dependency on Ukraine (approximately 80 per cent of the gas from Russia to the EU market transits Ukraine) in order to secure its main gas market in Europe.

Having this balance in mind, Crandall argues three main points. First, Moscow has a clear strategy which is basically to maintain its dominance in European gas markets and to prevent any attempt to bypass it. Put in other words, Russia intends to market its gas, to keep its traditional European customers via long-term contracts, to get new customers, and finally to go downstream in order to climb the gas industry's value chain. In parallel, Russia strives to preempt new gas flows that could challenge its gas markets in Europe. Second, Crandall highlights Europe's reduced ability to act as a key player compared to Russia, China and the US. These three countries seem the most influential ones when considering new pipelines in the future. Third, Crandall estimates that the war in Georgia affected the geopolitics of energy by casting doubts on the reliability of the Georgian corridor as a route of future gas supply to Europe.

Central Asian gas producers want to develop new routes to deliver gas to Europe and elsewhere, thus reducing their dependence on Russia. To clarify the stakes, Crandall recalls that there are currently four gas pipelines working in the Caspian area, and four main projects about which there is much speculation. Regarding the working pipelines, Crandall lists the SCP from Shah Deniz to Erzurum; the pipeline from Turkmenistan to Iran; the Central Asia–Centre system inherited from the USSR delivering gas from Turkmenistan, Kazakhstan and Uzbekistan to Russia; and lastly the pipeline from Turkmenistan to China inaugurated in December 2009. According to Crandall, CNPC expedited this project for strategic rather than economic reasons, that is to make inroads into Central Asia in order to reinforce its energy security. Regarding future projects, different factors deserve to be taken in consideration. The TransCaspian gas pipeline is reliant upon an agreement between the five littoral countries for delimitating respective zones. In addition, this project would require additional supplies. An unknown factor remains: the ability of Ashgabat to develop the Yolotan–Osman field in proper time. Without additional supplies, the TransCaspian would remain 'no more than an unrealistic hope'. The Nabucco project has been supported by both the EU and the US. However, it suffers from uncertainty in terms of supply, and remains a pure transportation play 'with no producer–company equity participation'. It is in competition with the South Stream, which has been strongly promoted by the Russian authorities. The last project is a pipeline to Pakistan and India via Afghanistan which has won US support as it would avoid Russia and Iran.

One thing appears very clearly in this discussion of existing pipelines and pipeline projects: Russia's resurgence since 2003. Compared to the EU, Crandall sees Russia's strategy as coherent. One of the four Russian proposals – North Stream – has become a reality with construction starting in April 2010. According to Crandall, Blue Stream–II, South Stream and the Pre-Caspian Pipeline represent a comprehensive programme to fix 'Europe in a natural-gas pincer'. She rightly notes that all the projects proposed by Russia would bypass both Azerbaijan and Georgia. These projects also demonstrate Gazprom's opposition to diversification in its future gas trade with Europe.

Crandall's chapter helps us to understand the strategic link Russia makes between policy towards Central Asian countries and policy towards Europe. As the closest, traditional and most profitable market, Europe is, and undoubtedly will remain, at the core of Moscow's strategy. Not only in the energy field. In this field, it is appropriate to consider the Caspian countries as a crucial third element in the EU–Russia relationship. Similarly, Europe is the third element in the relations between Caspian countries and Russia, just as the latter represents the third element in relations between the EU and Caspian countries. This book is aimed precisely at illuminating this type of geopolitical and energy interaction in a dynamic environment.

The book's second part focuses on the very particular relations between Russia and the net exporters of the NIS. It is not an easy task to explore these relations given the lack of reliable data. Vladimir Milov's chapter deals with Turkmenistan, the most isolated Eurasian country. To be more precise, Milov deals with Saparmurat Niyazov's Turkmenistan, and strives to describe the relationships existing between Ashgabat and Moscow. Even though Niyazov died and Gurbanguly Berdymukhammedov became president in December 2006, it is still difficult to see the impact of this political change on the Turkmen energy policy given the lack of information about this closed country. From a Russian point of view, policy-makers face the following problem: Russia has been increasingly dependent on imports of Turkmen gas in order to fill the gap between growing Russian and European gas demand and its own stagnant gas production.

Milov delivers an interpretation of the relations between Russia and Turkmenistan based on three main arguments. First, despite its unpredictability Turkmenistan has become 'a very important secondary player in the Eurasian energy arena', given its links to Russia and to China. Second, Russia's approach has been to limit Turkmenistan's opportunities to market its gas abroad. By maintaining a monopoly on Turkmen gas exportation, Russia has been in a position to profit from the significant price differential between gas purchased from Turkmenistan and European gas prices. Third, relations between the countries are unpredictable and lack trust. Added to this, Russia has failed to use a window of opportunity opened by 11 September 2001 to foster its relations with Ashgabat. Its traditionally domineering approach encourages the Turkmen leadership to seek alternative partnerships where possible, in particular with China. That is why Milov anticipates 'the progressive parting of the two countries rather than their developing partnership'.

Turkmenistan's international position deserves to be remembered. Its 'permanent neutrality' was recognized by the United Nations in 1995. However, as described by Milov, Turkmenistan was eager to cooperate with the Taliban regime after the capture of Kabul in September 1996. The project for a gas pipeline from Turkmenistan to Pakistan through Afghanistan is a prime example of this. Relations between Russia and Turkmenistan had started to warm up after 11 September, given Turkmenistan's need for support after the US intervention in Afghanistan and the increasing presence of the US in Central Asia. Niyazov may certainly have feared becoming a target of US military action. The most tangible outcome of this rapprochement with Russia was the signature in April 2003 of a 25-year agreement for a Russian–Turkmen gas supply. In August 2005, the announcement of Turkmenistan's withdrawal from the CIS illustrated the instability of this relationship.

This sort of oscillation is a telling demonstration of the unpredictability of the Turkmen leadership. Milov briefly describes the administrative system and state ownership practices, which traditionally restricted foreign investment, before underlining the influence of opaque intermediaries in the gas trade between Turkmenistan, Russia and Ukraine. There is no reliable information about the functioning of relations between Turkmenistan, China and Iran. This lack of objective information encourages scepticism about official estimates of gas reserves.

Milov is very sceptical of Turkmen ability to develop its gas potential on its own, but is much more sceptical of Russia's ability to establish a new energy relationship with Turkmenistan. This is partly due to the dire need to refurbish and expand the CAC system. According to him, this relationship will be reliable 'neither in terms of volumes of gas delivery, timeframe, nor prices'. Indeed, he expects further conflicts and supply interruptions. Milov's pessimism should be questioned in two directions. First, it remains unclear what consequences the relations between Ashgabat and Moscow will have on Russia's oft-mentioned 'gas deficit'. Second, the inauguration of the Turkmenistan–China pipeline illustrates what determined players can achieve in proper timescales in this region.

Martha Brill Olcott's chapter deals with the Kazakh–Russian relationship, which is at the core of the energy balance between Central Asia, Russia and Europe. In just one decade, Kazakhstan has become the pivotal country in Central

Asia, and has succeeded in implementing a consistent energy policy. Brill Olcott explains how important the relationship with Russia is among the competing options before the Kazakh leadership.

Brill Olcott argues that the Kazakh–Russian relationship is the single most important one for Astana because it includes not only energy relations but also security cooperation and economic ties. She draws three main conclusions. First, Astana has seriously improved its position vis-à-vis Moscow during the last decade. Becoming more articulate and assertive, Astana appears to be taking the role of Russia's 'younger brother' within the post-Soviet space. Second, Astana has succeeded in increasing its foreign policy options, especially by raising the issue of Russian pressure when dealing with the EU and the US. Third, regarding Kazakhstan, Russia's major concern today is Chinese competition rather than Western attempts to engage Astana. For all these reasons, Astana has secured its leadership position among the Caspian states.

Brill Olcott qualifies this final conclusion by raising the impact of the global crisis on the Kazakh economy, and consequently on its foreign policy choices. Given the depth of the crisis in 2009, the government was obliged to break into its National Fund. Yet the question remains: is the crisis an opportunity to escape from Russian influence by attracting alternative investments or, on the contrary, an opportunity for Russian investors to reinforce their positions by introducing liquidities? Whatever the answer, it is worth remembering how dependent the Kazakh economy will remain on both energy rent and its relationship with Russia. Nursultan Nazarbayev is completely aware of this dependency and consequently manages his personal relationships with both Putin and Medvedev very carefully, as he did with Yeltsin previously. Added to this, there has been a consistent play for the Kazakh elite to simultaneously seek to protect Kazakhstan's national interests, the interest of particular Kazakh enterprises and the personal assets of the Kazakh leaders themselves. In this regard, the leading energy company, KazMunaiGaz, has a pivotal position within the Kazakh system. Needless to say, regarding foreign investors, the Kazakh authorities have followed Russia's example by renationalizing its energy industry. Despite a new toughness perceivable since 2003, Kazakh officials maintain that the government has become more investor-friendly.

On the Russian side, the importance of Kazakhstan is very well understood. Astana was targeted by Putin as a top priority. According to Brill Olcott, the Kremlin under Putin began strongly encouraging Russian firms to take equity stakes in projects in Kazakhstan, and in Central Asia more broadly. This encouragement would have been made with the implicit assumption that if it was not commercially beneficial 'they could expect to receive compensations elsewhere'. At the political level, an important step was the meeting between Putin, Nazarbayev and Berdymukhammedov in May 2007 to support the project of a Russia–Kazakhstan–Turkmenistan pipeline, which would directly challenge the US-sponsored TransCaspian. The growing 'Chinese factor' has led Russia to reconsider the origins of the threat to its own influence. In terms of pricing especially, Russia started to propose adjustments to producers.

One further point would also deserve to be mentioned given its possible impact on energy transit. In June 2009, Putin declared that Russia will join the WTO only after having built up a customs union with Belarus and Kazakhstan. Medvedev slightly nuanced this approach afterwards, but there is clearly a wish in Russia to try to improve regional cooperation by deepening the relationship with Kazakhstan, which is seen as the most important and most promising partner with the post-Soviet space. From this point of view, the evolution of both Kazakh–Ukrainian relations and Russia–Ukrainian relations should be observed very closely.

Andrew Monaghan's chapter deals with Uzbekistan, which is an important, even if little known, component in the regional energy puzzle. It is not only a key transit country, but also a valuable producer. Without its production, the region's as well as Russia's gas supplies would be significantly constricted. Added to this, two key pipelines transit the country: the Central Asia–Centre and the Bukhara–Ural line. It is also important to remember that Turkmen gas reaches Russia through Uzbekistan. For all these reasons, Uzbekistan should be taken into consideration when assessing the energy regional balance and its weight in Russian energy policy towards the EU.

Monaghan's argument follows three main lines. First, he reports a return to Uzbekistan by Russia during Putin's second presidential term. There has been a growing partnership between both countries which is mutually beneficial. To some extent, this partnership contributes to Uzbekistan's fragile political situation. Second, Uzbekistan is host to difficulties and complexities which all investors in Central Asia face, including Russia. The nature of the regime is a critical obstacle to ambitious energy development programmes. Third, this regime is not always an easy partner for Moscow, which remains prudent, refusing to be too narrowly bound to Tashkent.

After underlining the lack of transparent official data, Monaghan astutely describes the main characteristics of the regime led since the independence by Islam Karimov. According to Monaghan, there is no doubt that the clannish nature of the regime has a dramatic impact upon its foreign relationships, especially that with Russia. For the presidency, energy rent represents the best way to secure the loyalty of key sections of the elite, in particular the security elite. The total concentration of power within the President's hands plays in two directions. On the one hand, it can be seen as a most efficient context in which to conclude bilateral energy agreements; Putin used to operate through direct links with his Uzbek counterpart. On the other hand, the future of Uzbekistan is too narrowly tied to the future of its president personally. Put in other words, the concentration of power fuels constant political uncertainty linked to the presidential succession, and implicitly the respect of foreign investments in the future too. The May 2005 events in Andijan illustrated the instability of the political situation given the seeds of Islamic radicalism. Because of the repression of anti-government protests, the EU imposed sanctions upon Uzbekistan. Five years later, relations remain tense.

Compared to its political situation, Tashkent's energy situation appears more sustainable. Uzbekistan did not suffer a serious decline in oil and gas production after 1991, and achieved energy self-sufficiency in 1995. Due to its industrial organization, it has still a high level of energy consumption. Uzbekistan also needs foreign investments to explore new fields, Karimov has tried to encourage investment through PSAs, but Western companies have shown little interest owing to the country's 'uncertain legislative atmosphere'. Partly because of this Western reluctance, Tashkent has not rejected Russian and Chinese attempts to invest into its energy industry. Putin clearly targeted Uzbekistan and complemented his energy proposals with security agreements. After Andijan, Putin and Karimov signed an agreement on allied relations between Uzbekistan and Russia in November 2005, and Russia succeeded in attracting Uzbekistan into the CSTO. As explained by Monaghan, Tashkent has been courted by Moscow and Beijing, and encouraged 'competition between Russia and China to enhance its own influence and room for manoeuvre'.

However, Monaghan delivers a nuanced conclusion. One the one hand, Moscow is undoubtedly more present and more active in the Uzbek energy field since 2005. On the other hand, he notes the lack of real substance in the energy links, due mainly to Gazprom's postponements of investments. Uzbekistan has stayed an unpredictable energy player, because of the nature of the regime and the fragmentation of its territory (Ferghana Valley remains a highly unstable area due to the situation in Kyrgyzstan and Afghanistan). The question of who will succeed Karimov is high in the minds of investors right now, Russians included.

Pavel Baev's chapter starts the exploration of bilateral relationships between Russia and the main producers of the former Soviet Union. Baev examines the key dimensions of the relationship between Baku and Moscow in order to assess its stability. Azerbaijan is situated in a highly valuable geopolitical location. As an energy producer, Azerbaijan is the point of departure for the BTC and the SCP, which bypass Russia. Azerbaijan is also Europe's gateway to Central Asia and Russia's to Iran. For all these reasons, relations with Baku are crucial to Russia's continuing influence in the Caucasus.

Baev identifies a paradox in Azerbaijani–Russian relations: the authorities on both sides are fixated on energy matters, but their direct interactions in this field are rather limited. He then goes on to argue two central points. First, he explains why Russia remains sanguine vis-à-vis Azerbaijan. Building up a strong relationship with Baku was one of Putin's main foreign policy targets. According to Baev, the apparent cordial relationship between Putin and Aliyev has masked the fundamentally imbalanced structure relations between both countries. Second, Baev suggests that Moscow has been both patient and careful vis-à-vis Azerbaijan. Russia did not try to block Azerbaijan's Westward drift during the first part of the 2000s. Having a longer term vision, Putin has invested in the consolidation of the authoritarian Azeri regime, which may be embraced in a new 'big-brotherly' approach.

According to Baev, Moscow misjudged the profitability of the BTC and SCP and their importance for the Azeri economy. Its attempts to block the TransCaspian have resulted from the desire not to repeat this mistake. Putin tried to correct this initial mistake by courting Aliyev Jr, who succeeded his father in October 2003.

Putin was partly successful, but personal relations cannot compensate for real political divergences based on security interests. In this regard, the inability to resolve the Nagorno-Karabakh conflict remains the main irritant for both countries. Armenia remains an important ally in Moscow's defence system, as it hosts Russian military bases on its soil.

Putin planned to exploit the fundamental difference between the Georgian and Azerbaijani regimes to draw the latter closer to Russia at the EU's expense. Georgia portrayed itself as a democratic regime, whereas Azerbaijan represented a system of patrimonial corruption. It was criticized more or less openly by the US and the EU for this fact. Putin expected that this divergence would have made things more difficult for the EU. As presented by Baev, relations between Moscow and Baku were limited by both internal and external factors. Internally, there was a lack of joint industrial projects despite pre-existing potential. Added to this, Azeri citizens living in Russia may have been treated as 'Caucasian criminals' on many occasions, creating tension between the governments. Externally, Iran constitutes an important variable in the energy field as in the military field. Baku carefully distances itself from the complex negotiation process between Tehran and the international community. Turkey is the other important variable in relations between the two countries. Having become a key energy partner of Russia, Turkey has increased its influence over Azerbaijan. Turkey has in fact become pivotal in every major energy project in the region, reinforcing its influence.

After having analysed the issue of the Gabala radar station, Baev insists on the consequences of the war in Georgia in terms of regional dominance. Moscow took great care not to inflict any damage to the BTC and SCP. At the same time, he makes clear that the war dissuaded Baku from experimenting with military force to regain control of Nagorno-Karabakh. After Georgia, Baku seems to have toned down its statements towards Armenia, especially on military matters. This is also certainly related to the start of a new process of negotiations between Turkey and Armenia. Relations between Moscow and Baku still lack substance given the potential for profitable cooperation. However, both countries seem to have reached a sort of balance in their relations, which has been accelerated by Russia's return to the Caucasus with the war in Georgia. Militarily outclassed by Russia, during the 2000s Georgia promoted itself as a key energy player. Its ability to play a stable energy role has been damaged dramatically. There is no doubt that Baku has drawn this conclusion too.

The book's third and final part explores the role of transit countries. Leonid Grigoriev's chapter focuses on Russian export policy and relations with transit countries, raising an often-overlooked premise: Russia too is a transit country, not only a producer and a consumer. Grigoriev explains the Russian point of view by distinguishing the Kremlin's interests as well as Gazprom's ones, mainly in terms of gas transit from the CACR. Western expert opinion is dramatically mistaken in refusing to recognize that the Russian authorities and Gazprom defend their own distinct interests. Western obsessions notably with EU gas market liberalization lead to limited understanding of each player's interests. Simply put by Grigoriev: 'Russian policy is always under scrutiny for non-commercial objectives. Yet in reality, both Russian companies and political actors are trying to protect their respective interests'.

Grigoriev argues three main points. First, from 2003 to 2008, the Kremlin imposed low gas prices for volumes imported from CIS countries 'leaving Gazprom hostage to Russian foreign policy'. Second, there is paradoxically no explicit policy from either the Kremlin or Gazprom towards the CACR and the CAC system. Lastly, there is no doubt that the best route for new transit capacity in the CACR is through Russia along the old CAC, which will remain a focus of regional and international politics.

For Grigoriev the CAC was 'a free gift', inherited by the NIS countries from their Soviet past. These countries inherited capacities for exporting their gas with no investment cost or corporate debt. This legacy is sometimes considered by Western experts as an obstacle or an inconvenience for EU energy security, but Grigoriev reminds us that this complex system still transports over 700 bcm every year. Until 2004, NIS producers could not finance the system's maintenance and repair. The Ukrainian section of the pipeline is seen to be the best maintained. There are still many uncertainties about the quality of maintenance and repair, as well as the level of investment required to bring the system back to its original design capacity. The principal point to be made here is that despite having been divided into independent states, the system 'linking the various national pipelines is a near-single entity'. In other words, the CAC gives sense to the notion of post-Soviet space.

Unsurprisingly, Grigoriev explains why the Nabucco project makes no sense given this legacy and the predicted production capacities. The TransCaspian route to the SCP will also be delayed further. According to him, Western experts should understand that Russia will be served first from gas coming from the CACR, and China second. This might prompt tensions between China, India and the EU. The competitive advantage of the EU in this battle for supplies lies in the higher prices paid on this market.

It is a basic fact that CACR countries are less trusting of global market mechanisms than Western consumers. In addition to their political organization, this may explain why they are so keen for guarantees of delivery through special agreements, and why bilateral energy issues with Russia remain highly politicized. In this regard, the collapse of energy prices in the second half of 2008 gave a strong incentive to secure transit incomes.

After outlining the different priorities of consumers and producers, Grigoriev explains why Gazprom's priorities do not match the Kremlin's priorities consistently, which is a way for him to underline the constant weight of political factors. To conclude, he is much more optimistic about Gazprom's future than about the EU's energy policy which is 'a very complicated affair'. Undoubtedly, Gazprom will retain its position as the major carrier of CACR gas, as the CAC will 'remain the backbone for transiting CACR gas to the West'.

Reading Grigoriev raises the question whether any uncertainty and unpredictability in the gas market takes its origins more on the EU side than on the Russian side. According to him, the natural flow of investments coming from Russia, CACR countries and consumers should be driven to the maintenance and modernization of the CAC, which has a major competitive advantage: it already exists. This argument does not deal with the anticipated increase in EU gas demand, as well as the increasing gap between the real value of Gazprom and its perceived value by its European customers after the gas crisis in Ukraine.

John Roberts' chapter then takes a look at transit issues in the South Caucasus after the Russia–Georgia war of August 2008. This war highlighted the vulnerability of the Southern Corridor very often presented as a secure route avoiding both Russia and Iran. One year later, the Russia–Ukraine gas crisis in January 2009 dramatically revived European concerns about the reliability of Russia, spurring renewed interest in the Southern Corridor. In any case, the Southern Corridor is the most sensitive geopolitical issue when dealing with EU gas supplies, and will continue to be so in the near future. According to Roberts, governments will be more and more involved in this area, which cannot be only driven by the energy companies' concerns.

Roberts' argument follows three main directions. First, in order hope to boost their political as well as their economic independence, both Georgia and Azerbaijan over-invested in promoting the Southern Corridor to the EU. From 2003 to 2008, they tried to involve Europe deeply in South Caucasus. The success of this policy is debatable. Second, in this region, Russia remains nervous and married to a zero-sum game approach. Third, the combination of the BTC and the SCP are the first main achievements of the Southern Corridor. However, its stability is still questionable given 'the risk of frozen conflicts erupting again', namely Nagorno-Karabakh, South Ossetia and Abkhazia.

The description of the main sequences of the 2008 war is useful to understand that 'the war was not about energy, but it had profound implications for energy security'. The war reflects the Southern Corridor's lack of stability, and also Russia's influence upon it. Roberts reminds us that the BTC line was mysteriously attacked two days before the outbreak of war. Physical safety of pipelines has been a topic of concern for both companies and governments. During the war, BP preferred to shut down the SCP. Physical safety of pipelines has been used by both Azeri and Georgian authorities to justify their wish to join NATO. There is no doubt that the main political consequence of the war is to have dramatically postponed this possibility.

Roberts strives to understand the EU policy towards this complicated region. According to him, the European Commission is moving in two directions. On the one hand, it has encouraged Azerbaijan to develop its gas resources quickly and to commit them for export to the European market. On the other hand, it has encouraged a Caspian Development Corporation working as an umbrella group for European companies to develop and to export the gas resources from the area. Roberts makes an analogy between this approach and the Chinese approach in courting Turkmenistan, Kazakhstan and Uzbekistan for integrated exploration and production projects. Added to this, Roberts delivers his own interpretation of the debate about Nabucco and South Stream.

Regarding Russia, Roberts explains Moscow's reluctance to support (politically or economically) any type of project reinforcing the Southern Corridor: 'the more effective the corridor is in attracting Central Asia gas away from Russia, the more it forces Russia to address its own gas imbalance'. The combination of the war and the financial crisis has created a tougher business environment for the entire range of actors. It leads to a situation in which governmental guarantees on volumes will be seen as absolutely essential.

Roberts' chapter illustrates the balance of opinion regarding the stability of the Southern Corridor. There is no doubt that Mikheil Saakashvili's initiative was counterproductive in terms of enhancing the security of the corridor and in encouraging NATO to welcome Georgia. This war has exacerbated two conflicts, Abkhazia and South Ossetia, and it is not particularly good news in terms of the stability of energy supplies. On the other hand, the gas crisis between Russia and Ukraine in January 2009 has revived the EU concerns to diversify the routes of their gas supplies to avoid total dependence on transit through Ukraine. In any case, the South Caucasus will continue to attract various influences coming from not only the EU, but also Turkey, Iran and certainly Russia.

The final chapter by Rainer Linder, Felix Hett and Felix Garbe concludes the debate on transit issues by examining often-overlooked and misunderstood Belarus–Russia relations. There are regular crises in the framework of these relations. Belarus is an interesting case study given its particular position in the post-Soviet space. Belarus is very often presented, primitively, as the most loyal country to Moscow. The reality is much more complicated, and energy relations are part of the complex relationship between Moscow and Minsk. Belarus' experience represents the antithesis of the 'energy weapon' argument, as bluntly described in the mass media. Indeed, the authors wonder why Belarus should be 'punished' given its loyalty to Moscow?

The authors resolve this conundrum by elaborating three arguments. First, under Putin Russia carried out a dramatic reappraisal of Belarus. It has gradually dropped its value-driven policy toward Belarus for a more economic rationale. Second, given its location and its regime, Belarus remains an object of potential rivalry between Russia and the West. Rejecting NATO and shunned by the EU, Belarus has been saved from complete isolation only through its partnership with Russia, and bilateral relations with distant countries such as Venezuela and Libya. Third, energy relations should be examined within the context of the two countries' broader economic and political relations. Belarus directly benefited from Russia's growth throughout the period, but was also keen to preserve its independence where possible.

The authors describe the historical background of the relations between both countries. Putin initially continued Yeltsin's Belarus policy; in January 2000, the Treaty of the Union State was ratified. However, Putin insisted in giving more consideration to Russia's economic, geographic and political preponderance, and strove to improve integration within parts of the post-Soviet space. Lukashenko also had a policy: to maintain his personal power. The authors explain how this policy has been most consistent, and successful to some extent. Added to this, Belarus has balanced its approach towards Russia. On the one hand, Lukashenko courts Russian diplomatic support for the Belarusian regime to avoid complete international isolation, having been blacklisted by the West. At the same time, the regime profited from economic privileges granted by Russia, which contributed to political stability and to secure Lukashenko's personal power. On the other hand, he could be critical towards Moscow, and did not hesitate to lambast Russia publicly. For instance, Lukashenko has attributed Minsk's failure to recognize South Ossetia and Abkhazia to the fact that Russia refuses to help mitigate any damage Belarus might suffer should it do so.

The Belarusian business model is directly connected to the Russian economy even if Minsk has tried to diversify its exports (mainly refined and petrochemical goods) to markets in the EU. From 2000 to 2008, the Belarusian economy benefited strongly from Russian growth and from its exports to the EU. Part of this success can be explained by the fact that Belarus was turned into an 'offshore' tax haven for Russian businesses. Being highly energy intensive, the Belarusian economy remains dependent upon low gas prices. For a long time, Belarus benefited from gas prices equal to those of Russia's internal market. According to the authors, Moscow has given these privileges to Minsk mainly to reduce its own dependence on the Ukrainian transit. Like Ukraine, Belarus is a transit state, upon which Gazprom relies to fulfill its long-term obligations. This is why Gazprom has used various crises to gradually gain control of Beltransgaz, the Belarusian transport utility.

The authors describe in detail the consequences of the different crises between both countries. They conclude by saying that this relationship reflects Moscow's attempt to reconstruct relations with the states of the post-Soviet region 'on pragmatism and devoid of ideology'. Moscow has been partly successful, having gained access to the distribution networks. However, the maximization of its profits may destabilize a regime which remains pivotal for its strategic stability on its Western flank.

The basic point is that Belarus has already played its best cards during the last decade. In fact, the completion of the North Stream will dramatically reduce its importance in terms of gas transit in the coming years. The trend is a dwindling bargaining power for Minsk, quite aside from the uncertainties related to the unclear succession of Lukashenko. Integration projects between Russia, Kazakhstan and Belarus need to be followed very carefully to try to anticipate the coherence of Russian energy policy towards the European market.

There is no chapter dedicated to Ukraine. There is a practical explanation for this. Due to the crisis between Russia and Ukraine, it was difficult to reach experts able to involve themselves in a medium-term editorial project. Moreover, as the main transit country, Ukraine is in itself a topic already well researched.[11]

Part I

Energy's link to domestic and foreign policies

1 Energy security in Eurasia

Clashing interests

Frank Umbach

The consumer's predicament

> The rise of resource nationalism and the global search for energy resources is forcing countries to rethink their foreign policies and their alliances. Nowhere is this impact more evident than in the European Union, which has struggled to meet the twin demands of energy security and climate change, while contending with a resurgent Russia.[1]

In November 2000, the European Commission warned in its first *Green Paper* that in the next 20 to 30 years up to 70 per cent of the Union's energy demand (presently 50 per cent) will need to be imported. With regard to oil, the EU's dependence could reach as much as 90 per cent, and 100 per cent for coal.[2] In 2006, the EU-27's total primary energy supply was generated by oil (37 per cent), gas (24 per cent), solid fuels (18 per cent), nuclear energy (14 per cent) and renewables (7 per cent). Future new capacity will be predominantly generated by fossil resources, especially natural gas.[3]

The expansion of natural gas as an environmentally clean energy source is widely seen as the most problematic factor in the next two decades for the supply security of the EU member states.[4] Europe is already the largest market for natural gas import and will continue to be the world's top gas importers until 2030. In 2006, nearly half of the EU's gas consumption was imported from only three countries: Russia (23 per cent), Norway (14 per cent) and Algeria (10 per cent). Together, these three countries alone cover approximately 75 per cent of EU gas imports in 2006 (Russia alone representing 42 per cent).[5] Some new EU members, as former allies of the Soviet Union, are considerably more dependent on gas imports from Russia's monopolistic gas exporter Gazprom, with some importing the entirety of their gas via Gazprom.

In order to cope with rising energy insecurity and global challenges, the EU has adopted an ambitious 'integrated climate and energy policy' in March 2007 under the German presidency, with a 20–20–20 per cent formula in its 'Energy Action Plan' (EAP) that aims to reduce greenhouse gas emissions, raise the share of renewables and improve energy efficiency and conservation.[6] It is worth noting that these decisions, taken in March 2007, were largely a consequence of the Russian-Ukrainian gas conflict of January 2006 and that they had been negotiated within the EU just in a year.[7]

The EU's future security of supply is also supposed to be strengthened through active diversification of oil and gas imports. Hence, for instance, the EU's proclaimed new Central Asia strategy (also a product of the German EU presidency, declared on 22 June) aimed at intensifying energy cooperation with Central Asia and the Caspian Region (CACR).[8] Nevertheless, in this the EU will have to compete with Russia, China, Iran, India and the US.

While the EU has strived to liberalize its internal energy market, Russia has moved in the opposite direction. These opposing policies have complicated and even hindered substantial progress since a 'strategic energy partnership' between these partners was declared in 2000. They could also come to a head in light of the EU's declared energy and foreign policy interests in the CACR, which may become another theatre of competing interests between the EU and Russia.[9] While bearing this in mind, it is also important to note that Russia's more assertive energy and foreign policy since 2000 is not the only reason why energy security has become a top priority of the European and international policy agendas.

Between 2000 and 2008, international oil prices quintupled to US$ 147 – until crashing brutally at the end of the summer. In November 2008, the International Energy Agency (IEA) warned in its new annual *World Energy Outlook* that the world's energy system 'is at a crossroads' and that current worldwide trends in energy supply and consumption 'are patently unsustainable' for environmental, economic and social reasons. As a consequence, the IEA demanded nothing less than a 'global energy revolution'.[10]

For the first time in history, the rising demand for energy in emerging economies such as China and India coincided with mounting uncertainties concerning long-term oil and gas reserves, and thus how many resources would really be available on the future global market. Hence the global energy price and supply crisis has been very different from those in the past, and further complicated by a number of other factors.[11]

First, the economic rise of Asia (above all, China) has not only created enormous regional demand, but has also raised numerous foreign and security issues concerning both regional and global stability. In the view of many Western experts, China's appetite for energy, and its resulting energy foreign policies in Africa and Iran, have complicated, if not outright undermined, Western strategies for conflict prevention, management and sustainable development funding in Africa and the Middle East.[12]

Second, recent research has indicated that rising oil prices have triggered a new wave of nationalizations in the oil and gas sector, particularly in 'petro-authoritarian' countries with weak political institutions (see Chapter 2). The more valuable oil and gas assets are, the stronger the incentives seem to be to expropriate them. Nationalization and cartelization trends have in turn contributed to higher oil prices, since state control of production almost inevitably leads to inefficiency and corruption which raise export prices. Instead of investing in the exploration, production and refining sectors, the revenues from state-owned companies are more often than not diverted from social welfare programmes and government operating budgets to arms purchases and foreign aid. In this twenty-first-century-style nationalization, governments mandate that oil and gas projects turn over majority stakes to the state-owned company, though they tend to allow foreign investment in minority stakes and with no real control over the business.[13] While a new 'Great Game' over energy resources is no longer confined to Central Asia as was the original Great Game of the nineteenth century between the British and Russian empires, the game's theatre and the numbers of players have increased at a time when new rules of the game are still being written.[14] As a result, the gap between energy supply and demand may widen. Political factors are also increasingly determining the access to oilfields in Africa, the Caspian Basin and the Middle East.

The various definitions of energy security

Since the end of the 1990s, international energy experts have stressed the increasing strategic importance of supply security in the triad of goals to which economic efficiency and environmental compatibility also belong. However, 'energy security' has long had a different meaning depending on the perspective; the concept of energy security varies between the producer, consumer and transit states. Whereas consumer nations are primarily interested in security of supply, producer countries are more focused on security of demand from foreign markets. Transit states, for their part, are often equally interested in their own national security of supply *and* security of demand from neighbouring markets in order to benefit from stable and ever higher transit fees. To complicate matters further, the final concept of 'national energy security' also depends on the individual countries' geographical location and domestic policies, as well as the traditional state, economic and business ties it entertains with its partners.

The challenge and opportunity in the new and global 'Great Game' lies in avoiding misunderstanding and in recognizing mutual interdependence between producer and consumer nations as well as between their governments and companies.[15] However, it remains doubtful whether the presently difficult consumer–producer relationship is solely or primarily a problem of misunderstanding and insufficient communication as well as lack of information and not one of a fundamental conflict of power and competing strategic interests, influenced heavily by domestic policies, foreign policy and geopolitical ambitions.

This chapter will focus on the EU's projected energy security in the face of competition from Russia's and China's own concepts. Concretely, this will focus on these actors' energy policies toward the CACR, since the region represents a rare opportunity to diversify oil and gas imports. Several reasons account for this: (1) the CACR contains large quantities of both oil and gas; (2) Russia's increasingly eastward geopolitical and energy orientation and political instability in the Persian Gulf encourage diversification of supplies; (3) although the main global consumer countries are struggling for influence over the Caspian's resources, they are still not under the domination of a single major power; (4) the newly emerged Caspian states are eager to export their resources in exchange for the capital necessary for economic development; and (5) the EU does not invoke the same fears in the CACR as does encroaching Russia, the US or China.

Bearing this in mind, this chapter will examine the role of the CACR in global energy demand in light of the various definitions of 'energy security' that exist between producer, consumer and transit states as a way to understand the increase in clashes between the EU, Russia and transit states. It will also seek to shed light on the EU's own difficulty in developing an actual and effective common foreign energy policy.

Clashing interests in the CACR

Recognizing that it has no real alternatives in the short- and mid-term other than to cooperate with energy producer and exporter states – even authoritarian ones – in the CACR,[16] the EU has identified the region as an important source of diversification for its future oil and gas imports (particularly the latter). Though Central Asia cannot replace the Persian Gulf in satisfying the world's total crude oil demand, regional production is not negligible: the CACR's combined proven reserves of oil and gas are estimated at up to 3.9 per cent (see Table 1.1), or 4.3 per cent of the world total.

Global gas reserves are even more concentrated than are oil reserves. Thus Russia, Iran and Qatar – often seen as a potential 'gas cartel' – together control more than 55 per cent of global gas reserves.[17] Even excluding Iranian and

Table 1.1 Proven oil and natural gas reserves in Central Asia and the Caspian Region (CACR) in comparison with the EU-27, US and the Middle East, 2008

Country	*Proven oil reserves, thousand million barrels (% share of global reserves)*	*Proven natural gas reserves, tcm (% share of global reserves)*
CACR		
Azerbaijan	7.0 (0.6%)	1.28 (0.7%)
Kazakhstan	39.8 (3.2%)	1.90 (1.1%)
Turkmenistan	0.6 (>0.05%)	2.67 (1.5%)
Uzbekistan	0.6 (>0.05%)	1.74 (1.0%)
Total	48.0 (>3.9%)	7.59 (4.3%)
Iran	138.4 (11.2%)	27.80 (15.7%)
Russia	79.4 (6.4%)	44.65 (25.2%)
EU-27	6.8 (0.5%)	2.84 (1.6%)
US	29.4 (2.4%)	5.98 (3.4%)
Total Middle East (incl. Iran)	755.3 (61%)	73.21 (41.3%)

Source: British Petroleum. BP Statistical Review of World Energy 2008. June 2008

Russian gas reserves, the CACR's proven gas reserves, at 7.59 trillion cubic metres, are more important than those of Saudi Arabia. In 2007, the region's natural gas production totalled 163.5 bcm (billion cubic metres), or 5.5 per cent of the world total. While Kazakhstan and Azerbaijan lead in oil production (and together account for 92 per cent of the region's total proven oil reserves), Turkmenistan and Uzbekistan are the major natural gas producers in the region (representing respectively 40 per cent and 27 per cent of the region's proven reserves). As far as Europe is concerned, the CACR cannot replace Russia as an energy partner, though it could be an important *supplementary* supplier and provider of diversified resources, especially as far as natural gas is concerned.

Although the five Caspian littoral states have not yet settled the legal division of the Caspian Sea (and therefore of its sub-sea resources), a trilateral agreement on sub-surface boundaries and collective administration was reached between Russia, Azerbaijan and Kazakhstan in May 2003. The full development of the region's oil and in particular gas reserves is nevertheless hampered by inadequate export infrastructure, disagreements over new export routes (mainly with Russia), unresolved border disputes between the littoral states and regional instabilities.

With regard to future regional stability, the five Central Asian states (Tajikistan, Uzbekistan, Kyrgyzstan, Turkmenistan and Kazakhstan) and three Caucasian states (Azerbaijan, Armenia and Georgia) alike face the threat of religious extremism, authoritarian rule that neglects human rights and mismanaged economies. Kyrgyzstan's revolution in March 2005 and the violent unrest in Uzbekistan in May 2005 underlined growing discontent at economic problems, the continued suppression of dissent and a return to authoritarian tendencies. Further upheaval in the region is even more likely; the region also remains geopolitically incoherent, given the states' 'multi-vectored' energy and foreign policy orientations.[18]

Although the EU's 'Baku Initiative' of November 2004 envisaged enhanced political dialogue for energy cooperation between the EU and the countries neighbouring the Black Sea and Caspian Basin,[19] the EU has remained hesitant toward the CACR until 2007. Its policies have been criticized as being too fragmented, project-driven, focusing mostly on technical assistance instead of seeking a more strategic and long-term perspective of cooperation with the region's countries.[20] Furthermore, the EU's focus on strategy based on regional rather than bilateral cooperation needs to be modified, if only because cooperation between CACR countries has not been forthcoming, given their own divergent economic, energy, foreign and security policies and strategic orientations. The EU's demands on human rights and political and economic reforms have also been considered by these regional governments as an infringement of their sovereignty. As a result, the EU enjoys only limited influence compared to Russia, China, India, Iran and other powers competing in the region.[21]

Despite its failing policy, the EU's energy security has benefited – mostly thanks to US policy in the region rather than its own – from the opening of the Baku–Tbilisi–Ceyhan (BTC) oil pipeline in 2006, designed to carry 1 million barrels per day (a little over one per cent of the world's daily consumption) and a parallel gas pipeline from Baku through Tbilisi to Erzurum in Turkey (the South

Caucasus Pipeline, or SCP).[22] An east–west axis of new pipelines to Europe also constitutes an important aim for Central Asian states themselves, as it would reduce their dependence on Russia while strengthening their own energy security – that is, enlarging the number of potential buyers.

Despite this, until 2007 the EU has played only a small role in CACR through its TACIS and INOGATE programmes. Hence Russia has been able to hold on to a mostly unchallenged monopoly on transit routes to consumer countries. Europe, as a consumer, has in fact been slipping into a dual dependence on Russia: it is dependent both on the country's supplies and on Russia as a transit country for CACR oil and gas. Indeed, almost a third of the EU's total gas imports already come from the CACR, via Russian gas pipelines or as a result of Russia's gas swap deals with countries of CACR. In 2005 and 2006 Gazprom purchased 45 billion cubic metres (bcm) of Central Asian gas rather than developing its own Arctic and Siberian fields.[23] In 2007, it purchased 60 bcm.[24] Russia aims to increase these imports from CACR countries to 90 bcm in the coming years.

In November 2006, the EU finally began strengthening its energy relations with Kazakhstan. Two agreements provided for exploring possible energy supplies to Europe as part of efforts to diversify European supplies and cooperation in the peaceful use of nuclear energy, with Kazakhstan as the world's third-largest uranium producer. Hitherto, Kazakhstan had accounted for just three per cent of uranium deliveries to EU nuclear facilities.[25] In mid-November 2006, the EU had signed new neighbourhood agreements with the three South Caucasus states (Georgia, Armenia and Azerbaijan), while Poland and Ukraine agreed to extend the Odessa–Brody pipeline to the Polish city of Plock and carry Caspian oil to Poland and Europe. Finally, in preparation for Germany's EU presidency in the first half of 2007, German Foreign Minister Steinmeier became the first foreign minister of any EU member state to visit all five CACR countries (Kazakhstan, Uzbekistan, Turkmenistan, Kyrgyzstan and Tajikistan), which he did from 30 October to 4 November 2006, with the aim of identifying common strategic interests as well as expanding German and European energy ties to the region.

On 22 June 2007, the European Council declared a new Central Asia strategy that strives to intensify the EU's energy cooperation with the CACR.[26] However, EU member states were left facing a dilemma: the EU's foreign energy policy will further complicate either EU–Russia relations or its diversification strategy as embodied by the proclaimed aim of importing oil and gas from the CACR.

Several factors, however, mean that the EU cannot abandon its goal of diversifying its imports of natural gas from CACR. First, the Kremlin is exploiting energy dependencies as a tool of foreign policy. Second, Russia alone cannot meet the EU's forecasted natural gas demand until 2030.[27] Finally, in light of the EU's wider foreign and security interests in the CACR, Russia has little alternative but to pursue increased ties with the region's countries. Besides the region's energy resources, CACR is also a conduit for drug-smuggling and organized crime. What is more, militant Islamic fundamentalism is threatening the region's most secular states, and there cannot be a lasting peace in Afghanistan without regional stability. As the core and hub of economic integration of the Eurasian landmass, the CACR will inevitably have a key role to play in the future development of a transcontinental transport infrastructure linking Europe with Russia and Asia for the rapidly expanding transcontinental Eurasian trade and capital flows.[28] In other words, the EU has many strategic interests in the region other than energy that may increasingly affect its own security and economic future.

Yet the EU's growing energy as well as other economic and political ties with CACR are difficult to maintain for several reasons. First, the EU remains unwilling to confront Russian policies in the region due to the individual interests of its member states. Second, it has no choice but to take into consideration its crucial energy ties with Russia. Third, European public opinion is critical of the region's overall floundering respect for human rights, democratization and good governance. Fourth, the understanding of both security and energy policies within the CACR is very much state-centred in contrast to the EU and its believe in 'market policies' and the separation of economics from politics.[29]

Facing China

China, as the world's most populous country, will require ever more energy for its economy. This in turn will have a long-term influence on Beijing's own foreign and security policy as well as on regional and global energy supplies. Energy security has always been a very sensitive subject in China, where self-sufficiency became a mantra during Mao Zedong's rule. Still today, the Chinese government regards its rising energy imports as a 'strategically vulnerable resource':[30]

> After all, with a widening gap between domestic supply and demand, energy is being recognized as a core national interest among China's national security apparatus. As such, energy security is not only economically vital, but also has political, diplomatic and military implications. The legitimacy of the Chinese Communist Party is largely based on rapid and sustained economic growth. That is why China's top leaders have been paying full attention to this issue and becoming actively involved in energy diplomacy toward Russia and states in the Middle East, Central Asia, West Africa and even Latin America.[31]

As Chinese economic experts explain, China needs an annual GDP growth of at least six per cent to solve the rising social-economic problems inside its country. In order to achieve this rate of growth, however, China requires a sufficient and uninterrupted flow of oil and, increasingly, natural gas. Accordingly, China's definition of energy security is closely linked with domestic stability and, ultimately, the survival of its authoritarian one-party regime.

Given the global energy demand and the potential vulnerability of shipping routes through the Indian Ocean to the US Navy, the Central Asian region with its eight successor states of the former Soviet Union has gained particular importance for Beijing as a strategic land bridge between the Middle East and the PRC. This is true not only for its energy policy but for its broader security policy as well.[32]

Furthermore, despite increasing energy cooperation with Russia, Beijing does not wish to become excessively dependent on Moscow. Beijing remembers all too well how, after the break with China in the late 1950s and 1960s, Russia not only withdrew its nuclear weapons engineers but also sent home its oil and energy experts and ceased providing technical support, which at the time led to serious problems for China's energy supply.[33]

Given China's increased energy and security ties in Central Asia, Russia fears that the constantly declining strength of its military since the mid-1990s, combined with the CACR's sparse population and underdeveloped economies, will make the region vulnerable to rising Chinese hegemony – this in Russia's backyard. Yet both China and Russia still give priority to counteracting US hegemony. Thus China has been careful not to challenge Russia's regional leverage with too much zeal. It has tried to counterbalance the negative impacts of its increased ties to CACR by cooperating with Russia in the Shanghai Cooperation Organization (SCO) and other regional organizations as well as bilaterally with Moscow.

Russian experts can nevertheless hardly ignore China's increased energy and foreign policy ties to the region, in particular with Kazakhstan, since the latter follows 'multi-vectored' energy, foreign and security policies by maintaining good relations with Russia, China and the US simultaneously. Indeed, bilateral trade between China and Kazakhstan tripled in the period 2002–5, now representing two-thirds of all Central Asia–China trade.[34]

As a relative latecomer, China has had to invest in Kazakhstan's more isolated fields, particularly in the Aktobe region and along the route of the China–Kazakhstan oil pipeline presently under construction. In less than a decade, China has quietly managed to produce approximately 24 per cent of the Kazakh oil output. When the third and last section of this transnational oil pipeline is completed, it will secure about five per cent of China's total oil imports. This could easily double in the years ahead by expanding capacity.

Furthermore, a 2,000-plus km and US$ 7 billion gas pipeline, originating in Turkmenistan, crossing both Uzbekistan and Kazakhstan and ending in Xinjiang, is also being built with a capacity of 30 bcm per year. It is scheduled to be operational at the end of 2009. It is expected that the pipeline's capacity along its Kazakhstan–China section will be increased beyond its presently agreed 10 bcm per year, which has caused concern at Gazprom (see Chapter 6).[35] The whole Turkmenistan–China gas pipeline was inaugurated on 14 December 2009 during China's President Hu Jintao's visit to Turkmenistan with the leaders of Turkmenistan, Uzbekistan and Kazakhstan. The Turkmenistan–China pipeline is the first pipe to bring Central Asian natural gas to China and highlights China's quest for Central Asian energy exports. Furthermore, China is also interested in negotiating uranium imports from Kazakhstan and exploiting its hydroelectric potential. In April 2008, both sides signed an agreement to develop more than 40 projects for exploiting natural resources.[36]

China has also shown interest in reviving Unocal's 1997 project of pumping gas from Turkmenistan through Afghanistan and Pakistan to India (TAP or TAPI). Furthermore, China's construction of the Gwadar harbour in Pakistan, which could serve as a terminal for Iranian oil and gas exports to Pakistan and beyond, has also been linked to a northbound oil pipeline in Pakistan which could be extended to China.[37]

As both producers and transit states, Central Asian governments, for their part, seek to maintain good relations with all potential customers. This is a key component of both their energy and national security policies and is reflected in their broader foreign policy concepts. Enhanced energy cooperation with China represents for them a means of becoming less dependent on Russia and increasing their ability to diversify their Asian clients. This in turn enhances their leverage in dealing with Russia, China, the EU and the US, as they are able to play one against the other, and increases their influence in the region. This is particularly true of Kazakhstan.

The European theatre after the Russia–Georgia war

Given its tendency to think in zero-sum terms, Russia has regarded the EU's June 2007 strategy paper on Central Asia as a threat to its own geopolitical and energy interests in the region. After the March 2007 summit, therefore, Russia intensified energy cooperation and pipeline projects with individual EU and CACR states. This in turn threatened the EU's strategy for diversification and cooperation with Central Asia.[38]

From the Russian perspective, embodied by the Kremlin and Gazprom, the most dangerous EU project undermining Russia's pipeline monopoly from CACR to Europe is the Nabucco gas pipeline project, the EU's major project for a 'southern corridor' (see Chapter 10).[39] This 3,300-km pipeline, by running from Azerbaijan via Georgia to Turkey and from there to Austria via Bulgaria, Romania and Hungary, would bypass Russian territory entirely. According to the original plan, Nabucco ought to be built between 2009 and 2012 and is designed to have a capacity of 31 bcm per year, or approximately 10 per cent of all EU-27 gas imports.[40] The pipeline is expected to be operational by 2015.

In order to undermine the Nabucco project, however, the Kremlin and Gazprom have proposed an alternative pipeline called South Stream. It was first announced on 23 June 2007, with the participation of Italy's Eni – only three months after the EU's Caspian summit in March and the same month in which the EU proclaimed its new Central Asia strategy.[41] In contrast to Nabucco, which is to be an overland pipeline, a large section of South Stream is to be built below the Black Sea (considerably deeper than the Baltic Sea, where the Nord Stream pipeline is being built) to Greece. From there the pipeline would split, one spur going to Italy and the other northward to Austria, parallel to Nabucco. Due to the high costs of building South Stream's sub-sea section, the project's overall costs (officially €12.8 billion but possibly up to €20 billion) are believed to be twice those of Nabucco (up to €7.9 billion).[42]

Austria's state-dominated OMV, also a member of the Nabucco project, signed agreements with Gazprom in May 2007 and January 2008 to take part in South Stream, simultaneously pledging to turn the Baumgarten gas transmission centre near Vienna (the third-largest in continental Europe with a capacity of 16–18 bcm of mostly Russian gas) into a joint venture with Gazprom. If this project came through, it could deprive Nabucco of a final outlet, thereby threatening Nabucco's strategic rationale.[43] Nevertheless, Nabucco is the only EU pipeline project to have received the status of high priority infrastructure project by all of its 27 member states, including Austria, in the EU's Energy Action Plan of March 2007.[44]

Yet the fate of South Stream, like Nabucco, is also uncertain because it also lacks gas.[45] Given Russia's gas shortage, Gazprom may be able to supply only 10 bcm of South Stream's annual capacity, while 20 bcm would need to come from additional gas supplies, either from CACR countries or from the diversion of existing transit routes.[46]

Furthermore, Romania has twice explicitly ruled out participating in Gazprom's South Stream pipeline project, in January and November 2008,[47] despite increased pressure from both Russia and Gazprom.[48] Instead, the Romanian government has reaffirmed its full commitment to the Western-backed Nabucco. Even if Gazprom disregards Romania's inclusion in the South Stream pipeline, the construction of the pipeline on its presently designated seabed route from Russia to Bulgaria would require the consent of Ukraine and Romania or Turkey due to the route crossing their Exclusive Economic Zones (EEZ). Although coastal countries such as Ukraine, Romania and Turkey have no official veto to block the pipeline project, they can nevertheless require that extensive studies be conducted on the impact of the pipeline on environment, fishing, shipping and maritime safety. Following Gazprom's experience with Poland, the Baltic states, Sweden and Finland over Nord Stream, it is clear that coastal and transit states have the potential for considerably delaying a pipeline's construction and for raising its overall construction costs. Many European diplomats as well as representatives of those energy companies involved in the South Stream and Nabucco projects claim that both pipelines will be required to meet the EU's energy needs. This seems more of a political/diplomatic response than one based on sound economic planning, however, given new forecasts of the EU's impending gas demand. Furthermore, both projects remain uncertain, since it is unclear where the gas should come from (see Chapter 2). Russia, already confronted with a shortage and possibly a gas crisis,[49] does not alone have enough resources available to guarantee throughput for both South Stream and Nord Stream – neither in the short- nor in the mid-term future.

Collateral damage

The August 2008 Russia–Georgia war[50] also raised new questions concerning the EU's aim of closer energy cooperation with the CACR, since the war clearly highlighted the vulnerability of Western-funded Caspian export pipelines avoiding Russian and Iranian territory.[51] An important and often overlooked aim for the Kremlin during and even before the war was to discredit Georgia's role as an important transit state, thereby reinforcing its own supply and transit monopoly from the CACR. Even if the pipelines were not openly targeted, the war nevertheless forced BP to shut down oil and gas export routes from Azerbaijan to Georgia

(Baku–Supsa and SCP). Though the Baku–Tbilissi–Ceyhan (BTC) oil pipeline itself was not hit by Russia, it was also closed shortly before the war began due to a mysterious explosion on its Turkey section (5 August). Should Russia have purposely targeted the BTC, the West would have had the clear proof that Russia's war aims went beyond the official argument of self-defence.

Although Russian President Medvedev has claimed the CACR its 'privileged' sphere of influence, the overall effect of Russia's military intervention on Western plans for pipelines crossing Georgia – and the overall perception of Georgia as an increasingly important transit state – were not dimmed for long, and Russia's success was, in fact, short-lived.[52]

This has been visible, first of all, with regard to the 'European theatre' and energy rivalry. For example, the European Council, showing rare unity and resolve at its emergency meeting on 1 September, called on the EU Commission to undertake new initiatives to diversify its energy supply and transport routes in cooperation with the European Commission.[53]

The Russia–Georgia war, like the Russia–Ukraine 'gas wars', has in fact helped to promote the EU's declared common energy and energy foreign policy. While Russian experts argued that the war further undermined the Nabucco project, due to increased doubts as to the safety of investments in the CACR, the EU has in fact increased its efforts at securing gas sources bypassing Russia. In September, a 'strategic-cooperation conference' in Baku, backed by the US, the EU, Turkey, Azerbaijan, Turkmenistan and Kazakhstan, again declared their commitment to the pipeline.[54] In addition, the European companies involved, including RWE on the German side, have also made clear that while commercial risks have indeed increased, they remain no different from most encountered in energy projects throughout the world.

Since Russia's attempts to preserve its claimed 'privileged' sphere of influence now include war as a policy tool, entirely new concerns of regional security are now bundled in European – and, indeed, Eurasian – opposition to Russia. As far as Western investors are concerned, the war has placed greater emphasis on pipeline and other energy infrastructure safety and security, thus increasing their costs. Both factors will play heavily in future energy and foreign relations. On a side note, it is worth pointing out that many experts, particularly from Azerbaijani, maintain Russia's real purpose in Georgia was not in directly controlling Georgia, but rather Azerbaijan, since it is the West's and Europe's main source of oil and gas supplies from the CACR.[55] In other words, the conflict's consequences far from stopped at Georgia's borders.

More worryingly for Russia in terms of security issues, while Moscow was hoping to gain support from its Shanghai Cooperation Organization (SCO) and Collective Security Treaty Organization (CSTO) allies, neither China nor the regional states of CACR declared official support for Russia's military intervention, nor did they recognize Abkhazia and South Ossetia's independence.[56] Such a lack of diplomatic support was also the consequence of Russia's selective use of the CSTO, SCO and other regional organizations: '... while sending its troops to Georgia without consulting its member states, it expected their approval after the fact'.[57] Russia became largely isolated diplomatically. Its only remaining allies have been the big European companies with large stakes in Russia's expanding consumer market.[58]

Even Russia's closest strategic ally in Central Asia, Kazakhstan, refrained from supporting Russia's stance, since this would threaten its own multi-vectored foreign and energy policies. Although Kazakhstan has abandoned its plans to build an oil and gas terminal in Georgia for the time being, the approximately US$ 2 billion investments it has made in Georgia's oil, gas and other industries over the past three years (as well as the huge investments and logistics opportunities the EU-backed Transport Corridor Europe–Asia represents for Astana) means that Kazakhstan will not easily give up its energy cooperation with either Georgia or Azerbaijan.[59] Beginning 1 November 2008, Kazakhstan in fact increased oil shipments from its super-giant Tengiz oil field to Baku and onward, via the BTC, to international markets. In 2009, Kazakhstan's oil input into the BTC will be raised from 70,000 barrels per day to 98,000 barrels per day in 2008. This signifies that the country's confidence in the Azerbaijan–Georgia transit corridor has not waned.[60]

Although Russia has managed to increase its energy cooperation with Uzbekistan, Kyrgyzstan and Serbia on a bilateral basis,[61] regional cooperation in energy and trade between Turkey, Azerbaijan and Turkmenistan – key actors in the Southern Corridor (see Chapter 10), already promoted before the Russia–Georgia war,[62] has only been enhanced after the war. Reportedly, the focus of the last trilateral cooperation at the end of November 2008 was even natural gas supplies for Europe and Nabucco. The talks also included the possibility of a TransCaspian pipeline, strongly opposed by Russia (see Chapter 4). In November, Azerbaijan had also signed a new five-year agreement to supply Georgia with gas. By doing so, it de facto rejected Russia's offer to buy all of its available gas volumes for exports to Europe.[63]

The trilateral meeting between Turkey, Azerbaijan and Turkmenistan at the end of November and new initiatives by the Hungarian government and the private MOL energy company, including plans for a Nabucco summit in January 2009, have also signalled that Russia's hopes for killing the Nabucco project with its war are more than premature and may have had the opposite effect. By promoting closer economic trade with Kazakhstan, which expects to produce some 70 bcm of gas by 2015, Astana has promised Hungary as the current main project driver future gas supplies for the Nabucco project, albeit conditional on the opening of a TransCaspian pipeline to Azerbaijan.[64]

Finally, the Russia–Georgia war has also enhanced Iran's role in the region.[65] In spite of EU and, especially, US opposition to further energy cooperation with Teheran, Ankara has heightened its cooperation with Iran by developing some of its gas fields and transporting production to Europe. Iran is in fact quite keen to supply gas for Nabucco or any other pipeline to Europe. Although giving Iran an increased role and influence in the region is presently not in the West's interests, allowing Iran to contribute to European gas supplies would help thwart Russia's ambition in the CACR. Such a move would also bring about increased rivalry between Russia and Iran, possibly straining their relations and hindering the creation of the announced 'gas troika' that includes them and Qatar (December 2008).

This has also all occurred while the 2008 financial crisis hit Russia, threatening Gazprom and other Russian energy companies' entire investment programmes, while severely limiting the Russian government's ability to support these projects financially.

A context changed

New international concerns, including global energy supplies and climate change, have progressively driven a wedge between the agendas of producers and consumers, fuelling resource nationalism, increasing competition for resources, stimulating public and private investment in alternative energies, and stoking disagreements over the post-Kyoto world.

Such a divide became apparent as oil prices climbed from 2003. The world's seemingly insatiable demand for oil and gas, combined with increased political instability in many producer countries, as well as the possibility of 'peak oil'[66] approaching, then tilted the overall balance of power between energy producers and consumers toward the former. Thus, while consumers tried to shift their energy mix by reducing future growth in demand, increasing energy efficiency and conservation, promoting clean fuels and alternatives, boosting domestic production and diversifying imports, producers focused on downstream activities, acquiring market shares in order to lock up future demand, in order to further influence prices. Yet the emergence of a 'seller's market' on a global level led to a profound change in the nature of competition between producer and consumer states as well as among consumers themselves.

In this context, the once sharp dividing line between foreign, domestic and economic policies is increasingly blurred. Global economic stability, fuelled largely by oil and gas, has become much more dependent on the domestic and regional stability of producer regions and trade routes. In the age of globalization, any policies that ignore detailed analyses of various domestic and regional stability factors in these regions will prove extremely shortsighted and will be paid with high financial, economic and political costs in the mid- and long-term future. The necessary symbiosis between energy security, in all its forms, and regional stability in fact requires new thinking.

Thus, for example, Russia's various oil and gas conflicts with its neighbours – themselves fuelled by domestically designed concepts of energy security – have focused the EU's attention on its own energy security – that is, security of supply.

In the face of global energy security issues, but also in the face of Russia's actions, the EU-27 for example announced the world's most ambitious 'integrated climate and energy policy' in March 2007. According to this document, the EU's energy security is broadly defined through a balance of three objectives: security of supply, economic competition and environmental/climate protection. Its 'Energy Action Plan' of 2007 agreed on three 20 per cent targets for increasing energy efficiency and conservation, the share of renewables and reducing greenhouse gas emissions by 2020. Furthermore, the EU's concept of energy security seeks to foster liberalized energy markets, broadening Europe's energy mix as much as possible by including the civilian use of nuclear power and clean coal consumption as domestic energy sources, and diversifying energy imports (in particular gas) in order not to become too dependent on a single supplier.

With every passing energy crisis of European countries with Russia, the EU has taken further steps toward defining and executing a common energy policy, as well as speaking to Russia with one voice. Although such progress is as yet insufficient, and the rather slow speed on this path has been and still needs to be criticized, Europe has begun learning from its mistakes. This, it must be added, is in no small part due to Russia's own actions. Indeed, in the light of the global challenges, the EU's responsibilities and the need to preserve its own strategic interests, the EU needs an active strategy and to be able to speak with one voice; few actors – within the EU or outside it – have contributed as much as Russia has to defining a common European interest, since Gazprom cut off gas deliveries to Ukraine in 2006.

Russia, for its part, has never seen its energy sources and pipeline policies as just a normal commodity – as did the EU until 2006. Traditionally, energy has always been considered, and wielded, as a key tool of Moscow's foreign policy, especially in Soviet times when several Soviet republics and former Warsaw Pact allies were entirely dependent on Moscow. This history also highlights the fact that Russia mostly lacks more benign instruments of influence or soft power, leaving it to follow more aggressive policies that damage its reputation. Whereas its understanding of energy security often seems defensive and reactive by nature, the result of it instrumentalizing energy policies and dependencies can only be considered by its neighbours in Eurasia as offensive, since it threatens their own energy security as well as economic and foreign policy sovereignty.

With such a prism, Russia has considered the EU's liberalization strategy, its diversification policy of energy imports as well as the new Central Asia concept of June 2007, as a major threat to its own understanding of energy security, and thus to its foreign policy and geopolitical ambitions, particularly in the CACR. This has been manifest in the region, where it seeks to preserve or expand its monopolistic status with regard to energy pipelines and other critical infrastructure, especially in the gas sector. Pushed further, Russia's prism and concept energy security is based on the (unrealistic) assumption that its regained power can lead to a new Yalta-type sphere of influence (between the EU, the US, China and others). This of course requires a strong Russia, while the sovereignty and independence of the CACR states are downgraded. Accordingly, Russia must defend its energy security interests abroad by all means, including coercive ones, and sees no interest in transparency. Within Russia, this means nationalizing its energy sector. In the CACR, it means exploiting transit dependencies. In Europe, it means playing nations against each other, taking advantage of the EU's lack of political unity. However, whether such a strategy is tenable in the long run is another question. Indeed, with its present mindset and concepts, Russia is focusing on short-term economic and diplomatic gains, to the detriment of its longer-term strategic interests. The situation is even more complicated in the CACR, since the region is also a focal point of Chinese energy security interests and foreign energy policy.

In November 2008, the European Commission's '2nd Strategic Energy Review' and its new 'EU Energy Security and Solidarity Action Plan' identified major weaknesses and problems which need to be overcome on the way to a real common energy (foreign) policy and by enhancing the energy supply security of its 27 member states.[67] It has proposed five key areas for joint cooperation and projects in the forthcoming years: (1) infrastructure needs and the diversification of energy supplies; (2) external energy relations; (3) oil and gas stocks and crisis response mechanisms; (4) energy efficiency, and (5) making the best use of the EU's indigenous energy resources. Should the EU implement its March 2007 decisions and its 2nd Strategic Energy Review package, it may drastically decrease its gas import demand, and Russia's energy policy, partly responsible for this, will have proven self-defeating. Rather, both the EU and Russia should recognize the mutual interdependence inherent in their relationship, and for this both sides need better information, understanding and communication channels to solve their conflicts and problems.

Prospects

Russia, however, is confronted with even greater immediate challenges, though these may ironically offer new perspectives for closer EU–Russian energy cooperation in the years ahead. Indeed, with oil prices falling in 2008 and the global economic slowdown of 2009, Russia's very economic and social stability is at stake. The state budget for 2009 was originally based on a crude price of US$ 95 per barrel – while prices had stabilized at around $US 30–40 at the end of 2008. What is more, the current gas price of $US 450–500 per mcm – indexed on oil prices at their peak – is estimated to decrease to around $US 280 per mcm by mid-2009. Should gas prices remain nearer $US 360–400 per mcm, Gazprom's profits would still drop by at least $US 20 billion in 2009. What is more, the Kremlin's original plans of aligning domestic gas prices with those of Europe (presently 15–20 per cent of what Germany pays wholesale for Russian gas) may spark civil unrest and even mass protests.[68] Worse, Gazprom actually expects its natural gas production for 2008 to be three per cent (or 15–18 bcm) lower than previously forecast. Much of the shortfall will have to be covered by imports from CACR, which Russia has already promised to purchase at market prices.

Without government intervention, Gazprom, already $US 55 billion in debt, will have major problems securing funds for investing in costly pipelines and exploration projects, such as Nord Stream and South Stream gas pipelines and the new gas fields in the Yamal region and Shtokman in the Barents Sea.[69] Gazprom has already begun construction of the Bovanenkovo–Ukhta gas pipeline, designed to transport gas from the Yamal peninsula, though much work remains. Gazprom estimates Yamal reserves at 16 trillion cubic metres, and plans to produce up to

360 bcm per year from the region. These new gas fields are to become the core of Russia's gas supply system in the next years, as well as the main throughput of the Nord Stream pipeline. Yet the investment required to put this region on stream – and rapidly, to compensate for the fall in its traditional fields – is massive, and the global credit context quite unfavourable.

Russia's energy investments are therefore likely to be met only through more governmental support.[70] Given the various industry bailouts, and combined with declining oil and gas revenues as well as the deep tapping into its stabilization fund to stabilize the ruble, will make it very difficult if not impossible to carry out the plans to modernize Russia's aging infrastructure, and promote its geopolitical ambitions abroad to reassert the regional dominance it lost with the Soviet Union's demise. In addition, Russia's looming demographic problems and the cost of the vast network of patronage required to maintain loyalty across Russia's vast regions and republics cast doubt as to whether the Russian state can afford such profligacy.

These problems are compounded by OPEC's December 2008 decision to cut its production by 2.2 million barrels per day. While Russia is not a member of OPEC, it nevertheless tends to coordinate its oil output with that of OPEC; as OPEC pledged to reduce its oil exports by 4.2 mb/d in 2008, Russia has announced that it will cut its own exports by around 300,000 b/d. Whether these decisions will help nudge oil prices upward to US$ 75 per barrel, as oil producers hope, remains to be seen, though given slumping demand this seems unlikely to happen in 2009.

Given such a conjecture, it will be decisive for both the EU's and Russia's future energy security whether the Kremlin and Gazprom will define their long-term strategic interests based on the recognition of a mutual interdependency and according cooperation strategies. They need to be focused on building trust, confidence and transparency in the views of their foreign consumers instead of questionable short-term economic and diplomatic gains. So far, however, few signs point to such a convergence of minds or interests.

2 Hydrocarbon production and exports in Central Asia

The impact of institutions and policies

William Tompson

This chapter offers a brief overview of the political economy of oil and gas production and export from Central Asia and Azerbaijan. It begins by situating the non-Russian Caspian hydrocarbon producers in the larger global market, before proceeding to examine the ways in which the Azerbaijani, Turkmen and Kazakh energy sectors have been organized and governed since the collapse of the USSR. The chapter then examines the likely consequences of recent policy shifts in Kazakhstan, the region's most important oil producer. This is followed by a brief conclusion.

The chapter's central argument is that there are good reasons to doubt whether the region's hydrocarbon potential will be developed in a timely, economically efficient way. While the impact of geology, geography and international price movements can hardly be ignored, policy-makers can do much to raise or lower the long-term elasticity of CIS supply. On present trends, policy in the region seems more likely to lower supply elasticity than to raise it. While slower development of Azerbaijani and Central Asian oil resources would have relatively little impact on Russia, Gazprom's reliance on Central Asian gas supplies to help balance its internal and external commitments means that Central Asian gas developments are a significant concern for Russia.

Caspian oil and gas in an international context

The Caspian states' long-term oil and gas supply potential must be assessed against the backdrop of growing global dependence on OPEC for oil and on Russia for natural gas in Eurasia. The IEA sees OPEC's global market share rising from somewhat under 40 per cent in 2002 to around 49 per cent in 2030, slightly below the historical peak recorded in 1973.[1] Non-OPEC reserves are being depleted faster than those of OPEC, and global oil reserves are more and more concentrated in a handful of OPEC states, chiefly those in and around the Persian Gulf. World oil supply will therefore depend increasingly on developments in countries where investment is not currently allocated by the market and where long-term political stability is far from assured. This, in turn, has implications for natural gas prices, which are linked to oil prices owing both to contracting practices in the sector and oil's potential as a gas substitute. For Eurasian consumers, moreover, reliance on natural gas supplies produced in, or transited through, Russia means that their energy futures will be heavily dependent on Russia's policies as well.

The Commonwealth of Independent States (CIS) is both the largest oil-producing region outside OPEC and, at present, the region with the largest concentration of non-OPEC reserves. It is also an increasingly important source of natural gas for Eurasian consumers (see Table 2.1). The Caspian CIS states will in no sense emerge as a real rival or alternative to OPEC in oil or to Russia in natural gas, but their supply potential remains important, for at least three reasons. First, oil and gas consumers will wish to maintain diverse supply channels. Second, the potential for instability or threats to supply within much of the OPEC area is considerable, as is concern in Europe about Russia's ability to use its position vis-à-vis natural gas supply and transit as a source of political leverage. Third, what OPEC and, with respect to Eurasian gas, Russia do as reliance on their resources increases will depend in large part on the elasticity of both non-OPEC/non-Russian supply and international demand.

Yet if CIS oil production matters far less than that of OPEC in absolute terms, the same cannot be said of its contribution to *incremental* supply. During the nine years to the end of 2007, CIS producers accounted for roughly 62 per cent of the increase in world output and all of the increase in non-OPEC output (Figure 2.1). The CIS share of global production climbed by almost half over the period, from 10 per cent to just under 16 per cent, while the former Soviet Union's (FSU) share of world

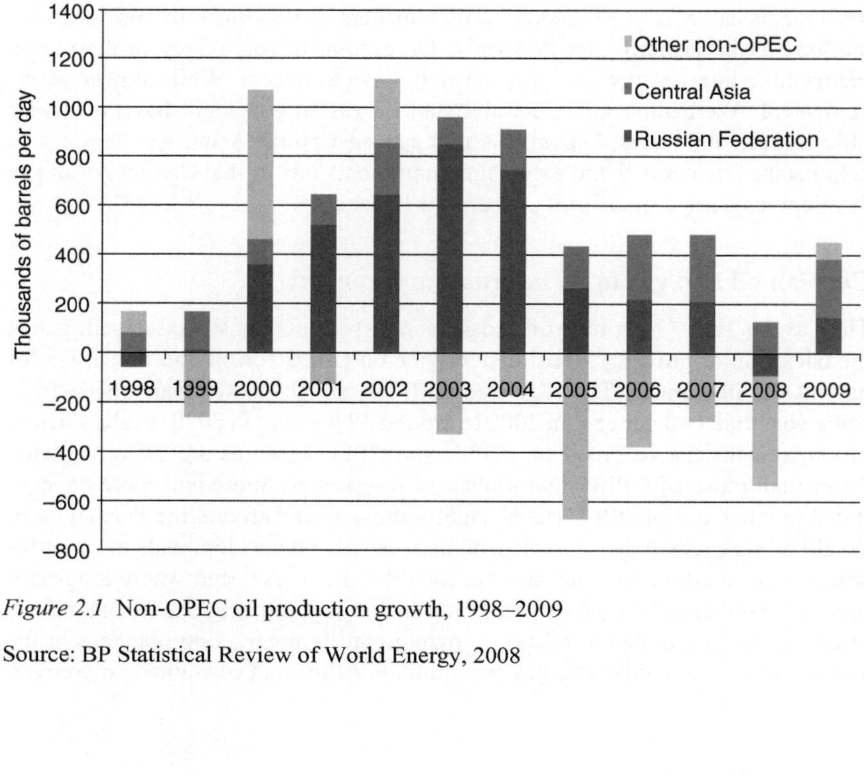

Figure 2.1 Non-OPEC oil production growth, 1998–2009

Source: BP Statistical Review of World Energy, 2008

Table 2.1 CIS oil and gas sectors, 2007

	Crude oil					*Natural gas*			
	Proven reserves		*Production*		*Exports*	*Proven reserves*		*Production*	
	bn bbl	*% of world*	*000 bbl/day*	*% of world*	*000 bbl/day*	*tcm*	*% of world*	*bcm/day*	*% of world*
Azerbaijan	7	0.6	868	1.1	721	1.28	0.7	10.3	0.4
Kazakhstan	39.8	3.2	1490	1.8	1200	1.9	1.1	27.3	0.9
Russian Federation	79.4	6.4	9978	12.2	7000	44.7	25.2	607.4	20.7
Turkmenistan	0.6	0.05	198	0.24	79	2.7	1.5	67.4	2.3
Uzbekistan	0.6	0.05	114	0.14	-56	1.7	1.0	58.5	2.0
Total CIS[1]	128.1	10.4	12804	16.0	8334	53.53	30.2	790.2	26.8

Source: BP. USEIA. OECD calculations

[1] Including other CIS producers. National export totals include intra-CIS trade but CIS total does not.

exports rose from 8.9 per cent to around 15.2 per cent. This contribution was critical at a time of rapid demand growth. Increases in CIS output during 1998–2005 came close to matching the combined consumption growth of the United States and China (roughly 4.2 million barrels per day), by far the two largest sources of demand growth over the period. The IEA, moreover, expects the non-OPEC share of world supply to continue growing to the end of the present decade, with Kazakhstan and Azerbaijan accounting for an increasing proportion of incremental CIS supply.

It is worth noting that Kazakhstan, alone among the CIS hydrocarbon producers, has a significantly larger share of estimated global oil reserves than of current production. These data, moreover, take no account of recent assessments suggesting that Kazakhstan's enormous Kashagan field may be roughly 25 per cent larger than previously thought.[2] Kazakhstan's relative weight in global oil supply is therefore likely to grow substantially. By contrast, unless reserves growth in Azerbaijan and Russia accelerates substantially over the coming years, their role as oil suppliers appears likely to diminish. The biggest question mark in the region hangs over Turkmenistan: there are grounds for believing that Turkmenistan's oil potential is considerably greater than current reserve estimates suggest. The same appears to be true of both Turkmenistan and Azerbaijan with respect to natural gas (see below).

The rapid recovery in CIS oil output in recent years, driven chiefly by the turnaround in Russian production, has been one of the most important developments affecting world oil markets (see Figure 2.2). From a peak of almost 12.7 mbd in 1987,[3] the region's oil output fell below 7.2 mbd in the mid-1990s. In 1996, however, a recovery began in Kazakhstan, followed by Russia, Turkmenistan and Azerbaijan.[4] By 2007, CIS output had exceeded the Soviet-era record. Russia accounted for about three-quarters of this increase (over 3.5 mbd), but Kazakhstan recorded the most impressive growth rates, and Azerbaijani output and exports skyrocketed in 2005 after the opening of the Baku–Tbilisi–Ceyhan (BTC) pipeline. The rapid expansion of Caspian output was not unexpected, in view of the ongoing development of substantial new fields by foreign consortia, but the Russian recovery was a tremendous surprise to most observers.[5]

Assessments of Central Asia's future supply potential vary widely. There are considerable differences in the methods and definitions used to estimate the volume of petroleum physically in place. Judgments about how much of this are commercially recoverable depend on assumptions about technological development, oil price trends and other variables. BP (2008) has estimated CIS proven reserves[6] at 128.1 billion barrels. Russia accounts for 60 per cent of the total, with 72.3 billion. It is followed by Kazakhstan with 39.6 billion.[7] This leaves the CIS with somewhat over 10 per cent of world reserves and more than 40 per cent of non-OPEC reserves. This compares with 60 per cent for the Middle East, including Saudi Arabia's 22 per cent. The CIS has, in recent years, been the only major region to record strong reserves growth, although this partly offsets very slow growth during the late 1980s and early 1990s (Figure 2.3). The situation with natural gas reserves, by contrast, has been roughly stable over the last decade, though there is reason to believe this could change in the coming years (see below).

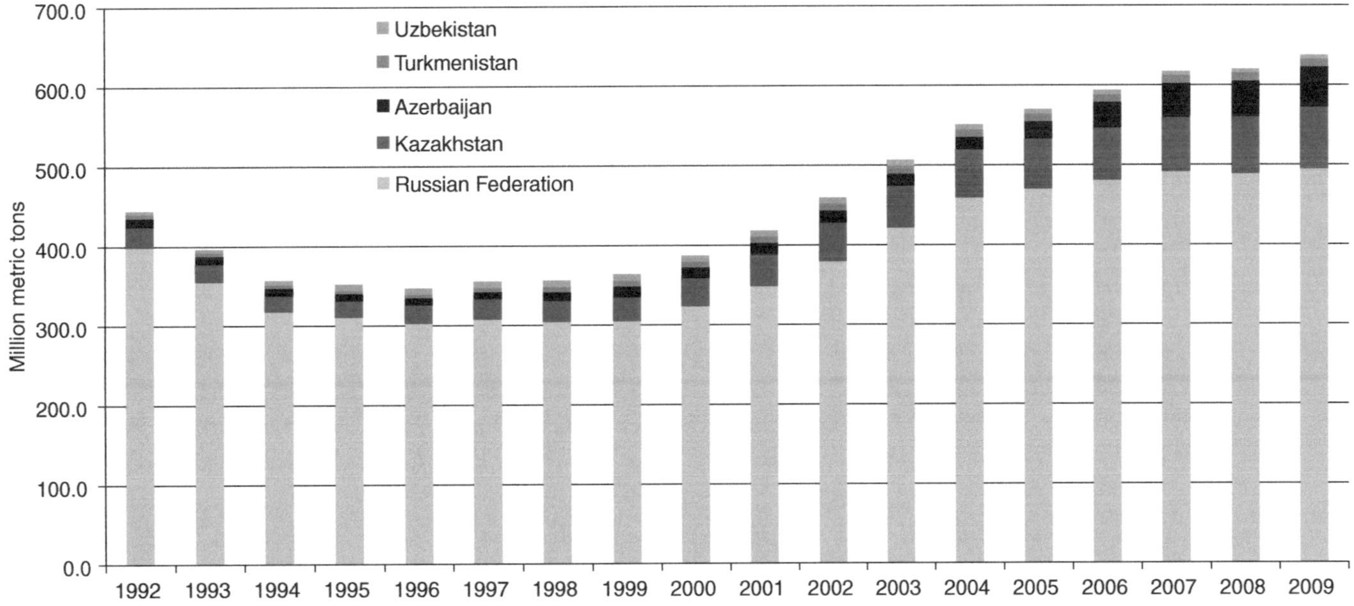

Figure 2.2 Production of crude oil and NGL, millions metric tons

Source: BP Statistical Review of World Energy, 2007

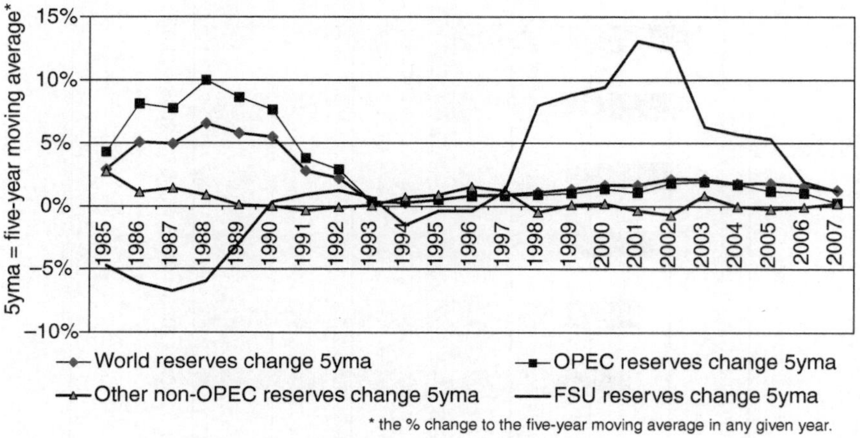

Figure 2.3 Change in oil reserves

Source: BP Statistical Review of World Energy, 2008, OECD calculations

The Central Asian share of both reserves and output could well rise further. Much of Central Asia, including promising regions of Kazakhstan, is still underexplored. In 2000, the US Geological Survey (USGS) estimated that the CIS states held 17.9 per cent of as yet undiscovered, technically recoverable oil in the world and 34.5 per cent of undiscovered natural gas.[8] These assessments were based on the mean probability for the USGS estimates, which cover a range of probabilities, from 5 per cent to 95 per cent. Significantly, the estimated potential CIS share is substantially larger than its actual 2007 share across the whole range of estimates for different probabilities.[9]

Turkmenistan may be the most interesting case. It is still underexplored, and the US Geological Survey (USGS) estimates it to be exceptionally promising with respect to as yet undiscovered oil and gas reserves. While the country's *proven* reserves are currently estimated at around 500 million barrels, the USGS mean probability estimate for *undiscovered* oil is 6.8 billion barrels.[10] At the very least, Turkmenistan could have oil production potential comparable to that of Azerbaijan, which, while far smaller than Russia or Kazakhstan, is nevertheless an important second-tier producer. Similarly, Turkmenistan's proven natural gas reserves, estimated at 2.7 thousand cubic metres (tcm) in 2007, are far smaller than the USGS mean probability figure of 5.9 tcm still to be discovered.

The USGS identifies Azerbaijan and Kazakhstan as the other Caspian states with significant untapped gas potential: its mean probability estimate for undiscovered gas reserves in 2000 was just under 2.5 tcm for Azerbaijan, enough to make it a gas producer on a scale comparable to Turkmenistan. The corresponding figure for Kazakhstan was roughly 2 tcm, slightly more than that country's existing proven reserves. The USGS assessment for Azerbaijan, moreover, looks increasingly plausible in light of industry estimates suggesting that the Shah Deniz gas field in the Azerbaijani sector of the Caspian may be more than double the size originally estimated by the project's operator, BP.[11] (Both Azerbaijan and Turkmenistan's estimated potential reserves were nevertheless dwarfed by Russia's estimated 32.7 tcm in undiscovered gas reserves.)

Ownership, control and taxation

During the 1990s, the hydrocarbon producers among the CIS successor states adopted a range of different strategies for governing and developing their fuel sectors. Patterns of ownership and control, as well as tax regimes, and attitudes toward both the extent and modalities of foreign involvement, varied widely.[12] There are, of course, certain problems common to the investment environments of all the Central Asian producers. These include a generally weak institutional framework, pervasive corruption, opaque and often changeable policy-making on the part of the authorities, and relatively high levels of political and economic uncertainty. Such problems are hardly unique to Central Asia. Much of the world's energy production originates in countries afflicted by similar problems. However, there is some evidence to suggest that, even by the standards of oil-producing countries, the Caspian region is a particularly difficult place in which to operate, particularly with respect to corruption.[13] In addition, the legacies of the Soviet past – including both physical infrastructure and institutions – and the problems of post-communist transition confront local and foreign oil companies with a unique set of challenges. Nevertheless, an awareness of the different paths taken by CIS producers is critical to understanding their development over the past decade and the challenges they now face.

At the beginning of the transition, the two most promising Caspian producers, Azerbaijan and Kazakhstan, needed foreign capital and technology to develop their oil and gas resources. In the immediate post-Soviet period, they were inclined to welcome a high degree of Western involvement, for political reasons as well as commercial ones – reducing their dependence on Russia was an important goal in itself. Subsequently, they also developed closer commercial links to Russia.

Azerbaijan

The newly independent Azerbaijan, the historic cradle of Soviet oil production, inherited a mature oil industry that was already in decline, and it required substantial new investment in order to refurbish existing fields and to develop large-scale new projects offshore. While maintaining full state ownership over its energy companies, Azerbaijan was quick to invite foreign investors to assume a direct role in the development of its hydrocarbon reserves. Most of the state's existing hydrocarbon assets were consolidated in September 1992 in the State Oil Company of the Azerbaijan Republic (SOCAR). While government ministries handle exploration and production agreements with foreign companies, SOCAR is party to all international consortia developing new oil and gas projects in Azerbaijan. SOCAR was created precisely in order to attract foreign players without compromising state control.

Azerbaijan has generally been the most consistently accommodating CIS state in its dealings with foreign investors, relying more than its neighbours on project-specific arrangements tailored to meet investor needs – major deals have virtually all been structured as production-sharing agreements (PSAs).[14] This investor-friendly approach attracted an influx of foreign investment since independence that has revitalized Azerbaijan's oil sector. To date, Azerbaijan has signed over 20 major field agreements with approximately 30 companies from 15 countries. Between 1995 and 2004 Azerbaijan attracted more than $11 bn in foreign investment, roughly 90 per cent of which went to its oil and gas sector. Crude oil output rose by just over 80 per cent, from a post-Soviet low of 180 thousand barrels per day (kbd) in 1996 to 868 kbd in 2007. The spectacular growth recorded in 2005–7 reflects the fact that production and export growth was constrained until the BTC pipeline began operations in the spring of 2005. Once it did, oil production and exports soared. Gas exports also began to grow rapidly after the South Caspian Pipeline (SCP) began operations; running parallel to the BTC for most of its route, the SCP has allowed Azerbaijan to begin exporting gas to Turkey.

Crude production growth since 1997 has mainly come from the international consortium known as the Azerbaijan International Operating Company (AIOC).[15] The AIOC operates the offshore Azeri Chirag and deep-water Gunashli (ACG) mega-structure, which is estimated to contain proven crude oil reserves of 5.4bn bbl. Not all foreign investment projects have been so successful, however. Several have announced disappointing drilling results in recent years, and a number of joint ventures and PSAs have shut down after failing to find the anticipated volumes of commercially recoverable reserves.[16] Indeed, there has been only one major new oil discovery since the 1994 agreement with the AIOC was signed. While production from the ACG structure is set to grow rapidly between now and 2010, it is also expected to have a relatively brief production plateau; in the absence of substantial new discoveries, Azerbaijan's oil production could peak as early as 2010, at about 1.5 mbd, before going into decline later in the decade.[17]

Kazakhstan

The existence of Kazakhstan's considerable hydrocarbon resources was well known in Soviet times, but the Soviet authorities were slow to develop them. After 1991, the newly independent Kazakhstan quickly recognized the need for foreign capital and expertise, in view of the relatively under-developed state of its most important deposits, the technical problems posed by some of those developments and the desire to reduce the country's reliance on Russia. It began courting foreign investors more or less immediately after independence.[18] Like Azerbaijan, Kazakhstan opted to rely heavily on tax and regulatory arrangements tailored to meet the needs of investors in large projects, whether in the context of PSAs, joint ventures or concessions. However, Kazakhstan proved less willing than Azerbaijan to operate on a case-by-case basis. While PSAs and concessions are negotiated as a matter of course in Kazakhstan, Kazakh contracts over time have come to be based increasingly on a body of law applicable to all projects. For investors in the 1990s, this was not a major impediment. The crucial factor was Kazakhstan's readiness to conclude contracts that would protect investors from subsequent changes in taxation or other policies. The tax regime was to be stable for the life of the contract. The question of contract sanctity and stability was central to Kazakhstan's drive to attract foreign investors.

Unlike Azerbaijan, Kazakhstan privatized most of its existing oil and gas enterprises – mainly to foreign owners and often with significant social commitments included in the terms of the sales.[19] Many small, onshore fields were taken over by private companies under various contracts. The fact that these private companies were often controlled by interests close to the government or the state-owned companies involved did raise questions about the terms of these transactions (which looked like a form of 'insider privatization' in some cases). However, the production performance of the small onshore privates has generally been good. The pipeline network constituted the major exception to the authorities' readiness to privatize – while Kazakhstan has accepted the construction of some new private pipelines, the network inherited from Soviet times has remained in state ownership, which may well make sense in view of its natural monopoly character, and the same is true of most major new export pipelines being built.

Like Azerbaijan, Kazakhstan was extremely successful in attracting Western capital to the oil sector. Indeed, the country became the largest recipient of foreign direct investment (FDI) per capita in the CIS. During 2001–5, net FDI inflows averaged around 10 per cent of GDP.[20] As successive projects have come onstream, Kazakhstan's crude oil production has grown rapidly, reaching approximately 1.49 mbd in 2007, from a post-Soviet low of 414.8 kbd in 1995, raising net exports to 1.2 mbd. The government hopes to increase production levels to around 3.5 mbd by 2015, a target that appears entirely feasible on the basis of projects already under way. Some Western oil executives believe that Kazakh production could reach 4 mbd if the authorities wish it so and are prepared to offer more attractive fiscal terms to investors interested in developing the remaining North Caspian blocs. As will be seen, however, there are questions as to whether or not the authorities really want even faster output growth, given the problems this could create for macroeconomic management.

Although Kazakhstan's Soviet-era oil and gas enterprises were largely privatized in the mid-1990s, the authorities created a national oil and gas company, Kazakhoil, in 1997. Its role was to manage the state's remaining oil-sector enterprises and its interest in PSAs, which had until then been held by the Ministry of Energy and Mineral Resources. In February 2002, a vertically integrated state oil and gas company, Kazmunaigaz, was created via the merger of Kazakhoil and Transneftegaz, the country's state-owned trunk pipeline operator. Kazmunaigaz was to ensure a united and coherent state policy on using the country's hydrocarbon resources, and it was also given the task of overseeing a major licensing round, which began in 2003, involving over 100 blocks in the Kazakh sector of the Caspian shelf. Kazakhstan ultimately intends for Kazmunaigaz to compete with foreign energy companies.

The state's ambitions for Kazmunaigaz are linked to a broader trend toward greater state involvement in, and direct control over, hydrocarbon developments in Kazakhstan. This has led to increasing tension with foreign investors in the country and, at times, raised questions about the Kazakh authorities' commitment to property rights and contract stability. Kazakh officials frankly acknowledge that the authorities' attitude toward foreign investors has grown tougher, insisting that the state allowed investors to take advantage of it in the early post-independence period and that it has since been reasserting its legitimate interests, trying to enforce rules and contract provisions that it had previously been too weak or inexperienced to uphold.[21] The shift in the authorities' stance vis-à-vis the industry has found expression in both formal and informal ways. Much of the conflict has centered on the interpretation of the major contracts concluded between investors and the state – both PSAs and the Tengiz contract. Since the contracts themselves remain secret, it is difficult for outsiders to judge the merits of any given dispute, but there is little doubt about the overall trend towards interpreting contested provisions in favor of the state. Tax administration, in particular, has become both tougher and more arbitrary.

The most prominent conflict during 2002–3 concerned ChevronTexaco's management of the Tengiz project. The main issue concerned the switch to an accelerated depreciation schedule and the financing of the investment programme for the next stage of the project. The matter was settled after a dispute that lasted for over a year when ChevronTexaco agreed to pay a substantial advance on its royalties. There have been hints that the authorities might revise the agreement again before the end of the three-year period. Canadian-based PetroKazakhstan, which was bought out by China's CNPC in 2005, found itself in a series of long-running conflicts with the authorities in connection with its development of the Kumkol field. Conflicts arose over the schedule for developing the 7–9 billion barrel Kashagan offshore project. The authorities indicated that they might want to redraft the contract if the agreed schedule was not met. Even after the two sides reached agreement on the delay in late 2003,[22] leading politicians suggested that original contractual obligations should be revised.[23] Some Western oil executives fear another 'renegotiation' once the infrastructure for the project is built.

Concerns about the Kazakh authorities' commitment to the sanctity and stability of contracts were further aroused by the government's pursuit of a stake in the Kashagan project. In 2003, UK-based BG agreed to sell its 16.7 per cent stake in the field. Under the contract governing the project, the other members of the consortium had right of first refusal to the stake, which they wished to exercise. However, the government wished to secure the stake for Kazmunaigaz. When the investors proved unwilling to sell the stake to the state company, Kazakhstan adopted a law giving the state the right of first refusal over stakes in existing projects, asserting that this right, once enshrined in statute, trumped the terms of the contract itself. In the end, the two sides reached a compromise under which the Republic settled for half the BG stake, with the investors buying up the other half.[24] The conflict nevertheless marked an important turning point. While there had been many disputes concerning the application of various contract provisions, this was the first time the authorities had openly violated the principle that PSA provisions were protected from subsequent changes in national legislation.

The Kashagan row now appears to have been a harbinger of things to come: in October 2005, Kazakhstan adopted legislation stipulating that the transfer of rights and obligation under a contract on subsoil use or the transfer of a stake in the legal entity which has such a contract can be made only with the written sanction of the state if one contractor leaves and another takes its place. The measure granted the state pre-emptive rights to stakes in legal entities possessing the right to subsoil use.[25] The amendments also provided for the authorities to make decisions concerning the transfer of subsoil use rights on fairly wide-ranging – and ill-defined – 'national security' grounds.

Small frictions notwithstanding, Kazakh authorities have largely honoured the contracts concluded with investors in the first post-Soviet decade. Indeed, after several years in which senior officials had frequently mooted the possibility of significant contract revisions, the Kazakh authorities recently stated very explicitly that existing contracts would *not* be renegotiated.[26] However, the conditions under which *new* projects can be launched have been substantially revised. Since 2002, Kazmunaigaz has been legally entitled to a 50 per cent stake in all new projects.[27] This does not mean that all projects must be executed on a parity basis, because the law does not prohibit Kazmunaigaz from selling some or all of this stake, but it does have the right to nominate the operating company (or to demand this role for itself) even if its share in the project is far below 50 per cent. Moreover, new projects no longer enjoy the assurance of contract stability that had been previously available: the 2003 law 'On investments' states explicitly that contracts signed after it entered into force would be subject to changes in legislation and obligations under international treaties that 'change the procedure and conditions for the import, manufacture and sale of goods subject to excise duties'.

Some of the most dramatic changes have taken place in the field of taxation. However, investors taking over a field from another company still inherit the tax regime under which the field was being developed, so the changes will only affect the country's ability to attract investment in *new* projects. When the new tax regime was introduced in 2004, critics argued that it would effectively freeze new developments in Kazakhstan, and it now appears that this was indeed the government's aim.[28] There is no doubt about the Kazakh authorities' determination to capture a larger share of oil rents, but it is significant that the tax changes were adopted against the backdrop of growing concern about the possibly damaging impact of overly rapid development of the oil sector on the rest of the economy. While oil consumers might want Kazakhstan to press ahead with oil developments as rapidly as possible, there are indeed reasonable grounds for concern about the implications of too much oil-sector growth for macroeconomic management – specifically, the threat of 'Dutch disease' and the prospect of overheating. Modifications to the new regime in 2005 brought some improvement for investors, although Anderson has estimated that these changes raise the investor's internal rate of return by only about one percentage point. The tax take under the new regime is still heavier than that found in most other oil-producing states, including many with far lower lifting and evacuation costs.[29] They nevertheless appear to have been enough to satisfy some investors: while the major Western oil companies remained sharply critical of the revised regime, Russia's Lukoil agreed in early 2005 to join Kazmunaigaz in developing the Khvalynskoye field in the North Caspian, which holds an estimated 322bcm of natural gas and 53mt of oil and gas condensate.

While the past few years have undoubtedly witnessed growing friction between the Kazakh authorities and major oil companies, the situation should not be over-dramatized. Kazakhstan has not witnessed anything like the large-scale assault on property rights mounted against Yukos in Russia. The fact that the major private investors are foreign companies would in any case constitute a significant deterrent to any such attack on private property. With the exception of the state's exercise of pre-emption rights to the BG Kashagan stake, the authorities have so far resisted the temptation to revise unilaterally the contracts concluded with investors – though they have sometimes pushed questions of contract interpretation as far as they could. The much tougher tax and regulatory regime created in recent years will undoubtedly slow the development of new projects in Kazakhstan, but, given the trajectories of projects like Tengiz and Kashagan, and the sums already committed by investors, Kazakhstan is still well placed to deliver continued strong growth in oil production over the years ahead.

Turkmenistan

Turkmenistan has pursued a quasi-Soviet approach to economic policy throughout the post-independence period. The oil and gas sector is no exception to this rule. The Turkmen system is characterized by the predominance of state ownership over the means of production, restrictions on foreign economic activity, widespread subsidies and an approach to 'planned' development on the basis of import-substituting industrialization. While some privatization has taken place, the authorities remain committed to keeping the major productive sectors in state hands. This has not prevented Turkmenistan from making some efforts to attract foreign investors to its hydrocarbons sectors. It has recently been particularly concerned to bring them into oil production and petroleum refining in an effort to reduce its dependence on gas exports. In practice, however, Turkmenistan has proved an extremely difficult place in which to do business and has attracted little foreign investment. Arbitrary state action, high levels of corruption, poor infrastructure and a lack of export routes, together with the general unattractiveness of the heavily regulated business environment, have combined to deter investors. So has Turkmenistan's refusal to sign up to the agreement on dividing the Caspian Sea, which has stalled any significant offshore development. This is a particularly significant opportunity foregone, given that 40 per cent of its proven reserves are reckoned to be offshore. It is thus hardly surprising that Turkmenistan has so far concluded only small-scale deals with outside investors and these have not

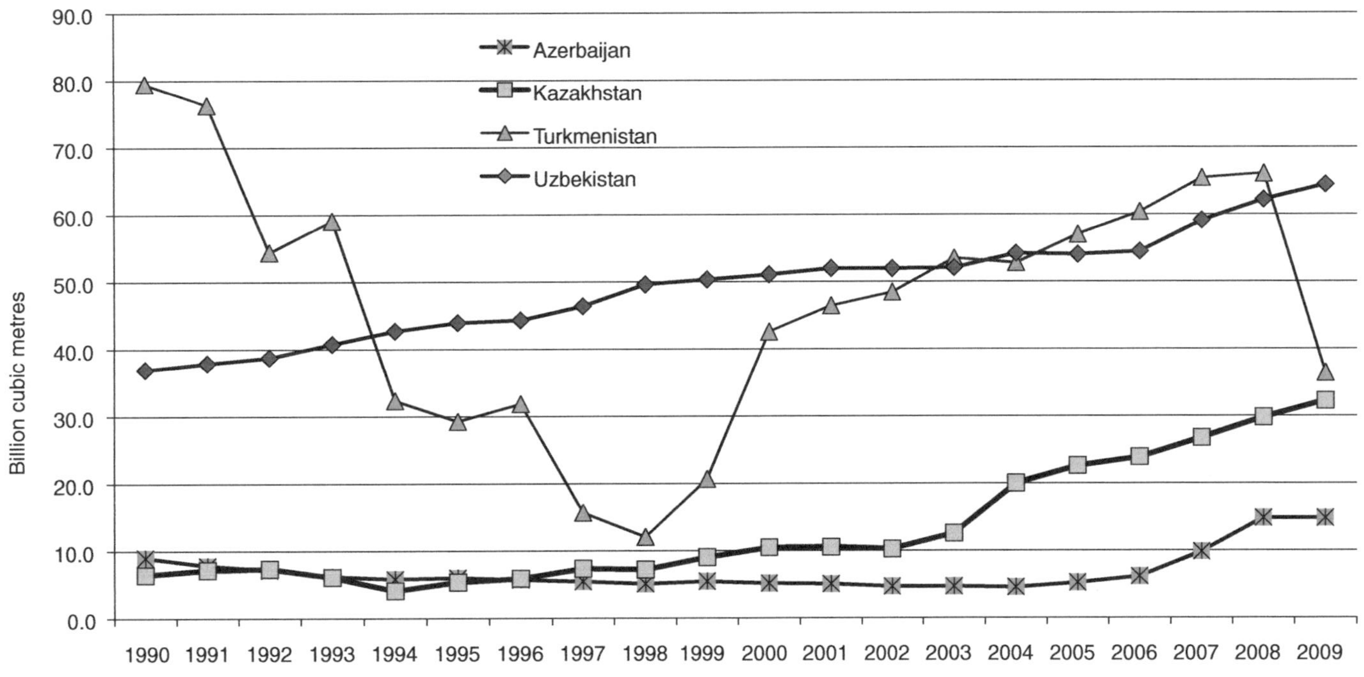

Figure 2.4 Natural gas production in Azerbaijan and Central Asia

Source: BP Statistical Review of World Energy, 2008

always been handled in such a manner as to encourage others. Foreign investors have complained of extremely arbitrary treatment, including the unilateral, *ex post facto* revision of major contract terms.

In terms of output, Turkmenistan's performance was initially fair. Its resistance to structural reform and its readiness to invest heavily in raising oil and gas output enabled it to weather the early transition better than some other CIS states and, after an initial fall in output, it managed to increase crude oil and gas production rapidly. Gas production rose more than fivefold in the decade to 2007 (Figure 2.4), though it remained at late-Soviet era levels, and oil production recovered to a level of around 200 kbd in 2004. Since then, however, oil output has stagnated, and there have been questions concerning Turkmenistan's ability to sustain production over the longer term in the absence of substantial new investment. So far, the fall in production has been limited by rapid increases in the output of Burren, Dragon and other foreign producers involved in PSA projects; the production of the state-owned Turkmenneft has been dropping. Barring a dramatic shift in policy, it seems unlikely that Turkmenistan will realize its production potential.

Uzbekistan

Like Turkmenistan, Uzbekistan has been reluctant to cede much control to outsiders. Rather than privatizing enterprises, or putting fields up to tender, Uzbekistan has tended to prefer loan-financed projects and contracts with foreign services companies. In 1992, a national Uzbek oil and gas company, Uzbekneftegaz, was established by merging the existing enterprises in charge of upstream activities, oil refining and product distribution, gas transmission and distribution, and petroleum facility construction. Later, Uzbekneftegaz also took over enterprises responsible for petroleum exploration, which had previously reported to the Uzbek Committee on Geology. In global terms, the Uzbek hydrocarbon regime matters little, as Uzbekistan is by far the least important of the Central Asian hydrocarbon producers: in 2007, it produced around 114 kbd of oil and 53.5bcm of natural gas. However, the country did manage, for much of the 1990s, to meet its own consumption needs, borrowing at least $1bn in the 1990s in an effort to achieve self-sufficiency in oil and to sustain self-sufficiency in gas. Indeed, Uzbekistan has managed to sustain steadily rising gas production throughout the post-independence period, raising production from 39.9bcm in 1991 to 58.5bcm in 2007, allowing for modest but growing net exports, which reached 12.9bcm in 2007.

Foreign oil and gas companies contributed little to the development of Uzbekistan's hydrocarbon resources during the 1990s: only Malaysia's Probady secured exploration and production rights and was carrying out field development work. Lukoil formed a joint venture with Uzbekneftegaz to develop some fields. Enron, Texaco, Pertamina and Unocal signed letters of intent with the Uzbek company at various times, but they faced long delays in dealing with a government that was clearly unenthusiastic about foreign involvement in the sector, and little concrete work was actually undertaken. Uzbekneftegaz signed its first production sharing agreement (PSA) in April 2001, with Britain's Trinity Energy, through a specially formed subsidiary known as UzPEC Ltd.[30] In 2004, control of UzPEC passed from Trinity Energy to the Moscow-based Soyuzneftegaz, and in February 2005, the Uzbek government launched an investigation to determine whether UzPEC had properly fulfilled the licence conditions for the fields. Uzbekneftegaz broke the PSA in mid-2005, and the government indicated that it was minded to withdraw UzPEC's licences. UzPEC and its new owners, Soyuzneftegaz, contested these decisions, and a new PSA was signed in 2007. Russia's Gazprom has been the dominant foreign player in the Uzbek gas sector, though Lukoil, China's CNPC, Petronas and South Korea's KNOC have also been involved; Gazprom signed a strategic cooperation agreement with Uzbekneftegaz in 2002, providing for Gazprom to purchase Uzbek gas exports on a long-term basis. This was followed by a PSA to develop the Shakhpati field in 2004 and the acquisition of a number of export licences in 2006.

For a time, at least, Uzbekistan could, like Turkmenistan, argue that its formula had prevented disruption to the sector and thus ensured continuity of supply. Production grew rapidly during the 1990s, from 69 kbd at the end of the Soviet period to a peak of 191 kbd in 1998–9. This enabled the country to achieve rough self-sufficiency in oil, an important goal of the Uzbek authorities. However, domestic regulation of the market meant that the hydrocarbon sector suffered from considerable distortions and inefficiencies – artificially low domestic prices for crude and refined products generated substantial smuggling of oil and products to neighbouring Kazakhstan.[31] Moreover, the production levels reached in the late 1990s soon proved unsustainable. Output fell by more than a third during 2000–5. As no significant new discoveries have been reported, there are questions about Uzbekistan's ability to sustain its self-sufficiency in oil, at least in the absence of outside assistance. Gas production has held up better, but the main Uzbek fields developed in the Soviet period are now in decline, and the country is struggling to develop new fields fast enough to sustain output.

Indeed, poor recent performance appears to have prompted the Uzbek authorities to reconsider their approach to the hydrocarbons sector. They have been trying since 2003 to privatize a minority stake in Uzbekneftegaz, but their determination to retain majority state control has so far deterred investors. A stake in the company may well end up in the hands of a state-owned company from one of the major consuming countries, most likely China. The authorities have also signed a number of agreements with Chinese and Russian companies concerning investment in the sector. Malaysia's Petronas hopes to develop oil and gas deposits in the Aral Sea basin together with Uzbekneftegaz, and there are also plans to borrow from Sinopec and the Chinese government to modernize the pipeline network and to rehabilitate exhausted oilfields over the next five years.

Policy change in Kazakhstan: the return of the 'obsolescing bargain'?

According to USEIA estimates, which are not particularly bullish with respect to Russian output, Russia will still account for well over half of CIS supply even in

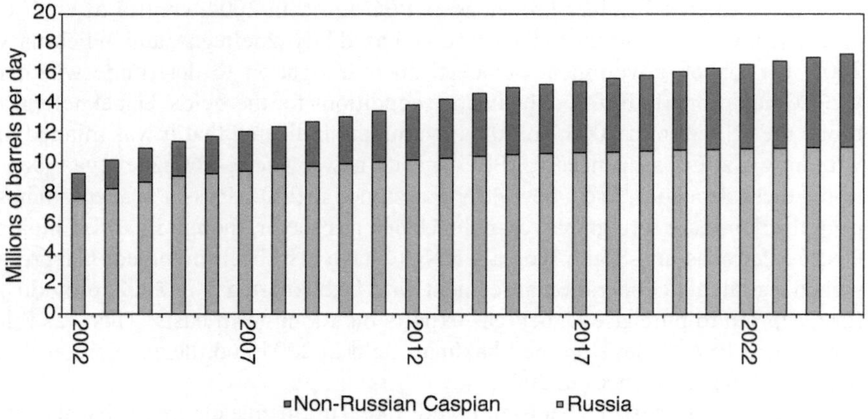

Figure 2.5 CIS oil production forecast 2002–25
Source: USEIA (2005)

2025. However, Russia's role in incremental supply will shrink dramatically. It is Central Asia that is will constitute the major generator of supply growth in the region (Figure 2.5). How issues of ownership, governance and stability of contract are handled in Kazakhstan – by far the region's largest and most promising producer – is therefore likely to be of considerable importance not only to the region itself, but to global oil consumers.

In many ways, recent developments in Kazakhstan echo the history of the oil industry in other parts of the world, as well as in neighbouring Russia in recent years. Until the 1960s, a handful of international oil companies dominated the market, and oil-producing states had little option but to accept a substantial degree of foreign ownership and control over their reserves. Over time, however, local elites' bargaining power increased. This resulted from a combination of increased knowledge within the state administration and the emergence of competitors to the international majors who were willing to cede more revenue and greater managerial control to host governments in order to capture market share from their larger rivals. They also profited from the rise of Western oil services companies, which provided yet another way to access the kind of technology and expertise that the majors provided. In these circumstances, developing countries were able to conclude more favourable contracts with foreign investors, and in many cases, to outright nationalize their respective oil sectors.[32] The capital-intensive nature of oil extraction also helped, since it imposed high barriers to exit when the state began to revise the bargain.[33] Initially, the states' need to attract large-scale investment gave the oil companies tremendous leverage over taxation, regulatory policies and questions of institutional design. However, once investments were made, and costs were sunk, it was relatively easy for states to revise the terms of their interaction with investors.

One can identify similar patterns in Kazakhstan. Its political leaders believe – not without reason – that the state's weak bargaining position in the very uncertain environment of the early 1990s led it to make concessions to investors that it would not otherwise have made. These concessions have become ever less palatable as oil prices have risen.[34] The question of how oil rents are to be allocated has grown more, not less, acute as those rents have grown. At the same time, the strengthening of the state's extractive and coercive capacities has made it much more assertive toward the oil companies. The accumulation of local experience in the oil industry has probably also contributed to a tougher stance vis-à-vis foreign investors. As in the developing world in the 1960s and 1970s, the fact that investors are already heavily committed clearly affects the parties' bargaining power. The Kazakh authorities are counting on their reluctance to exit. This calculation has so far proved correct: while there have been numerous complaints about the authorities' behaviour, no major player has opted to withdraw.[35] Finally, if the majors were undercut in the late twentieth century by emerging competition from smaller rivals and oil services companies, Kazakhstan now also stands to benefit from the increased activism of national oil companies (NOCs) from large, rapidly growing consumer countries like India and China.

To be sure, the expansion of the state's role in the hydrocarbons sector is hardly a novel phenomenon. On the contrary, Kazakhstan is arguably just moving somewhat closer to the international norm in the oil business. Most of the world's major oil exporters have state-controlled oil industries, and the extent to which the private sector dominated oil production in Kazakhstan in the 1990s was anomalous. However, it appears to have been a *positive* anomaly, at least in terms of investment, efficiency and growth. The country stands to lose much by moving toward something more akin to the international 'norm'. The shift toward a more 'hands-on' approach, relying on direct control over assets and intervention in markets, is likely to contribute to poorer performance. Subsoil resources in Kazakhstan belong to the state, and it is certainly not difficult to comprehend the state's desire to exercise its ownership rights more effectively, yet the economic costs imposed by the new approach may not be fully appreciated by policy-makers. The problem is that the cost of this shift toward greater dirigisme are largely *opportunity* costs – welfare losses arising from investment that does *not* take place and wealth that is *not* created. Since opportunity costs are by definition 'invisible', they are often insufficient to deter the pursuit of economically inefficient policies.

At issue is not merely the question of state versus private ownership in general but the capacity (or lack thereof) of state administrations in Kazakhstan and other Caspian states to manage large state-owned companies in technically complex sectors. While private entrepreneurship is generally preferable to state ownership and control, the implications of the latter depend on the institutional context. A country where the democratic accountability of the rulers to the ruled is well established and the rule of law is relatively strong has a far better chance of ensuring that state-owned companies are reasonably well run. Where political accountability and the rule of law are weak, the creation of large state companies in the most lucrative sectors is likely to be associated with greater opacity, corruption and rent-seeking by insiders. Moreover, those in control of such companies will face strong incentives to resist steps to increase transparency and accountability.[36] The performance of large state-owned companies throughout the CIS suggests that such poor governance will contribute to higher costs, lower productivity and slower growth of output. Kazakhstan is now proceeding further down this path.

This point is extremely important in the larger debate over the so-called 'resource curse'. A number of scholars have suggested that state ownership rather than resource wealth per se may lie at the root of resource exporters' apparently chronic underperformance.[37] The expansion of state ownership in major minerals sectors may thus be detrimental to overall economic performance and not merely to the performance of the sectors in question. This line of argument would be even more persuasive if, as suggested above, underdeveloped state capacities increased the incentives for states to nationalize. Underlying all of this is a basic contradiction in the actions of the Kazakh authorities. The government has been seeking to capture a substantially larger share of oil rents, which probably makes some sense in itself, provided it does not overshoot the mark and discourage investment in future development. At the same time, however, the authorities have acted in ways that increase the risks investors face and thus raise the rate of return those investors require in order to justify investment. In other words, investors are being asked to accept a lower rate of return in an environment of increased risk.

Policy shifts notwithstanding, the major projects now under way should be sufficient to deliver continued production growth for some time to come. However, there are already indications that growth is slowing and any further large projects will almost certainly take longer to develop. This may be what the Kazakh authorities want – to slow down the further development of their hydrocarbon resources in the interests of macroeconomic management. Those projects that do go forward will involve Kazmunaigaz in cooperation with other state-owned companies – chiefly the NOCs of major consumer countries – or at least with companies subject to less capital discipline than the international majors. While the eagerness of the NOCs to participate in Kazakh projects may enable the authorities to get new projects under way without making big concessions to the international majors, the NOCs are likely to bring less efficient management to the projects, and their role increases the likelihood that oil-sector development will be shaped by political factors rather than market forces.[38]

Conclusion

The impressive growth in CIS hydrocarbons production since the mid-1990s has been overwhelmingly driven by the private sector – even in Turkmenistan, oil (though not gas) production has been sustained in recent years by fast growth among the handful of small private investors who are active there. Despite this success, there has been a marked trend in Kazakhstan towards a greater direct role for the state in hydrocarbon production. This bodes ill for the future. Azerbaijan appears to remain far more investor-friendly, despite the large role assigned to SOCAR, but its oil production potential looks far less promising than does that of

Kazakhstan. Turkmenistan appears determined to cling to tight state control, even the price of forgoing a potentially enormous opportunity. The long-term consequences of further dirigiste policies are potentially very serious. Given the sums already committed to existing projects, Azerbaijan and Kazakhstan are still well placed to deliver strong growth in the years ahead. However, major new projects are unlikely under the present tax and regulatory regime in Kazakhstan, and it is far from certain that the country will triple output by 2015 as planned, not least because of conflict with investors over the development of export infrastructure.

For Russia, this state of affairs presents both opportunities and problems. On the one hand, the drift toward more *étatiste* policies and more political decision-making implies an operating environment that will be less problematic for Russian companies – whether state-owned or privately owned – than for the private international players. In any case, Russia will not lose if its Caspian neighbours fail to realize their full potential as oil producers. The situation in gas, by contrast, is more problematic for Moscow. Gazprom has no interest in seeing the other Caspian gas industries' stagnate, since it has been trying to lock up a large part of their exported output as part of its own strategy for meeting domestic and export demand.

3 How to get a pipeline built

Myth and reality

Jérôme Guillet

Introduction

The debate about Russia's relations with the West, and its relations with several of its formerly Soviet neighbours, is dominated by the topic of energy, and in particular by that of the 'energy weapon' supposedly created by the need for gas imports from Russia. In that context, discussions of alternative pipeline concepts to bring gas to Europe are a favourite pastime of pundits and armchair geopoliticians, yet these often consider the diplomatic and balance of power context in mostly abstract terms, and the economic and financial drivers of pipeline construction, as well as the profound interdependencies that existing pipelines create (and which new projects have to willingly create in order to become reality) are too easily ignored.

This article will try to explain what the practical requirements for a pipeline to be built actually are, and to demonstrate that pipelines do not happen for geopolitical reasons alone; indeed, it will argue that while politics and diplomacy can usually kill pipeline projects and, at best, help make a choice between two viable projects, they cannot make happen projects that do not have a strong underlying financial and business rationale. As we will see, the conditions to meet that test are rather strict.

Proposals for new pipelines usually occur in three cases: when a large enough supply of gas needs to be brought to the market; when a large enough market/customer is insufficiently or not at all supplied with gas; or when a large supply of gas and a large market are close enough that it might be worth linking them.

The emerging geopolitical consensus in the West is that Europe is too dependent on Russian gas, and that this dependence is set to grow, thereby representing a menace for the continent (with the Russia–Ukraine disputes in 2006, 2007 and 2009 representing a blatant case of Russia using the 'energy weapon'). Europe thus fits the second category above: a market in need of new supplies[1], and one of the nearest sources of alternative gas is in the area around the Caspian Sea. While Iran is a special case (it has large gas reserves but, because of its suspected nuclear programme, is seen at least by Americans as an unsuitable supplier for now[2]), Turkmenistan, Kazakhstan and Azerbaijan, with their significant (and Western-major controlled[3]) reserves[4] and their current lack of access to international markets,[5] are seen as a perfect match for Europe's needs; and pipelines to connect the two regions – Nabucco[6] and the TransCaspian pipeline in particular - are now heralded as a perfect solution, with Turkey, an important Western ally, to play a vital role both as a regional gas hub and as a transit country.

While some pipeline concepts indeed turn into reality, as SCP or the Blue Stream pipeline from Russia to Turkey[7] attest, the mere existence of a plausible-sounding idea, even if put forward by large companies or governments, certainly does not guarantee that this will indeed happen. The long-running saga of the Turkmenistan–Azerbaijan–Turkey TransCaspian pipeline, the Iran–Pakistan–India proposal,[8] the Turkmenistan–Afghanistan–Pakistan (TAP) pipeline,[9] or the ongoing unsuccessful promotion of Nabucco suggest that some concepts remain just that, i.e. something is missing to make them happen. In fact, one must be very suspicious of pipeline projects that seem to make sense but never get anywhere. If any planned route frequently mentioned in the media or decision-making circles is not happening, they can often be considered little more than 'PR concepts'.

Before going into more detail on what makes a pipeline possible, it is worth flagging that in most instances, the profound interdependency and the fundamental commonality of interest between gas sellers, buyers and transit countries is downplayed. A pipeline, and the attendant framework of explicit long-term contractual obligations of the parties involved, physically and economically links these parties in ways that they cannot ignore and cannot break. Natural gas, and natural gas pipelines, make for very unwieldy 'weapons'. The producer cannot easily do without its dedicated market, and the buyer cannot easily do without its supplier. The fact is that nobody can win a gas war: all parties lose – supplies, income, reputations, trust – while remaining unavoidably linked by the pipeline. The only case when gas deliveries can conceivably be used as a weapon is in highly asymmetrical relationships, for instance when a seller is the sole provider of a supplier, *and* non-commercial prices are used. Russia's relationship with Armenia may be categorized as such, and Armenia is rightly seen as a Russian vassal state. When there is a balanced relationship, such as that between Russia and Europe (which is, when including Turkey, the only possible destination of Russian gas exports), or between Russia and Ukraine (which, while depending on Russian deliveries, has absolute control over a large portion of Russia's gas exports[10]), the 'energy weapon' is just like nuclear ones: it can be useful for mutual deterrence, because it can cause massive damage – but it cannot actually be used, because the damage will ultimately be on both sides. In fact, brushing up on the theory of MAD (mutual assured destruction) might be worth while when talking about gas pipelines.

A common flaw in the analysis of Eurasian pipelines is that they are viewed first and foremost through the prism of what they mean to European importing countries, i.e. Russia's actions are seen purely through that prism and not as determined by its own interests and needs. One of Russia's main requirements 'and a stark reminder that pipelines happen only if they benefit all parties involved' is to keep in place the contractual relationships that provide long-term stability of *demand* and allow it to invest in the infrastructure needed to fulfil Europe's needs.

Large European buyers know this, but seem to be demonized for it all too easily[11] – and yet it is they who are more aware of what it takes to build a pipeline, and bring the gas we all clamour for to our homes.

Building a pipeline – or: Getting it financed

The easiest way to understand pipeline construction is to break up the task into its smaller components, and determine how these are to be put in place and fitted together.

This discussion will focus on natural gas pipelines, which are somewhat more complex to get built than their oil 'cousins' and allow for a better discussion of all interrelated issues.

A gas pipeline is like a chain: it requires every single link to be in place in order to work. The required components of a pipeline are as follows:

- a supply of gas (Reserves)
- a supplier of gas (Seller)
- a market for gas (Market)
- a purchaser of gas (Buyer)
- a route and technical design (Design and Engineering)
- an entity operating the pipeline (Operator)
- government authorizations for pipelines crossing their territory – for each country (Permits)
- a price for gas (Gas Price)
- a price for gas transport (Tariff)
- an entity (or more) building the pipeline (Constructor)
- an entity (or more) paying for the pipeline (Investor).

The fundamental point is that all of the relevant components and parties need to be present at precisely the same time for the project to exist. Here 'being present' means 'irrevocably and credibly making binding commitments for a very long period, representing large sums of money'. It is a surprisingly difficult job to bring all the parties at the table to commit in that way and at the right moment, which is why fewer pipelines are built than one would expect, and why few entities are actually able to pull it off.

What is more, being able to pay for the construction of the pipeline (often thought to be the most obvious obstacle) is actually not the main issue. Indeed, there is a difference between paying for construction and actually *financing* a pipeline. Financing, the prerequisite to building any pipeline, requires much larger commitment than the pure construction cost, which is but one link in the chain. All the other links need to be in place also. A pipeline that costs $3 billion will carry, over the life of its original contracts, something like $20–100 billion worth of gas (depending on prices): when financing the pipeline, the relevant amount that needs to be committed is not the $3 billion figure, it is the $20 billion at the bottom of the gas price range, out of which the $3 billion will be paid over 20 or 25 years.

Pipeline concepts

Such subtleties do not prevent those involved from making sensational announcements about their business-altering projects. A good example is provided in the ENI press release about the South Stream project, which includes the assertion that the project 'aims at strengthening Europe's energy security'; in fact, the text mentions this no less than three times in a few paragraphs. What this in fact states is that, at that point in time at least, the pipeline does not exist yet.[12] The aim is to try to make the project seem a reality, to build momentum around it in order to get other players interested and attract the necessary partners to get it done. Such announcements will be all the more grandiloquent and upbeat if the purported pipeline route is competing with others for a given gas resource or a given gas market, in which case they will be part of a sustained campaign to make one project appear inevitable, or preferable to any other, so that the commitment of the third parties whose involvement is required may be secured. However, this is again but one link in the chain.

It is rather safe to assume that the more one talks about a given project, the less likely it is to happen: in fact, it may instead be said that the more numerous such public announcements are, the more it implies that the project is unable to move toward the next steps (described below), and is stuck in what may be dubbed the 'diplomacy-and-PR-heavy, partner-seeking phase'.

The various pipelines noted above (TAP, IRI, Nabucco, TransCaspian) appear to be in this situation. They are being discussed endlessly, are subject to various announcements and declarations of intent, and are supposed to be built in a few years' time. Yet something always seems to be missing. The case of South Stream is a bit different in that the proposal is still relatively recent, and thus its future prospects are not as obvious.

Engineering stage

If the concept looks potentially attractive (basic economics seem plausible; the resource base is sufficient; there exists a need for transport capacity), more detailed studies will get funded to look more closely at the engineering side and get a better handle on technical issues and actual costs. Such studies cost almost nothing (a few million dollars) compared to the overall likely investment amount, and conducting one of these means little in terms of the likelihood of a project happening. But as they provide useful information, in particular route maps and quotable numbers (kilometres, pipeline diameter, volumes to be transported), they can be used very effectively to make a public case for a project, by getting the media and politicians interested. Such studies create a base of underlying technical facts that are sufficient for politicians to get involved in signing 'memoranda of understanding' with the project promoters, or with neighbouring countries that may also be potentially involved.[13] Again, the goal here is to generate publicity and momentum: at this point, the pipeline is still nowhere closer to being built, and has not been subject to any kind of investment decision. At best, it can be said that technical obstacles have been identified, but these are almost always the easiest to solve.

We can now go into more detail on the various objective factors that determine if a pipeline will be built.

Securing the resource base

Enough gas must be available from the production area to fill the pipeline. 'Filling' a pipeline means using up its capacity for at least 20 years. A relatively small 10 bcm/y[14] pipeline will thus require a resource of at least 200 bcm, which needs to be extracted and then transported. That number must also refer to *proven* reserves, not just probable or potential reserves.

This is no trivial point: many gas fields are labelled 'stranded reserves' because they are not big enough to justify economically the construction of a pipeline to bring them to market. The alternative to sourcing gas from an identified gas field is to procure it from a gas producer that controls several fields and agrees to allocate the relevant production volume (with or without specifying its origins) to the pipeline.

As Table 3.1 suggests, the entities controlling the required reserves for a 10 bcm/y pipeline in the Caspian area are not that numerous.[15]

With the exception of the Shah Deniz and Karachaganak fields, which are controlled by Western majors and can be expected to reliably provide the required volumes of gas (and indeed, the gas production capacity of which requires the corresponding transport capacity, as their owners will be keen to monetize their assets on the international markets via exports), the only sources of gas in the region are the national companies. Kazakhstan's gas sourcing capacity is limited by its domestic consumption, which until 2005 was equal to or higher than its production capacity. Iran's net potential for export is hard to determine, given its large reserves, but is not very significant right now, and its current focus is to use gas domestically, whether for industry (petrochemicals), power generation or oil production enhancement via injections.

Table 3.1 Entities controlling significant reserves in the Caspian area

Country	*Gas field or entity*
Russia	Gazprom – various fields Gazprom has an export monopoly
Turkmenistan	Turkmengaz (national gas company, 2000–5000 bcm overall reserves) Dauletabad gas field (1000–1600 bcm remaining) Shatlyk gas field (400–1200 bcm remaining) Iolotan gas field (200–4000 bcm)
Azerbaijan	Shah Deniz (BP operator, 1200 bcm)
Kazakhstan	Karachaganak field (ENI/BG, 900 bcm) Kazmunaigas (national oil and gas company, 2000–3000 bcm) Production is barely more than domestic consumption
Iran	NIOC/ NIGC (20,000 bcm overall reserves) Production is barely more than domestic consumption

This is a major issue for the Nabucco pipeline. The only entity able in the foreseeable future to commit the required volumes of gas from the Black Sea/Caspian Sea region is… Gazprom, precisely the entity which the Nabucco project seeks to avoid. Neither Iran nor Azerbaijan will be able to support the project, Azerbaijan because it does not have enough gas (existing volumes from Shah Deniz are already sold to Turkey and Georgia and would not be sufficient anyway), and Iran because its own export capacity to Turkey is insufficient, and it seems unable to deliver even that volume to Turkey reliably (and adding a Iranian component to the scope of the project would further increase its complexity).

The need for transport capacity

The gas resource must not have any other existing or obviously cheaper transport alternatives to be brought to market. Any project where the gas is already using some other route, or *could* use another route, is unlikely to happen. Similarly, buyers in a market which is already oversupplied are unlikely to commit to buying gas only via a new pipeline that would be built to service that market.

For instance, all the gas pipeline projects from Central Asia (including the recently announced Russian one, the 'Pricaspii'[16]) make no sense if a pipeline already exists and is not full. The mere existence of that available capacity is enough to undercut any alternative project and kill it, because existing capacity can always be offered at a much lower price than capacity that needs to pay for its construction cost. Given how transportation costs over the typical distances required in the region (say, 2,500 km from Central Asia to Europe) represent 20–40 per cent of the price of gas,[17] and how such transportation costs will inevitably lower the income received by the seller of gas, no seller will commit to providing volumes to a transportation option that significantly reduces its expected income.

That argument on its own strongly suggests that no new gas pipeline will be built from Turkmenistan as long as there is spare capacity in the existing lines going to Russia. Not to Pakistan, not under the Caspian, and not to China. The Turkmen authorities could conceivably decide to commit to sending all their gas in another direction, for the sake of finally creating an alternative to their one-on-one relationship with Russia, but who will take the bet that they will stick to that commitment for the required 20 years when Russia is able to pay them more net money for the same gas at any time?

Similarly, Russia will not build a new pipeline just to replace existing transit through Ukraine: that makes no economic sense as it will always be cheaper to pay off the Ukrainians (with free gas, for instance). But as demand keeps on growing in Europe, and new capacity is needed, then it makes sense to look for alternative routes for the *additional* volumes, because they will then fulfill a real need and cannot be undercut as easily.

This suggests that the most interesting situation is when there is an actual need for transportation capacity, and competing projects, which are in each case economically viable on a self-standing basis and cannot be killed by an existing alternative. In that case, it is not necessarily the most economic that will happen, but the one *among those that would be economic* which first fulfils the condition stated earlier of bringing all the necessary parties together.

This is when PR campaigns and diplomacy make sense, as they become part of an elaborate game to convince the undecided parties whose participation is required (all the other links needed to have a full chain from gas field to gas user) to come on board and support one project rather than another, by committing to it, politically and, more importantly, contractually.

As long as no firm commitments have been made to permit a project, invest in it, deliver gas and purchase gas, no project is actually inevitable, however much public support it may appear to have. But a strong campaign will help a project happen.

A good example of that is that of the BTC (Baku–Tbilisi–Ceyhan) oil pipeline form Georgia to the Mediterranean coast in southern Turkey. This pipeline is a perfect illustration of how a sustained effort to privilege one project over another can succeed. The oil companies (led by BP) developing the ACG fields in the Caspian Sea near Baku needed an large capacity (1 mb/d) export route and had various options, via Russia, Iran or Georgia. Iran was seen as the easiest and most profitable route, but was ruled out quickly because of US sanctions (showing that 'campaigns' can take more than diplomacy and include the use of blunt instruments and demonstrating that politics can indeed kill projects); a Russian option was technically feasible, using the route of the smaller existing Baku–Novorossisk pipeline, but was soon excluded by the oil companies given the unreliability of Russia in previous years, and the risks associated with crossing Chechnya and/or Dagestan. That left the route to Georgia, which could end either in Supsa on the Black Sea (using the route taken by another existing small pipeline already built by the same consortium), or a much longer route going through Turkey to the Mediterranean. The shorter route was definitely cheaper, but was opposed by Turkey which wanted to avoid more oil going through the Bosphorus and could potentially slow the traffic through the Straits for safety or environmental reasons, thus making it less attractive. Turkey also wanted some of the transit income. With the USA supporting Turkey's stance, as a way to anchor the close relationship between the two countries, and to reinforce Turkey's influence in the Caucasus region by giving it a seat at the table, the oil companies were under a lot of pressure to choose the longer option. In the end, in addition to sustained diplomatic action by the USA,[18] which convinced the three transit countries to agree to the required legal framework for the pipeline, financial contributions by the US government (in the form of loan facilities provided by OPIC and the USEXIM export credit agency, which supports exports of US equipment) and the Turkish authorities (in the form of a commitment to build the long Turkish section of the pipeline in exchange for reasonably low transit fees) helped trigger a decision which was by no means guaranteed in view of the expected costs of the alternatives.[19]

Calling it purely a PR campaign is a bit unfair in this case, as a lot of diplomatic work went into convincing the various countries involved in the project not only to agree to its principle and support it politically, but also to commit to more concrete things like tariffs, permits and a specific long-term legal and regulatory framework for the project.

But it has to be noted that essentially similar work was conducted at the same time to support the Shell- and Enron-supported TransCaspian gas pipeline project,[20] using the very same instruments of public and private diplomacy, regular speech-making underlining the US position and supporting the project, and promoting intergovernmental agreements. The same amount of diplomatic effort led nowhere in this case, because that project made no sense in view of the overwhelming reality created by an *existing* alternative. Thus, it is essential to note that a sustained diplomatic campaign does play a role in building or reinforcing some of the links required in the gas chain, but it cannot help in the absence of some of them, in particular, the lack of a business case for the project.

The demand side: a market and buyers

Finally, the gas must have a final destination: a large enough market, or end-user, that actually needs the gas volumes that would be brought by the pipeline, and would be prepared to pay for it. Additional capacity to a large market, like the continental European market, to keep up with economic growth makes sense, as does adding new capacity to a developing market (for instance, a country building up its gas-fired power generation, such as Turkey). Unless there are large end-users like aluminium smelters, power plants or petrochemical factories, the difficulty is that market growth is a fairly continuous process, whereas pipeline capacity can only increase in large sudden fixed numbers, so targeting a market is always going to be a difficult bet for a seller or for any pipeline project. That market risk will need to be borne by the buyer of the gas.

One of the most fascinating stories in the region in the past decade has been the 'race for Turkey': with economic growth and a strong need for more electricity generation, the country embarked on an ambitious plan to develop its gas infrastructure including urban networks and power plants. Demand was expected to at least quadruple between 2000 and 2020 (from 10 bcm/y to 40 bcm/y), and gas-fired plants to grow from 3,500 MW in 1997 to 19,000 MW in 2010.[21] All neighbouring countries were jockeying to supply the required gas which, in turn, required the building of new pipelines. Thus contracts signed with Iran (9 bcm/y signed in 1996), Azerbaijan (up to 20 bcm/y signed in 1999), Turkmenistan (up to 16 bcm/y signed in 1999) and Russia (up to 16 bcm/y for Blue Stream signed in 1997, in addition to the existing 14 bcm/y contracts using the old route via Bulgaria) for various volumes.[22] An economic crisis in 2000–2001 severely cut the gas demand growth in the country, however, making Turkey unable to honour most of its gas purchase contracts.[23] But by then, the Blue Stream pipeline had secured financing,[24] and its construction was well under way, whereas the other projects were delayed or cancelled.

What made the difference is that Gazprom's contract with Botas was seen as more likely to be honoured, because of the existing pipeline from Russia to

Turkey via Bulgaria and Greece: volumes could be adjusted a lot more easily between the two routes, and ensure that the Blue Stream was used as necessary to back the long-term contracts underpinning it. In addition, Gazprom 'cheated' by supporting its sales commitments to Turkey with the revenues of existing independent gas sales contracts going to Italy; in effect, it provided a highly credible commitment to use the pipeline by offering matching recurrent revenues from an existing source. Furthermore, the Italian connection was underpinned by the involvement of ENI, the Italian oil and gas group, in the project, as the counterparty to the existing gas contracts, as an investor in the project, and as the main constructor of the pipeline via its Saipem engineering subsidiary.

This allowed the pipeline to be financed, and built, and it allowed Gazprom to capture most the Turkish market – because once its pipeline capacity was in place to fulfil existing demand, no new pipeline could be supported. The pipeline from Azerbaijan did get built, but on a smaller scale than envisioned (7 bcm/y).

Another factor that made the difference is that, from Gazprom's perspective, Turkey was a hard-currency client paying market prices for gas, i.e. a much more attractive counter-party than countries such as Ukraine. For Western investors or banks, contracts supported by a Botas purchase commitment were not seen quite as favourably, which limited the ability to close that link, as the requirement is to have demand from a creditworthy entity able and willing to pay over the long term for the gas.

Project economics work

Project economics are determined by a combination of the price of the gas supply (production costs plus taxes) and the distance it needs to be transported, compared to the expected sales price on the destination market. That sales price itself depends on whether the pipeline connects to a liquid market/network, to a single client or to other transport facilities (LNG terminal, more pipeline transit), which impose additional costs before the gas is actually sold. A very rough estimate is that gas transportation costs about $2/100km/tcm.[25] Project economics are driven either by the buyer (looking at the cost of production, and adding transport costs to determine a potential net cost of supply) or by the supplier, looking the other way round (starting from the point of sale, identifying the likely market price for gas, deducting all transport and other costs to bring the gas there, and calculating the 'netback' price that such a market can provide at the point of production, i.e. the net income ending up in the gas producer's pocket). In the first case, the purchase price should be lower than existing available alternatives (if any), or low enough to make the use of gas at such price economically rational for the user;[26] conversely, the netback price must be higher than gas production costs to justify any investment in production capacity by the holder of gas reserves.

After the basic economic 'smell test' has been run, more detailed studies will give potential players a better grasp of the potential cost of the pipeline, and of the volumes of transit gas required to cover that cost in an economically sensible way, taking into account geography, likely tax regimes and market conditions. Such studies will provide a good indication of the requirements – in particular in terms of taxes and transit fees payable to the relevant public authorities – that will allow the project to be economically viable, or not.

Project leadership

What is needed at this point is an entity able to drive the project to fruition. A prerequisite is that such an entity has to have a direct interest in getting the project done (which explains why it's usually a gas producer or a large gas consumer), but something more is needed, which is a lot rarer, and which explains why so many pipeline projects don't become reality.

That additional vital requirement is that this entity needs to be able to credibly convince others that the project will happen and thus that such third parties can actually make their own commitment to it in the certainty that it's not one-sided. In other words, that entity needs to be an acceptable counter-party to all the other participants to the project, from both a financial and a contractual point of view.

The fundamental fact of pipeline economics is that pipelines have high initial investment costs, and low operating costs, and can pay for themselves only if the initial cost is spread over large volumes transported over long periods of time. Thus those who commit their money to build it, or their gas to fill it, or their demand to underpin it, or their restraint in not taxing it more than initially agreed, need to do so for a long time. And they will do so only in the comfort of knowing that the other commitments – those that materialize the existence of each link of the gas chain – have been made by the others. This requires satisfaction with both the willingness and the ability of each of the others to perform their part – and a complex framework of legal agreements formalizing the whole thing.

Each participant will say: 'I will commit to my part, but only if others simultaneously commit to theirs.' To get a pipeline built, you need all the commitments aligned, and all the conditions to each individual commitment fulfilled. This requires more than money: it requires the ability to bridge the gaps until all conditions are met, and to do so credibly in the eyes of everybody.

In theory, that argument would be enough to kill the notion that 'China has billions of dollars, it can pay to build a pipeline from Turkmenistan (or elsewhere).' China cannot credibly convince the Turkmen that it will pay for the gas even if Kazakhstan blocks transit for some reason. It cannot credibly promise the Kazakhs that it will pay the transit fees even if gas is not delivered. The amount at stake is not just the cost of the pipeline, it is the full value of gas sales over 20 years, which dwarfs by an order of magnitude the cost of the pipeline. And, more importantly, the Chinese themselves cannot trust the Turkmens to deliver the gas even if they have built the pipeline. From this point of view, the inauguration of a pipeline in December 2009 deserves further exploration regarding the future way of delivering.

So, in practice, the leader of the project is either a company or an entity that controls most of the chain (say, Gazprom, which has the gas, the technical know-how, the financial means and the ability to get approval for pipelines inside Russia), or a company that both has a stake in the pipeline and the track record to lead such projects - i.e. an oil major or, in a few cases, very large gas buyers such as the aluminum-smelting companies. Lacking such a project coordinator, a project is highly unlikely ever to happen.

The entity that operates the pipeline will be a central contractual counter-party to the other players in the project, dealing with the supplier of gas, the buyer of gas or both. In the simplest case, it is just the provider of a technical service (running the pipeline, providing maintenance and security, ensuring that the requisite capacity is available, transporting and measuring the gas coming in and out, etc.) with no involvement in the commercial arrangements of the project beyond a small fee for the services. In more complex situations, it can be a party to some of the commercial arrangements (such as the financing of the construction, and the corresponding transit tariffs it will receive). In either case, it is the entity that creates the commonality of interest between all the parties, and whoever does that job has to be representative of the entities with a stake in the overall project. It should be either an emanation of the project leader or a consortium including several project parties, and acknowledged by the public authorities. It is an entity that needs to be able to bring together all the players at the right time, and thus be sufficiently credible, even if it is just a shell created for this purpose (an 'SPV', or special purpose vehicle).

Whenever looking at pipeline announcements, identifying the entity that plays this role can go a long way in identifying the strength of a project. Amateurs out of their depth or political wishful thinking can easily be weeded out this way. Projects pushed by politicians (like the Iran–India pipeline) or led by smallish players (like Bridas, the Argentine oil company, trying to push the TAP pipeline back in the 1990s) will go nowhere.

Simply stated, politicians cannot get a pipeline built on their own.

The commitments needed

a) A supplier of gas

Identifying a potential gas-producing area is one thing. More specific questions need detailed answers when plans become more concrete: who will supply the gas? How much of it? For how long? Does that entity have the physical resources to do so? Does it have the legal rights to produce it and sell it? Has it committed all or part of that resource to anyone else? Does it make economic sense for it to bring the gas into the proposed pipeline? Are there, or will there be, more attractive (and likely) alternatives in the future?

In practice, that means identifying a particular gas field (or group of fields), or a particular player active in that production basin, and convincing the owner(s) of that field, or that player, to join the project. In fact, in many cases, pipelines are going to be driven by producers that are looking for an outlet, so that question is solved in the sense that the gas supplier is the initiator and is looking for cost-effective ways to sell its gas. But for projects talked up by others, this should be one of the first questions: *Who will fill up the pipeline, day in and day out, for the next 20 years?*

And it is that question that makes pipelines like TAP or the TransCaspian gas pipeline no more than pipe dreams – because the only entity that has the resources to conceivably provide for that requirement (Turkmengaz) cannot deliver it: (i) because commitments from that company over 15 years are not credible given its track record and the country's likely lack of political stability over such a period, and (ii) because they can already ship their production in an existing pipeline, to Russia, where the buyer, not having to support any construction or financing costs, can always undercut any alternative purchaser that does.

As noted before, Nabucco has not solved that question yet. Conversely, it is the strength of projects supported by Gazprom, such as the proposed Nord Stream pipeline currently being built under the Baltic Sea, or the more recently unveiled South Stream: the gas commitment comes from the largest producers and exporter of gas, with an unrivalled 40-year track record of reliable exports, including through the disappearance of the Soviet Union.

b) A purchaser of gas

Similar questions need to be asked at the other end of the line: who will actually purchase the gas to be transported, and pay for it? Does that entity need it? For the next 15 years? Can it afford it? Does it have the relevant infrastructure to use that gas?[27]

There are basically three kinds of buyers: utilities (those with retail gas distribution networks), power producers (operators of gas-fired power plants) and a few big industrial users (essentially in the metallurgic sector – steel and aluminium – and the petrochemicals industry). One or more of these three categories of players will imperatively need to be a party to the pipeline project. In a very real way, the buyer underpins the project – it is the payments it makes over the long term that will pay for the pipeline – as well as, of course, the gas itself.

Nothing underlines more the importance of the buyer than the almost systematic nature of the contracts they are asked to sign: the so-called 'take-or-pay' contracts.[28] Take-or-pay means that the buyer has to take the gas (and pay for it, of course), or pay for it *even if it does not take delivery*. Buyers are in effect asked to guarantee a minimum level of income, irrespective of their physical ability to deal with the gas, provided that it can be delivered by the proposed pipeline. That means, of course, that the buyer is willing and able to make such a commitment, and that such commitment is considered credible, both in that there is a binding contract and that it actually makes sense for the buyer to make it. The buyer should therefore be a large user (or distributor) of gas, with a strong balance sheet and, in all likelihood, a strong credit rating. It must need that gas, and not have a surplus of alterative cheaper supplies. The price it will pay must make sense for it (by being close to market conditions, or by providing some other economic advantage, such as a fixed price, or a maximum price).

As this requirement mirrors that above for the gas provider, it is obvious that both ends to the chain must be satisfied with one another, as they underpin what the other actually needs – a reliable supplier and a reliable buyer. In most cases, the core commercial contract underpinning the pipeline (the Gas Sales and Purchase Agreement) will be between these two parties alone, not including the other links in the chain, which will have their own contracts with one of them.

The joint venture between Gazprom on one side, and E.ON, Wintershall and Gasunie on the other, to build the Nord Stream, is a good example of the kinds of partnerships that are sought – and in this case the combination of the players suggested that this pipeline would be built as announced – as is indeed happening.

c) Government authorizations for pipelines crossing their territory for each country

Beyond geopolitical factors, plain politics play a role too. A pipeline is a major piece of infrastructure and will not happen without the explicit support (or, in the worst case, the mere consent) of all the local governments involved. And a pipeline, being a bit of equipment that stretches over large bits of territory, often straddles more than one government and/or local authority. All of those involved in any way will need to be brought on board – including the authorities regulating the gas provider and the gas buyer.

Every single one of them can block the project.

Each must be accommodated, and must provide its formal support, either via existing legal or regulatory procedures when they exist, or via ad hoc agreements. In the case of international cross-border projects, not only must each government be on board, but they must agree jointly and simultaneously on the package and its terms. The more parties involved, the more complex things are going to be.

The authorities' goals may vary, from simply getting income to protecting the environment to promoting the development of a particular region, to pleasing a neighbour or another powerful country, to promoting one's corporations.

There is an unbreakable rule here: the project must make sense for *strictly all* involved countries and authorities. The benefits may be unequally shared, but shared with all they must be. This requirement makes a project like Nabucco (involving Turkey, Austria, Hungary, Bulgaria, Romania, and potentially Azerbaijan or Iran) overwhelmingly complex and thus less likely to happen. The BTC oil pipeline shows the difficulty of multi-country projects, even with strong fundamentals (10 billion barrels of Western-controlled oil looking for an access to markets; BP as an undisputed project lead for both the oil production and the pipeline; massive diplomatic support from the US): it took almost 15 years from concept to first delivery.

This also explains the attraction, and the success, of bilateral projects such as Blue Stream (Russia–Turkey) and Nord Stream (Russia–Germany): there are fewer governments to bring on board – and fewer that can block the project.

d) Prices

Now that parties have been identified, and begin to discuss terms, the detailed economics of the project must be examined anew. The original 'smell test' simply yielded estimates for costs, and found that such estimates were compatible with the expected end price for gas on the targeted market, in view of technical production costs, tax levels and other endogenous price drivers. Now a more detailed case must be made that the pipeline will actually generate enough revenue for all in most foreseeable circumstances, taking into account the tax and accounting regimes. Those that bear the financial risk in the case of shortfall must be willing and able to do so.

Risks are pretty easy to identify: lower production or lower transit capacity than forecast, lower prices on the final market, changes in tax regimes, a party going bankrupt, delays in any part of the project. Who bears them will depend on the project and the parties involved, so it is hard to provide general rules, beyond a few simple ones:

As per contract, the 'take-or-pay' nature of the gas purchase dictates that gas volume risk is usually borne by the buyer – i.e. the seller has a guarantee that it can get cash for every molecule of gas it brings on the market;

Conversely, gas price risk is more often borne by the seller – i.e. it will only get the prevailing price for those molecules, unless it manages to get a fixed price or any other hedging mechanism, as part of the gas sales contract or as a separate transaction.

The pipeline company itself, which is in the middle, can see all configurations for the price of gas transportation. It may have to bear volume risk (via a payment structure which makes income proportional to volumes transported) or not (via a mostly fixed payment, irrespective of how much gas is actually transported – these are called 'capacity charges'). The expected income must in any case be sufficient to cover the cost of construction and financing of the pipeline, while not undermining the economics of the sale of gas. To ensure this, gas transport costs are usually paid first, before any income goes to the gas seller.

To add a few wrinkles, it must be noted that the pipeline company may bear volume risk even in the presence of a take-or-pay contract, because if the gas seller fails to provide the expected volumes, the buyer will not pay for unavailable gas. That risk may be covered by the gas seller via a capacity payment guarantee. This depends on the nature of the pipeline company, as a pure service entity (which will then bear very little risk), or as a standalone entity that must show a profit. This will usually be determined by the shareholding of the pipeline entity – dominated by the seller, the buyer or including third parties not otherwise present in the gas chain. If it is controlled by one party, the contracts will usually reflect that: for instance, if the gas seller runs the pipeline, the gas sales and purchase contract will usually incorporate a gas ownership transfer point at the exit of the pipeline. What matters here is that the contractual arrangements and price formulas are coherent with the risk borne by each party.

For instance, when one reads that India is unwilling to provide any take-or-pay clauses,[29] or that China is unwilling to accept market-driven prices for gas,[30] one can be certain that the IPI or, respectively, Russia–China pipelines are going nowhere, because there is a fundamental disagreement on the economics of the project and basic risk allocation.

A project leader rather than money

This brings us finally to the underlying question throughout this process: who will pay for the pipeline. It's actually an easy one, once the other issues have been solved: banks will usually be happy to finance a large part of the investment costs if the underlying commitments are in place.

So the payment issue brings us back to the project negotiation coordination.

As noted before, pipelines have very simple economics: they cost a lot to build, but are then very cheap to operate. So the main cost, at any time, will the amortization (or financing) of the initial investment, to be spread over a number of years, and whatever volume is transported over that period. The only thing needed to commit to a pipeline investment (and spend all that money upfront) is a reasonable certainty that the pipeline will be full enough for long enough.

The money at stake is not just the cost of the pipeline, it is the value of the gas transported by it over 20 years or more. Pipelines require contracts that include more or less binding guarantees to pay amounts equal to these full volumes (and that is implicitly recognized by the PR people, when they trumpet '$100 billion contracts'). That means these contracts have to be driven by at least one of the parties of the Gas Sales and Purchase Agreement – on the understanding that it is committing to its side of that contract as the anchor for the pipeline, and as an incentive for the other players to jump in with their own commitments.

So, to get a gas pipeline to China, what is needed is not a US$4 billion cheque, it is a signed contract by one of the big Chinese energy companies that states 'I will buy 10 bcm/y of gas at market prices (whatever the domestic price) for the next 20 years.' Without that, no gas pipeline will be built to China. And, as long as domestic gas prices in China are constrained by both price gaps and the competition from cheap coal, no commitment of the sort will be made (nor should be). As long as Pakistan is not seen as an acceptable credit risk for several billion dollars worth of gas per year, no pipeline will arrive to that country from Iran, Turkmenistan or anywhere else. And, as long as Nabucco supporters cannot credibly show where they intend to buy the gas needed to fill that pipeline, it just won't get built. No pipeline will cross the Caspian for as long as the existing pipeline going to Russia is not full. And Gazprom will continue to succeed in building its export pipelines underpinned by existing or future exports until Western buyers have doubts that the Russian giant actually has the reserves to fill these pipelines.

To commit parties to the intense negotiation effort needed to align the requirements and desires of all parties (and get them to allocate internal resources, hire legal and other advisors, and get internally approved negotiation guidelines) they must expect that the transaction has a reasonable chance of happening – which means that some entity must be there to ensure that it will indeed, an entity credible enough, motivated enough and strong enough to bridge the commitments of the various parties and bear the corresponding risk, even if temporarily.

The sustained coordination and management skills required to drive such an effort are beyond most entities – in practice, only oil and gas majors have them, and it is a large part of what they sell to host government these days: they are the only ones who can manage investments in both producing and consuming countries, and have the discipline, management depth and financial credibility to push these projects forward on all fronts at the same time.

Looking for the underlying requirements to get a pipeline built can help make sense of the jockeying and the public diplomacy that takes place around potential projects, and help one understand whether there are real stakes or if it is diplomatic theatre designed to distract public opinion or influence unrelated issues. The EU–Russia energy dialogue is dominated, at least on the visible side, by ghosts and unfounded fears of dependency rather than an appreciation of the real interdependency between the two entities; Europe, by ignoring its own strength in the existing gas chains, triggers hostile reactions from a Russia which sees itself, with some reason, as more vulnerable.

Conclusion

Pipelines are like kids in a couple: they create binding ties that cannot be dissolved by a fight or a divorce. And they are just as hard to bring to life: they cannot be willed into reality by parents or godparents, and they cannot happen without intimate relationships between willing parties who will have to live with the result for many years. They cannot be used to blackmail others without catastrophic consequences for oneself, and they are unlikely to grow up and thrive if not surrounded by a supporting environment. That metaphor should not be pushed too far, but it underlines the fact that pipelines are not a game – they represent long-lasting commitments by entities inextricably bound together, and they both require and foster such relationships.

4 The role of Central Asian gas

Is it possible to bypass Russia?

Maureen Crandall

Introduction

1 January 2006 was a key date in European–Russian natural gas trade. Moscow cut off deliveries to Ukraine over non-payment of a decade's worth of debt for past gas deliveries and Kiev's refusal to move to market prices. Kiev responded by continuing to siphon Russian gas transiting Ukraine, creating a problem for downstream European Union (EU) customers of Russian exports.[1] Russia reminded Ukraine in October of 2007 that unless it paid its outstanding 2007 gas debts of $1.3 billion, Gazprom would again reduce gas deliveries. Ukraine, briefly cut off in early 2008, paid within days. Russia's subsequent and prolonged 19-day gas cut-off in January 2009 to Ukraine and to customers downstream only served to underscore Europe's concern about its overdependence on Russian gas imports.

The net result of these changes was that Russia and the EU are seeking alternative export–import routes. This chapter analyses the viability of existing and new gas pipelines from the Caspian Sea and Central Asian region to Europe and Asia. Russia, on the other hand, is pursuing construction of new gas pipelines to Europe that avoid Ukraine and other troublesome countries. These projects are intended to retain Moscow's dominance in European gas markets and prevent any attempts to bypass it. Economic and geopolitical considerations loom large, as both Russia and China are key players in the development of future pipelines, and the US remains committed to projects bypassing both Russia and Iran. Russia's military defeat of Georgia in August 2008 cast doubt on the reliability of the Georgian corridor as a source of future gas supply to Europe. Additionally, the members of the Gas Exporting Countries Forum (GECF) are becoming more organized and may in the future pose a cartel-like force to be reckoned with by consuming countries. It is too early for the Europeans or the Russians, however, to assess the full measure of how the expected production of domestic European shale gas may affect the future volumes and directional flows of pipeline gas imports. Should domestic EU shale gas production become significant as the new technology becomes widespread, future gas pipeline import projects from Russia and elsewhere may be a casualty.

The state, role and ambition of Central Asian gas production

Central Asia wants to develop new routes to deliver gas to Europe and elsewhere as a means of reducing its dependence on Russian gas pipelines.[2] As EU gas production declines and gas simultaneously increases its share of the West European energy mix, the security concern pertaining to Russian gas imports has taken on great importance. Russia is the single largest external supplier of gas to Europe. In 2007, Europe imported just under 150 billion cubic metres (bcm) from Russia, or 50 per cent more than a decade ago.

Central Asian gas production amounted to 153 bcm in 2007, while domestic consumption totalled 87 bcm that year. The remainder was exported primarily to Russia, although some gas was used to maintain pressure in Central Asian oilfields. While production is growing, it has yet to make major inroads into either Western European markets or markets to the East. Turkmenistan in 2007 exported just over 6 bcm to Iran and about 40 bcm to Russia. Given Central Asian gas reserves – 6 trillion cubic metres in 2007 – these countries are looking for ways to build gas production and diversify exports away from Russia, possibly to link to the South Caucasus Pipeline (SCP), and to send as much as 30–40 bcm annually (bcm/y) to China via the pipeline currently under construction. The potential supply of gas from these countries is enormous, provided the development and transport investments are made and contracts with both Eastern and Western customers are signed. The ambition of the Central Asian resource owners is to develop and export their gas at market prices.

Three pipelines currently carry natural gas produced in the Caspian region, and a fourth is under construction. The first is the SCP that carries Azeri gas from the Shah Deniz field in the Caspian Sea to Baku, Azerbaijan and onward to Erzurum, Turkey, via Georgia. The second is the line from Turkmenistan to Iran. The third is the set of pipelines carrying Central Asian gas to Russia. A pipeline from Turkmenistan to China is under construction.

The 690-km SCP began carrying Shah Deniz gas in July 2007. Full Phase I capacity is 8.6 bcm/y. Future production levels have been placed at up to 50 bcm/y, should Shah Deniz reserves be sufficiently prolific and demand warrant, but the level of uncertainty is great.[3] Industry experts suggest that output could rise to 16–20 bcm/y by 2015 or earlier with relatively little additional cost. Contracts currently cover Turkey (6.6 bcm/y), with smaller amounts to Azerbaijan, Georgia and Greece. Future expansion of SCP will depend on new reserves as well as on both companies and governments concluding that it is in their best commercial interests to proceed.

With the 2007 completion of the 285-km Interconnection–Greece–Italy (IGI) pipeline linking Turkey to Greece, some Shah Deniz volumes have reached Greece, with IGI sized to carry 11–18 bcm/y from a variety of sources. Proposals call for the Greek line to be extended 550 km across the Adriatic to Italy to deliver as much as 8–10 bcm/y. The driving forces behind these proposals are those of energy diversification and security, as Greece, Turkey and Italy are already dependent on Russian gas, and all three countries view Azeri gas as a means of diversification.

The second gas pipeline runs from Turkmenistan to north-east Iran. Completed in 1997, it has a capacity of 8 bcm/y, but has never carried much over 6 bcm annually. It was initially viewed as the first step in a major project to deliver Turkmen and Iranian gas to Europe, but that goal was never realized. The third set of gas pipelines are those of the Soviet era, carrying gas from Turkmenistan, Kazakhstan and Uzbekistan to Russia. These pipelines, known as the Central Asia–Centre system, were built over the period 1960–88. They are in disrepair, and capacity is now less than half the original 90 bcm/y.

Bringing Central Asian gas to market: export options

In addition to the three existing gas pipelines, there are a variety of proposals to carry Central Asian gas to markets in Europe and China.

A TransCaspian gas pipeline

It is doubtful that a TransCaspian gas pipeline from Turkmenistan across the Caspian seabed to Azerbaijan is likely as a means of diversifying European gas supplies. Recent Caspian Summits have stated that there can be no TransCaspian pipeline projects without the consent of all five littoral countries, and Russian and Iranian objections are well known.

Current interest in a TransCaspian gas pipeline needs to be informed by history. In 1996, the idea was born: a twin set of pipelines, one for oil and one for gas, was to be built on the floor of the Caspian Sea between Turkmenistan and Azerbaijan ($3 billion estimated cost, 30 bcm/y capacity), with each fuel destined for Europe. The original proposal was shelved in 2000, due to no buyers, no suppliers, no financing, and no agreement on sea boundaries; those problems remain. Moreover, Azerbaijan's Shah Deniz gas development took centre stage and pushed TransCaspian pipelines aside. A decade later, the idea of a TransCaspian pipeline has resurfaced. Proposed capacity remains at up to 30 bcm/y, although costs have nearly quadrupled.[4] The change of Turkmen leadership in 2007 revived international interest. But political support is useless without commercial interest, and the issues affecting the line's prospects are similar to those that existed a decade ago.

A TransCaspian project certainly needs political support, but it also has to be in the commercial interests of both the participating companies and the participating governments. The President of Turkmenistan maintains that he has not ruled out a TransCaspian gas pipeline,[5] but he views that option as at the bottom of a list that includes the pipeline to China, an expanded pipeline to Iran, and a pipeline through Afghanistan to Pakistan and India, all of which seem either economically or politically doubtful at the present time.[6] What this indicates, however, is that the President is thinking primarily of routes to the east rather than to the west.

Commercial interests may yet force movement toward agreement on delimitation, and perhaps lay the groundwork for consensus. Only when the governments of the producing and transit countries agree on terms, and thereafter when business interests assess the risks and recognize the opportunities for profitable trade, will a project likely go forward. This means that without agreement on delimitation, dedicated supplies, identified buyers and markets, and project financing, a TransCaspian gas pipeline remains no more than an unrealistic hope.

Azerbaijan's offshore Shah Deniz field could supply 20 bcm/y to a TransCaspian project, after its Phase II completion, while meeting its commitments to Turkey.[7] But for a 30-bcm/y TransCaspian pipeline – even with approval by Russia and Iran – new and more gas would be needed to make such a line economic. It thus seems unlikely that enough gas sources will be identified to permit a TransCaspian gas pipeline for at least another 10 years, regardless of the political positions of Russia and Iran, or the European politics of diversification away from Russian gas.

There is, nonetheless, a new development which could change the calculus affecting the possibility and timing of a future TransCaspian gas pipeline. In 2006 the huge Turkmen gas field, Yolotan–Osman, was discovered. Gaffney Cline and Associates performed a reserves assessment in 2008; it estimated Yolotan gas at from 4 to 14 tcm, with a best estimate of 6 tcm. This estimate vaults Turkmenistan into the position of having the fourth or fifth largest gas field in the world.[8] It is so large that it could in time serve both the West and China, with some gas even going north to Russia. The discovery may be an extraordinarily positive development for the future of European energy security, but the gas could yet flow eastward to China. In the aftermath of the Georgia–Russia war, the political risk and uncertainty facing such a pipeline to the West have risen. If Yolotan feeds a TransCaspian line of about 30 bcm/y (or two lines at 60 bcm/y), it would likely make landfall in Azerbaijan or initially connect with Shah Deniz, thereafter travelling through Georgia, Turkey, Greece and Eastern Europe before reaching Western Europe. Alternative routes still involve Georgia, with one traversing the Black Sea from Georgia to Ukraine (the White Stream pipeline) and subsequently linking to existing/expanded West European gas delivery systems. The political risk of instability and interruptibility of such an immensely costly project, however, is a negative development for the future of European energy security.

An overland project: the Nabucco gas pipeline proposal

The Nabucco project is an attempt to bypass both Russia and Iran for future gas deliveries to Europe. Nabucco looks more dead than alive at the moment, despite continued political declarations of support from the EU and the US. While the US remains officially upbeat on the future of the project, as of mid-2010 only Azerbaijan has agreed to provide 10 bcm/y of gas; this is insufficient to justify investment in a $15–17 billion 30-bcm/y pipeline.

The 2000 Nabucco gas pipeline proposal appeared to be the latest version of Iran's and Turkmenistan's failed proposal of a decade ago to ship gas from both countries westward. In 2003 both the European Parliament and the European Council supported Nabucco, and the EU endorsed it as a top priority after the 2006 Ukraine gas cut-off incident. The EU earlier believed that Nabucco was the key to breaking Russia's control of Central Asian gas exports. Turkmenistan and China, however, have already broken Russia's hold by their agreement to build a gas pipeline between the two countries. An EU-supported feasibility study of Nabucco began in January 2006. The year 2009 was the target date for signing a document committed to moving ahead with Nabucco, although clearly this date has passed.[9] The EU's position, however, seems to have changed over time: while the EU supports Nabucco as a priority project, it does not by that support oppose South Stream, a Russian proposal to deliver Russian gas to the western shores of the Black Sea, and thereafter onward to Europe. The EU may *prefer* the Nabucco line, but is not averse to allowing individual member countries to choose between projects.

The Nabucco proposal is for the most part a pure transportation play, with no producer-company equity participation and only one major gas firm among its shareholders (Germany's RWE, which is also allegedly considering exiting Nabucco in favour of joining South Stream), and that may be a fatal flaw.[10] In general, transit countries or their member/national firms simply do not build pipelines; pipeline success depends on commitments from both large buyers and large producers.[11] Nabucco calls for bringing gas from the Caspian area and perhaps from Iran as well, to Europe, with the line terminating at Austria's Baumgarten pipeline juncture and hub. It is a partnership of companies from six nations, with Austria's state-owned OMV in the lead. The project originally included the state gas pipeline companies of Turkey, Bulgaria and Romania, along with the privatized Hungarian gas company, Mol. It now includes RWE of Germany as well, a large European seller of gas. Each participant has a 16.67 per cent share in the venture. Nabucco partners are thought to be willing to consider offering Gazprom an ownership share, which may in turn be linked to carrying Russian gas. Gas from Russia, as well as partial Russian pipeline ownership, would clearly dilute Nabucco's goal of diversifying away from Russian gas supplies.[12] With the new discovery in Turkmenistan, there is some speculation that that country will be invited either to join the project or to commit future gas to it.

Nabucco continues to battle a number of hurdles, including Turkish transit pricing, unidentified sources of gas to fill the line, no signed purchase contracts, and no secured financing. Currently an 'open season' for booking transport capacity is underway, to discover what supplies potential shippers might be willing to commit. To date, only Azerbaijan has expressed willingness to ship up to 10 bcm/y, which is insufficient alone to justify the investment. In view of the project's delays, Hungary's Prime Minister called in April 2007 for an extension of the Blue Stream pipeline (carrying Russian gas) as a preferred alternative to Nabucco.[13] After EU criticism, the Prime Minister re-declared his support for Nabucco.[14]

Nabucco and the Russian-proposed South Stream gas pipeline, to run from Russia across the Black Sea to Bulgaria, appear to be in direct competition with each other. This competition is likely to deter the interest of investors in Nabucco, if they perceive that South Stream offers not only serious competition but also a more rapid construction schedule and defined supplies of (Russian) gas, which

Nabucco does not have. Both pipelines are in the proposal status, and the supporters of each are trying to win supporters away from the other. Hungary is leaning toward South Stream, as are Serbia, Slovenia, Austria, Bulgaria and Greece. Gaz de France (now GdFSuez), rejected as a participant in Nabucco, now favours South Stream. Romania remains committed to Nabucco. If South Stream outpaces Nabucco, it will at the same time defeat EU interest in avoiding additional dependence on Russian gas supplies in the future. In March 2010 the CEO of ENI went further and proposed to merge the Nabucco and South Stream projects to 'reduce investments, operational costs and increase overall returns'.[15] This proposal was rejected by Energy Minister of Russia Sergey Shmatko saying that 'South Stream is more competitive than Nabucco' and that 'Nabucco and South Stream are far from being competitors'.[16]

Nabucco would run some 3,300 km, with an ultimate capacity of 25–31 bcm/y. The project calls for gas to enter through legs from Georgia and Iran, but no proposed volumes are yet on offer, other than the 10 bcm/y from Azerbaijan. Projected future gas sources include Egypt, Iran,[17] Azerbaijan and perhaps Turkmenistan and Kazakhstan, Iraq[18] and Russia. Estimated to cost €13 billion ($17 billion) or more, it is targeted for completion in 2015, although completion dates for big projects of this nature in this part of the world often slip.

As a diversification project that avoids transit of both Russia and Iran, it satisfies US stated policy objectives. The US, however, will probably frown on gas throughput from either Iran or Russia. The longer it takes Nabucco to come together, if in fact it does, the less likely that it will displace either the rival South Stream or the rival Blue Stream–2 (from Russia to Turkey) gas pipelines.

While most observers are sceptical of Nabucco's prospects, the landscape has now changed, what with the new large gas discovery in Turkmenistan and as a result of the Georgia–Russia war. Turkmenistan could almost single-handedly commit sufficient gas to this project to merit its going forward, provided the terms and price are agreed with customers and financiers. The most likely routing of Nabucco is through Azerbaijan and Georgia, and the politics of instability raise the same risks that face a TransCaspian seabed gas pipeline in the aftermath of the Georgia–Russia war. Reliability of deliveries takes on new meaning after the closure, albeit temporary, of the South Caucasus gas pipeline during the war.

A Central Asian gas pipeline to China

Turkmenistan and China signed a gas export agreement in mid-2007. Construction is now complete on a 30-bcm/y pipeline to China which will carry increasing volumes of Turkmen, Kazakh and Uzbek gas. The whole pipeline was inaugurated in December 2009 during Hu Jintao's visit to Turkmenistan with the leaders of Turkmenistan, Uzbekistan and Kazakhstan. This line has an option to expand to 40 bcm/y, and could easily be looped to double its capacity, potentially carrying future large amounts of Yolotan gas. The EU's chances of taking Turkmen gas may be hurt by the realization of the gas line to China.

China National Petroleum Corporation (CNPC) – owned largely by the state – expedited this project for strategic rather than economic reasons, and it is China that is incurring the risks of both transport and gas supply. The line is still a gamble on the economics, with the agreement calling for China to find and develop *new* fields to fill the line. The Turkmen government has reserved a number of undeveloped fields for CNPC in eastern Turkmenistan. Ashgabat is to provide 17 bcm/y, while China is to develop the untapped Bagtyyarlyk gas field to supply the other 13 bcm/y.[19] Once Yolotan gas is developed, it will clearly change the strategic gas export picture for Turkmenistan and its eastern and western customers. Future output from pre-existing fields (prior to the Yolotan discovery) of 80–90 bcm/y (well above current production of 67 bcm/y) is committed to Russia, but Yolotan output could reach China, Russia and/or Europe.

When the contract was signed in 2007, China agreed to pay Turkmenistan $90 per thousand cubic metres (tcm), less than the Russian price of $100/tcm, presumably because China was assuming both the exploratory and development risks.[20] Price negotiations continue, and other sources suggest that China would be willing to pay as much as $160–180/tcm,[21] still well below Russia's price to Europe as it resells Central Asian gas. With the line to China complete, Turkmenistan has the opportunity to sell to the highest bidder, either to the East or to the West.

A Central Asian gas pipeline to Pakistan and India via Afghanistan

A final proposed Central Asian gas export pipeline route involves crossing Afghanistan to reach customers in Pakistan and India. This idea has been around since the early 1990s, but was abandoned in 1998 due to war in Afghanistan, and continuing instability in both Afghanistan and Pakistan. In 2005 the Asian Development Bank breathed new life into the 15–30 bcm/y Trans-Afghan project. The bank completed a feasibility study of the proposed 1,680-km line. It found that the pipeline project, estimated to cost $3.5 billion at that time, is technically and economically feasible if financing is obtained. The project won US political support, as it passed through neither Russia nor Iran. Private investor interest was and is not high given the ongoing war in Afghanistan and instability in Pakistan. The future demand for Turkmen gas in Pakistan is also uncertain, since Pakistan has substantial domestic gas reserves thought sufficient to cover anticipated demand for some time. While India has been both for and against the project, by mid-2005 it was again interested in it.

Most observers think the line is unlikely to be built unless the volumes are sufficiently large – 30–40 bcm/y – to justify the investment, and that requires India's firm commitment. Competing with the Trans-Afghan gas line are other proposals, including one from Iran across Pakistan, and another from Bangladesh–Myanmar to India. Financing will be difficult to arrange without sufficient demand, greater political predictability and stability in both Afghanistan and Pakistan, and continued warming in Pakistani–Indian relations. Other than the substantial political and financial risk, a key question is whether there is sufficient gas demand to justify the project.

Russian resurgence and what it means for Central Asian gas

Russia has responded quickly and effectively to the various proposals to bypass it in the transport of Central Asian gas to Europe. Its strategy is well developed, coherent and clear. It seeks to market its gas, keep its existing European customers and win new ones, and integrate downstream into pipelines, storage and hubs. At the same time it seeks to pre-empt or block new gas flows that could challenge its gas markets in Europe. It will likely consider future interruptions of gas deliveries to assure payment by transit countries, and will explore the possible gains from participating in a 'gas OPEC' – the GECF.

As a result, the Russian pipeline proposals, with their first-mover advantage, look far more likely to reach completion than does Nabucco. While they are initially transit-avoidance pipelines, they may also serve to significantly increase Russian gas exports above presently anticipated levels. They do nothing, however, to allay EU fears of overdependence on Russian natural gas supplies – they offer alternate physical routes but not alternate supplies from different suppliers. This observation has been repeatedly noted, yet EU spokesmen are on record as saying that with the expected growth in EU gas demand, 'any new infrastructure that can match… demand… [is] a positive development.'[22] We now examine the four new Russian gas pipeline proposals.

Nord Stream – linking Russia to northern Europe

The first of these pre-emptive undertakings is the Nord Stream pipeline under the Baltic Sea. Under consideration since the 1990s and with German governmental agreement in 2005,[23] Gazprom is building from Portovaya Bay near Vyborg, Russia, while the Europeans are building from Greifswald, Germany. The 1,189-km line is specifically designed to enable Russia to bypass the troublesome/unruly countries of Ukraine, Belarus and Poland. It is an expensive insurance policy. Costs in 2006 were estimated at close to €4 billion for the first line (a total of two 27.5 bcm/y lines are planned), but were revised in 2007 to over €6 billion; costs may yet rise to at least €12 billion, as both offshore and onshore costs are rising.[24] More recent estimates place onshore construction costs at €6 billion, while offshore construction costs are likely to run at €8–9 billion.[25]

New Russian gas to feed Nord Stream will come from the Yuzhno-Russkoye field in Western Siberia, and from Shtokman gas in the Arctic offshore. Shipments were targeted to begin in 2010, but the date slipped to 2014–15 due to cost and environmental issues. Construction was officially launched at Portovaya Bay by President Medvedev in April 2010. Extensions include possible links to Denmark and elsewhere in Scandinavia, as well as to the Netherlands and the UK through existing gas pipeline networks.[26] The route is being altered to avoid concerns about WWII ordnance on the Baltic seabed. Environmental issues could delay the project, but are unlikely to scuttle it.

Two of the remaining Russian pipeline proposals parallel routes through Europe similar to that of the proposed Nabucco line, reaching Baumgarten, Austria in the north and Greece and Italy in the south.

Blue Stream–2 – linking Russia to Turkey and Europe

The second pipeline proposal is Blue Stream–2, an extension of the 16-bcm/y Blue Stream–1 gas pipeline, inaugurated in 2005, carrying Russian gas across the Black Sea to Turkey. This original Blue Stream was built as new capacity to counter any likelihood of completion of the original TransCaspian gas pipeline proposal a decade ago.[27] In that regard, it successfully fulfilled its purpose. Blue Stream–2 offers transit diversification (it avoids Ukraine and Belarus) as well as new capacity. It has a proposed capacity of 24–30 bcm/y, and would have a northern and southern leg serving Austria, and Greece and Italy, respectively. The proposal would parallel the Blue Stream–1 route across the Black Sea. The costs and technology are expected to be more easily manageable than those of either Nord Stream or South Stream, given the experience gained from building Blue Stream–1 and the advances in metallurgy and other related pipeline technologies.

South Stream – linking Russia to southern and northern Europe

The third of these Gazprom pre-emptive pipeline proposals is South Stream, which could displace the proposed Blue Stream–2. South Stream is widely viewed as Russia's effort to counter the rival Nabucco project, and the sooner South Stream moves ahead, the less likely Nabucco will move at all. The new 900-km pipeline is designed to run from Russia across the Black Sea to Bulgaria, thereafter splitting into northern and southern legs, and following routes westward similar to those proposed for Blue Stream–2 and Nabucco. The line would start at Russia's Beregovaya compressor station, which also pumps gas into the original Blue Stream pipeline.

South Stream proposes to bypass Ukraine, Belarus and Turkey. Capacity is targeted at 31–63 bcm/y. The line could be built in three years, with gas deliveries commencing in 2012–15.[28] Construction costs were estimated at $5.5–13 billion in June 2007, but by November that year had risen to nearly $15 billion.[29] Estimates made in 2008 place total costs at €12.8, or over $16 billion, more than double the original estimate.[30]

The pipeline is to be built and operated by several project companies. The offshore section would be built and operated by South Stream AG, a joint company of Gazprom and ENI. EDF's signature in June 2010 of a memorandum of understanding to acquire a minority stake (nine per cent) in South Stream AG gives the pipeline project extra credibility due to EDF's large role in European gas marketing.

Beyond the rivalry with the Nabucco pipeline project, a major political obstacle remaining is the problem of transit rights through Black Sea coastal countries' economic zones. The natural path for the pipeline passes through Ukrainian and Romanian waters, requiring their consent to be built.[31]

Pre-Caspian pipeline – linking Central Asia to Russia

In May 2007 Russia countered proposals to move Central Asian gas to Europe via routes that avoided Russia. Moscow signed an agreement with Turkmenistan and Kazakhstan calling for the repair of old gas lines and the building of a new line, called the Pre-Caspian line, on the eastern shore of the Caspian Sea, to carry Central Asian gas north to join the Russian gas network. The agreement covers up to 30 bcm/y of gas, at a price to be determined by the market. The 2007 agreement was widely interpreted at the time as a blow to Western interests and alternative non-Russian gas pipeline projects, but there is enough Central Asian gas to serve not only this pipeline but various others as well, if purchase agreements and financing are arranged. What the agreement does, however, is bind Turkmenistan and Kazakhstan to continuing gas shipments through Russia.

Perspectives on pipelines – making sense of it all

Russia's new gas pipelines, if completed, would 'encircle' Europe in the north by the Nord Stream pipeline (27.5–55 bcm/y), and in the south by the South Stream project (30 bcm/y) and/or by the possible rival Blue Stream–2 project (up to 30 bcm/y).[32] Russia appears to have Europe in a natural-gas pincer, giving Europe almost no opportunity to diversify away from dependence on its gas, since many contracts have been renewed to 2040.[33] All of the new proposals for Russian gas deliveries will serve only to increase European dependence on Russian gas, regardless of the fate of Nabucco.[34] In short, Europe's chances of breaking its reliance on Russian gas appear slim.

Moscow's policy responses to attempts to bypass it are to build alternative pipelines with sufficient capacity to serve its European customers, and realize these projects more rapidly than can the bypass proposals. It has the funds and the commitment of the state through Gazprom to move ahead in an expeditious manner. Potential lower energy prices will not derail Moscow's commitment to undermining attempts of other bypass projects, given its vast foreign currency reserves and the size of its oil fund. National oil and gas companies in Central Asia, despite their attempts to diversify their exports, are pleased with the Russian trade, especially as they have won ever-increasing prices. What is notable in all these proposals – South Stream, Nord Stream, Blue Stream and the Pre-Caspian shoreline pipeline project – is that not one of them passes through Azerbaijan or Georgia, and thus all of them avoid the risks of interruption due to potential new outbreaks of hostilities in the Caucasus region.

Russia appears to be using South Stream as a blatant attempt to undermine and derail the Nabucco project, and its efforts may yet succeed. At the same time, however, Moscow knows that it is incapable of preventing European purchases of Caspian or Middle East gas if such gas is available at competitive prices and can be securely delivered. Once Nabucco is truly abandoned, South Stream accomplishes its goal and may not necessarily go forward. Alternatively, South Stream, Nord Stream and Blue Stream–2 could all go forward, carrying 87–115 bcm/y to

Europe, thereby giving Russia export 'flexibility', but at some cost of building excess capacity. Moscow may intentionally be pursuing multiple pipelines and deliverability as 'flexibility' options in order to be able to switch future export volumes and destinations as a means of rewarding some customers over others.[35]

In all of these proposals, Gazprom is demonstrating its opposition to diversification in its future gas trade with Europe. It wishes to increase its vertical integration and extend its reach so as to market gas directly to customers, and this strategy is causing much concern in EU circles. Gazprom continues to seek ownership of downstream assets to ensure outlets for its gas. The EU fears that as Russia gains interests in downstream assets as well as controls upstream supplies, its market power grows at the expense of its customers. But as of October 2008, the EU permitted member states the freedom to decide whether to allow foreign ownership of downstream energy assets.[36]

Russia's behaviour is perfectly sensible, designed to guarantee it demand security and continued European gas sales. To sweeten the trade with its customers, Gazprom promised to assist in developing new, or enlarging existing, gas storage facilities in Austria, Hungary and Turkey. Hungary has now reached agreement to participate in South Stream.[37] Clearly Russia will not deviate from its strategy to keep and expand its gas presence in Europe.

Russia has additionally signalled that it will play hardball in the future with its European gas customers. It is now a founding member of a 'troika' group (including Iran and Qatar as well) of a 'gas OPEC'. This group will meet quarterly to coordinate future gas policy and discuss pricing and supplies. While the troika has not said so publicly, its future intent may be to control world gas prices, mirroring OPEC's control of oil prices. Clearly the idea of a gas producers' cartel is not well-received by the International Energy Agency (IEA), the US and/or the EU. The concern of each is that global energy security would suffer as a result, and that prices could be influenced or controlled. The IEA has underscored the idea that such a group's formation would send a signal to gas users to find alternatives, and thus undermine the cartel's ability to sustain prices in the longer run.[38] At present, however, the system of long-term gas contracts in place does not lend itself to cartel-determined price changes.

Conclusion

Central Asian gas supplies appear to offer Europe a means of diversification away from dependence on Russian gas. Yet proposals to bring Central Asian gas to Europe, either by an overland pipeline or by a TransCaspian Sea pipeline, are being scuttled by Russia's forceful responses to build Nord Stream, South Stream and perhaps a second Blue Stream, so as to keep its dominant position in Europe.

Russia will build these pipelines in order to deter Western bypass projects, and price its gas to retain its European markets. It knows Europe has few alternative sources for the large volumes of gas demand expected in the future. Any future non-Russian-sponsored gas pipeline projects will require careful scrutiny of Russia's pipeline capacity and flexible delivery options.

Moscow has reluctantly come to terms with the growing market power of its long-time Central Asian suppliers. It has granted them higher gas prices in the face of its own short-term needs for their gas. Its efforts to build the Pre-Caspian line linking Turkmenistan and Kazakhstan to the existing Russian pipeline network are a continuation of its role as a long-time purchaser. These efforts are now being undertaken to counter the power of China's rising gas demand and are an attempt to tighten the ties between Russia and its former republics, even if the price Moscow has to pay rises in direct response to competition from China.

In our view, Russia's strategy is clear and well-defined. The long-term trend of rising EU gas demand has correlated well with growth of Russian gas exports to Europe. Russia intends to keep its position in the EU market, which means rising future gas deliveries to Europe. In the face of the recession-dampened EU gas demand from all providers, and new EU sources of LNG, and EU shale gas production potentially displacing some Russian imports, Moscow will be forced to be more competitive on gas export prices Moscow plans to continue to act as the major gas supplier to the region. Its ways of doing this are to build new pipelines ahead of demand and to act quickly and decisively, while non-Russian-sponsored projects languish in uncertainty, without dedicated supplies, identified buyers and markets, and confirmed project financing.

Part II

Russian foreign and energy relations with NIS net exporters

5 Ups and downs of the Russia–Turkmenistan relationship

Vladimir Milov

Introduction

Turkmenistan is one of Eurasia's most isolated countries. It is one of the poorest countries in the post-Soviet space (according to EBRD, its poverty rate is estimated as 44 per cent) and one of the most criticized for its human rights and corruption record. Eccentric president Saparmurat Niyazov, who had been governing the country since 1990 and survived an assassination attempt in 2002 (only to die in December 2006), has become known for his systematic aversion to any international political alliances (backed by the country's declaration of 'permanent neutrality' formally recognized by the United Nations in 1995) and frequent position shifts, which allow some experts to characterize his political style as 'erratic'.

After gaining its independence from the Soviet Union in the early 1990s, Turkmenistan preserved a lot of policies distinctive of the administrative economy. These policies had featured the predominance of state ownership of the means of production, restrictions on foreign exchange activities, widespread subsidies and an approach to planned development on the basis of import-substituting industrialization. While some privatization has taken place, the authorities remain committed to maintaining the major productive sectors in state hands. Turkmenistan has proved an extremely difficult place to do business and has attracted little foreign investment. Arbitrary state action, high levels of corruption, and general unattractiveness of the heavily regulated business environment have combined to deter investors. According to the 2006 Transparency International Corruption Perceptions Index, Turkmenistan is one of the countries most plagued by corruption, occupying between the 142nd and 150th positions out of 163, together with Angola, Republic of Congo, Kenya, Kyrgyzstan, Nigeria, Pakistan, Sierra Leone and Tajikistan.

Nevertheless, Turkmenistan plays a remarkable role on the international gas markets. According to BP estimates, the country holds 1.6 per cent of global gas reserves and currently accounts for more than 5.5 per cent of gas production in Europe and Eurasia. Natural gas, according to the National Institute of State Statistics and Information, Turkmenmillihasabat, accounts for well over 50 per cent of the structure of the country's exports. Turkmenistan's gas resources are landlocked, and the country is heavily dependent on gas transit through Russia, since the traditional logistics of gas pipelines were built during the Soviet period and centered on Moscow. That situation is starting to change gradually with two main projects: the Turkmenistan–Iran gas pipeline and the planned Turkmenistan–China gas pipeline. Despite this, Russia – the largest Eurasian gas producer and exporter by far – has recently been increasingly dependent on imports of Turkmen gas in order to fill the gap between growing Russian and European gas demand and stagnating Russian gas production.

Turkmenistan is also one of the important players in the Caspian basin – Turkmenistan's position had been strongly influencing the negotiations of the five littoral states in the legal dispute on the division of the Caspian Sea. Resolution of this dispute is key to ensuring stable development of the Caspian region, particularly its oil and gas resources, since a lot of oil and gas exploration, development and infrastructure projects have been frozen due to uncertainty over the issue of the Caspian's division. Currently, entire offshore oil and gas fields in the Southern Caspian basin remain untapped due to the lack of clarity about ownership, the trans-Caspian pipelines are unlikely to be constructed without an agreement on ownership of the waters, seabed and resources, and even the regulation of tanker traffic on the sea remains controversial.

In fact, during all the negotiations on the Caspian dispute's resolution, it was Turkmenistan which (together with Iran) took one of the most inconsistent, selfish and unpredictable positions, actually contributing a lot to the failure to reach a final resolution (see Chapter 4).

In this respect, it is important to note that Turkmenistan, despite relatively moderate gas reserves and production (in absolute terms) and the landlocked nature of its gas resources, has become a very important 'secondary' player in the Eurasian energy arena. Taking into consideration the tight nature of the global and regional energy markets, driven by the continuing strong growth of energy consumption, which, in turn, is a result of strong economic growth in many parts of the world including the Asia Pacific region (China and India) in the first place, the importance of Turkmenistan's marginal supplies of gas, as well as of supplies of offshore oil and gas fields in the Southern Caspian basin, cannot be overestimated.

Latent tensions, emerging pragmatism

Over the last decade, political relations between Russia and Turkmenistan have been troubled by the unpredictability of Turkmenistan's political strategy and permanent political manoeuvres, first under Niyazov and then again under Gurbanguly Berdymukhammedov, causing strong Russian uneasiness and concerns. The most visible demonstration of Turkmenistan's independent, unpredictable and, as it is considered among Russian elite, 'anti-Russian' politics took place in October 1996, when Turkmenistan refused to attend the Central Asian summit gathered to discuss the risks of further advance of the Taliban in Afghanistan, shortly after Taliban forces took over Kabul in September 1996. The summit was attended by leaders of the Central Asian states and the Russian Prime Minister Viktor

Chernomyrdin, and was focused on the urgent development of a joint position and a possible plan of action to protect regional stability from the threat of increasing Taliban influence. Turkmenistan's leader Saparmurat Niyazov, however, refused to participate in the summit or send his representatives claiming that the country would 'maintain its neutrality' on the Taliban issue, in accordance with the country's 'permanent neutrality' status.

During the subsequent years, this Turkmen attitude evolved into a much broader cooperation with the Taliban regime, including alleged participation of Turkmenistan in drug trafficking, actual participation in the project to construct a gas pipeline from Turkmenistan to Pakistan through Afghanistan without Russian participation (and with participation of the US company Unocal), and, more broadly, political support of the Taliban regime. Turkmenistan was among the few countries opposed to a stronger international position against the Taliban regime, like the 1999 UN Security Council decision to impose sanctions on Taliban Afghanistan for support of terrorist activities.

That issue has been a major point of concern for the Russian elite with regard to Turkmenistan's politics, because Taliban influence in the region was traditionally considered as hostile. First, it was considered as the source of the Pakistani–US–Saudi influence in the region, taking into account that the Taliban forces initially had been established under the guidance of Pakistani intelligence, and supported by Saudi Arabia and, initially, the US. All of these countries have traditionally been considered, to say the least, as competing forces to Russian influence in Central Asia. Second, after the success of Taliban efforts to take control of Afghanistan in 1996–7, it became clear indeed that the Taliban's growing influence is itself a great source of instability in the Central Asian region, regardless of who is backing the Taliban, in terms of threat to state borders and the existing political regimes in Central Asia, increase in drug trafficking and terrorist activities, and the spread of Islamic fundamentalism. Third, some Taliban activities, such as the Trans-Afghani gas pipeline, were clearly targeted to reduce Russian influence in the region, diverting Turkmen gas supplies away from Russia and challenging the Russian gas transit monopoly.

Therefore, although Russia has seemingly been lacking clear general strategy toward Afghanistan after withdrawal of troops from the country in 1989 and the defeat of the Afghani communist government in 1992, resistance to the Taliban influence had always been considered a priority policy objective by at least a significant number of Russian policy-makers. In this light, broadly independent relations of the Turkmen leadership with the Taliban were always viewed in Russia as quite an unfriendly development.

Another issue of Russian concern was the dispute over the division of the Caspian Sea, a politically sensitive issue for Russia since the collapse of the Soviet Union. This issue is considered as key to ensuring stability in the Caspian area. However, Turkmenistan's position on the division of the Caspian has shifted during the course of the negotiations during 1990s and has probably been the most inconsistent of all parties. This arguably forced Russia to pursue the strategy of bilateral deals among Caspian littoral states: in May 2003, Russia, Azerbaijan and

Kazakhstan formalized this understanding in a trilateral agreement on sub-surface boundaries and collective administration of the sea's waters. They divided up the northern 64 per cent of the sea using a median line principle. Turkmenistan opposed this, refusing (together with Iran) to sign the agreement. Turkmenistan favours a modified technique for applying the median line principle, one that would give it possession of Azerbaijan's largest offshore deposits, while Iran, which would secure just 12 per cent of the seabed under the median line principle, wants the sea divided into equal shares. This issue, as well as Turkmenistan's permanently inconsistent position, has served as a major source of Russian disappointment with Turkmen policies.

But it was not only Russia that had reasons to be disappointed with Turkmenistan. There has also been strong discontent on the Turkmen side over Russia's position on the trade and transit of Turkmen gas. Although Russia did not oppose Turkmen gas being shipped and sold in principle, it has been from the beginning quite keen to limit Turkmenistan's capabilities for marketing of its gas abroad, in order to minimize competition with Russian gas on the international markets. Russia has traditionally considered the possible supplies of Turkmen gas to the European market as a serious competitive threat, and therefore has been reluctant to allow shipments of Turkmen gas to countries other than Ukraine. This had been possible due to the fact that, historically, all of the gas pipelines coming out of Turkmenistan ran through Russia, which allowed it to exert a complete transit monopoly over Turkmen gas. This has allowed Russia to maintain significant price differential between prices of gas purchased from Turkmenistan and European gas prices, particularly during 2000–2006, when European gas prices have been exceptionally high, and captured a large part of export rents that otherwise could have been skimmed by Turkmenistan.

Needless to say, the Russian position meets with a negative reception in Turkmenistan. It is important that Russia never actually tried to hide its intention to capture Turkmen gas through its transit monopoly, and has openly denied the right of European consumers to conclude direct gas purchase contracts with Turkmenistan. This has been the primary (and probably the only) cause of Russian's refusal to ratify the Energy Charter Treaty and the Transit Protocol –Russia has always defended its right to exert the transit monopoly over Turkmenistan and arbitrarily decide on Turkmenistan's access to international gas markets. It is quite natural that in this situation Turkmenistan has always been seeking opportunities to build new gas pipelines that would actually allow the gas to be exported bypassing Russian territory (see below). Needless to say, these attempts by Turkmenistan were always viewed in Moscow as less than friendly.

Relations started to warm up somewhat after 11 September 2001, when Saparmurat Niyazov, clearly afraid that he might be the next target of US military actions in response to his support of the Taliban regime. Turkmenistan started to make certain movements towards Russia in order to demonstrate a willingness to improve relations – a stance considered by many as simply seeking a defensive shield against the possibility of US action. This led to a temporary warming of relations, one of the results of which was the signing of the 25-year Russian–Turkmen gas supply agreement in April 2003, which raised high hopes among many Russian experts about the emerging solid partnership in gas supplies. The gas contract seemed so politically important to the Russian leadership that the Kremlin publicly closed its eyes to the subsequent cancellation of the Russian–Turkmen double citizenship rule by Niyazov, which proved to be harmful to the Russian population in Turkmenistan, but went publicly unremarked by the Kremlin.

In 2005–6 official announcements came from Ashgabat about Turkmenistan's withdrawal from the CIS. The first official announcement was made at the CIS summit in Kazan, Russia, on 26 August 2005. Although Turkmenistan was never an active member at the CIS, and Saparmurat Niyazov was a rare guest on CIS summits and had often publicly distanced himself from the Commonwealth, this announcement was very untimely for the Kremlin – as a sign of further deterioration of the post-Soviet landscape and the drift of more former Soviet republics away from Russia. After the emergence of clear pro-European orientation in Ukraine, Georgia, and Moldova (not to mention the Baltic states), it was very important for the Kremlin to present the case to the Russian public that Russian influence is still strong in Central Asia and it holds the countries in the region well within its orbit. In this environment, Turkmenistan's move to withdraw from the CIS (although not yet completed in legal terms) came as a very unpleasant public blow to the Russian authorities at the wrong time, despite the low profile of the CIS itself as an institution. The move was considered in Moscow as a sign of Turkmenistan's further separation from Russia's political orbit.

Also, it is important that, in spite of the many difficulties and instances of oppression experienced by the Russian population in Turkmenistan, the Kremlin was never particularly active in pushing for the improvement of Turkmenistan's human rights record. In the tradition of Russian politics (and particularly the political style of the very recent period), these issues were sacrificed for the sake of a more 'pragmatic' political approach, based on interests rather than values.

But, as recent history has proved, the 'pragmatic' approach with regard to Turkmenistan brought only partial results. Russia has been enjoying the gas price differential from re-exports of Turkmen gas to international markets, but this has proved to be only temporary and unsustainable, as Turkmenistan turned out to be quite effective in its efforts to push gas supply prices upwards during 2005–6 (see below). The capture of these pricing differentials mostly brought short-term benefits to some vague intermediaries, as will be illustrated, but in the longer term this only forced Turkmenistan to seek alternative gas delivery options, which have begun to develop in the form of the Korpezhe–Kurt-Kui gas pipeline to Iran, and new projects under discussion (see below).

At the end of the day, political relations between the two countries are currently characterized by a great lack of mutual trust and predictability. This can be attributed in large part to Turkmenistan's unpredictable politics, a significant part of which were taken as unfriendly in Russia (support of the Taliban, failure to develop a constructive position on the division of the Caspian, consistent drift away from various forms of post-Soviet integration including the CIS). But it should also be said that, despite the opening of the 'window of opportunity' for increasing Russian influence in Turkmenistan after 11 September 2001, Russia has failed to use this window and through its 'one-way street' policy approach with Turkmenistan has contributed a lot to push Turkmenistan to seek other partnerships and alternatives in terms of supply of its most important export commodity, natural gas, in other directions bypassing Russia.

Developments between Russia and Turkmenistan have not been particularly mutually positive, save for the remaining affiliations between influential stakeholders of mutual gas transit and re-export business in both countries (see below), and show strong signs of instability and uncertainty about the future: Russia–Turkmenistan relations are, generally speaking, continuing to deteriorate markedly. We are most likely witnessing the progressive parting of the two countries rather than their developing partnership, with Turkmenistan definitely drifting further away from Russia's political orbit. It should be said that both countries' policies have effectively contributed to this phenomenon.

Gas supplies and prices

Developments in Russian–Turkmen negotiations on gas supply prices were no less dramatic than the events of 2005–6, when Russia pressed for gas supply price increases from Ukraine, Belarus, the Baltic States, Georgia, Armenia and Moldova, although this received less public attention. In fact, some experts essentially consider the assertive Russian actions targeted at increasing gas supply prices for the westernmost post-Soviet states in 2005–6 as being at least partially initiated by the 'Turkmen factor', the pressure initially and consistently put on Russia by Turkmenistan in 2003–5 in order to increase prices for gas supplied from Turkmenistan to Russia, and the need to compensate additional expenses caused by more expensive Turkmen gas supplies. It should be noted that the connection here is only partially true, but it certainly exists to some extent. The upward pressure on prices initiated by Turkmenistan somewhere in 2004 was a serious factor that has influenced subsequent complex revision of the relations on the post-Soviet gas market.

During the late 1990s and early 2000s, the price of gas supplies from Turkmenistan to Russia[1] was relatively stable at a level of around US$ 42 per thousand cubic metres (tcm), regardless of the actual buyers on the Russian side. Russia was for a long time represented not by Gazprom but by Itera, and a fair amount of bartering was involved, making it impossible to accurately estimate the monetized price equivalent. The supply volumes to Russia at the same time were not large; most of the gas went to Ukraine (by that time, Russia's production deficit had not yet started to bite). At the same time, the actual cost of supply of Turkmen gas from the Uzbek border (the price setting point) was estimated as a maximum US$ 16–18/tcm, which gives an idea of the amount of rent actually levied by Turkmen authorities from gas exports to Russia.

A groundbreaking 25-year gas purchase agreement was signed between Gazprom and Turkmenneftegaz in April 2003. The contract marked a new era for Russian–Turkmen gas supply relations, marking a strong re-entry of Gazprom and a fading role for Itera, as well as defining Gazprom's new long-term strategy towards Turkmen gas purchases. Before 2007, a three-year gas supply contract between Turkmenistan and Ukraine that envisaged annual gas supply volumes from the former to the latter to the amount of 36 bcm per annum, remained valid, According to the contract, starting from 2007 Russia would purchase more than 60 bcm of gas (60–70 bcm in 2007, 63–73 bcm in 2008, and 70–80 bcm annually starting from 2009), or nearly all of the Turkmen gas produced. Before that, supplies to Russia had been set for gradual increase: from 5–6 bcm in 2004 to 6–7 bcm in 2005 and 10 bcm in 2006. This new agreement was praised by many in Russia, since a lot of experts had called for engaging Turkmenistan in a long-term gas supply agreement with Russia to pursue strategic goals, including the resolution of the problem of a temporary deficit in Russian gas supplies caused by the foreseen decline in production of major west Siberian gasfields, as well as preventing Turkmenistan from entering separate gas sale agreements with other competing parties, be it Europe (through pushing Russia to allow the free transit of Turkmen gas to the European market) or other countries like China, Pakistan or Iran (through construction of new alternative gas pipelines bypassing Russia). Despite the fact that the agreement between Russia and Turkmenistan beyond 2008 looks rather like a protocol of intent which is not legally binding, the agreement was cheered by many experts as 'final resolution' of the future of relations with Turkmenistan in the gas supply area, the 'closed deal' on the long-term capture of Turkmen gas for Russian interests.

However, at the end of 2004 Turkmenistan quite assertively demanded the reconsideration of gas supply conditions: Saparmurat Niyazov demanded of both Russia and Ukraine that they pay US$ 44/tcm in cash form, whereas before roughly half of payments came in barter form. Russia (Gazprom) resisted this move, which led to a temporary shutdown of gas supplies from Turkmenistan to Russia in January–March 2005. (It is important to note that the overall volume of gas supplies from Turkmenistan to Russia was not significant, the majority of gas was exported to Ukraine, and Russia did not try to interrupt the transit of Turkmen gas to Ukraine, primarily, as it appears, due to the fact that Russian commercial interests had been substantially involved in transit and resale of Turkmen gas to Ukraine through vague intermediaries – see below.)

Historically, this was not the first interruption of supplies – the first occurred in 1997–8, when Turkmenistan had near-completely shut down its gas exports for two years, but it is significant that such behaviour has recurred in the very recent past. Taking into consideration the series of such crises – gas supply interruptions from Russia to Belarus in January 2004, from Turkmenistan to Russia in January–April 2005, from Russia to Ukraine in January 2006 – it appears that such conflicts are becoming a systematic occurrence on the post-Soviet gas market, which presents a strong factor of concern for future Eurasian energy security.

In April 2006, however, gas supplies to Russia were restored, though the supply interruption had its effect – Russia was forced to agree to pay US$ 44/tcm in 100 per cent monetized form. Despite claims by Russian representatives that 'the solid price of US$ 44/tcm was set for 2006 and Gazprom would be ready to

Table 5.1 Gas supply prices from Turkmenistan to Russia on Turkmen–Uzbek border, 2003–7

Supply periods	*Apr 2003–Jan 2005*	*Apr1–Dec 2005*	*Jan–Sept 2006*	*Since 1 Oct 2006*
Gas supply price, US$/ tcm	44 (30)	44	65	100

Source: Gazprom

Notes:

In January–March 2005, intentional gas supply cuts from Turkmenistan to Russia had occurred due to lack of agreement on gas supply prices.

Nominal supply price was set as US$ 44/tcm, but part of payments were made in barter form (equipment supplies, etc.). Monetized equivalent of this nominal price is roughly estimated as US$ 30/tcm.

defend it in arbitration',[2] from the beginning of 2006, the price of Turkmen gas supplies to Russia had risen to US$ 65/tcm, fuelled by sharp increases in gas supply prices from Russia to the westernmost post-Soviet states since the beginning of 2006. Moreover, Saparmurat Niyazov had insisted that in Autumn 2006 the price of Turkmen gas supplies to Russia should further increase to US$ 100/tcm, which was finally agreed upon in September 2006.

It should be said that Russia's assertive efforts to increase gas supply prices to the westernmost post-Soviet states in 2005–6 had to a certain extent had the reverse effect, provoking Central Asian gas suppliers (Turkmenistan in the first place) to exert more pressure on Russia in order to increase prices on gas purchased by Russia from Central Asian countries.

More importantly, these recent developments have introduced further strong uncertainty about the level of gas supply prices. To some extent this is connected with the issue of continuing high gas prices in Europe, which has its net back effect. But it also looks like the key factor behind the growing Turkmen appetite for constant upward revision of prices when Russia attempts to raise gas supply prices to countries like Ukraine, Georgia, Moldova, Armenia and the Baltic States, as well as the success of Niyazov's effort to cut gas supplies to Russia in 2005 forcing Russia to agree to price reconsideration – which gave Niyazov an opportunity to blackmail Russia with the possible repeat of supply disruptions.

Such blackmail will now likely have a more serious effect, given that the volume of Turkmen gas purchased by Russia in 2006 will increase to 50 bcm/y. It is important to note that Turkmenistan's regime already has a strong experience of survival in the environment of nearly complete shutdown of gas exports over a long period during 1997–8 (and further in 1999), developing a strong resilience to the challenges this imposes.

Gauging Turkmenistan's opaque energy potential

Beyond the political smokescreen, however, lies more complexity: due to the lack of objective information, there has always been serious doubt as to the

Turkmen Government's official estimates of gas reserves, which put the actual implementation of the Russian–Turkmen 25-year gas supply agreement of 2003 at serious risk. Internationally, Turkmen gas reserves are estimated at 2,860 bcm, and production of Turkmen gas is expected to hit 120–130 bcm/y by 2010. Yet these reserves have never been subject to international audit and therefore lack credibility. In January 2005, Turkmenistan claimed its current recoverable gas resources to be as high as 20,420 bcm, but serious doubts persist about the veracity of Turkmenistan's claims for its total reserves.

In December 2005, the Ministry of Foreign Affairs of Turkmenistan issued a press release offering to disclose the results of an independent audit of its hydrocarbon resources if Gazprom agreed to start negotiations on the volumes and price of the natural gas it intends to buy from Turkmenistan. The press release stated that 'the audit of the hydrocarbon resources of Turkmenistan was carried out this year [2005] by independent international experts from an Anglo-American company and the Turkmen side does not want to hide its results. At the same time, in order to acquaint the Russian side of the results of the independent audit, it is necessary simultaneously to start negotiations on the price and volumes of natural gas, delivered from Turkmenistan'.[3]

The results of the reserves' audit have still not been released, and the conditional manner of the given promise to release them actually casts doubts on whether Turkmenistan is genuinely committed to publicly release real data about its gas reserves. After all, great uncertainty about the actual gas production potential of the country continues to exist, which increases the risk of relying on Turkmenistan's gas production in the future. The concerns are also associated with the fact that Turkmenistan's government is quite reluctant to allow foreign direct investment in the development of domestic gas resources, and strong concerns exist with regard to the efficiency of the state gas production company, Turkmengaz, and its ability to mobilize the necessary capital, technologies and human resources to develop resources in order to significantly increase gas production in the future.

Another issue about Turkmenistan's ability to deliver larger gas volumes, as envisaged by the Russian–Turkmen 25-year gas supply agreement of 2003, is the issue of transmission capacity availability to supply gas from Turkmenistan to Russia. The existing Central Asia–Centre (CAC) pipeline network, currently delivering the great majority of Turkmen gas to international markets, is made up of five different lines designed and built between 1966 and 1987 with an overall capacity of about 90 bcm/y. Four lines of the system originate in Turkmenistan and pass through Uzbekistan, with the fifth branch through Kazakhstan. The lack of maintenance and investment over time has almost halved the operational capacity of the system to about 50 bcm/y.

According to the IEA, if Russia intends to increase Turkmen exports to 80 bcm/y – not to mention the expected increase in exports from Kazakhstan (15 bcm) and Uzbekistan (10 bcm) – major refurbishment and expansion of the CAC system will be necessary. Gazprom has made its financial support of an expansion of the CAC contingent on the release of an independent audit of

Turkmenistan's gas reserves – but, as has been said above, this has still not been provided to Russia and the actual release of the audit remains questionable, particularly in the light of the new alternative developments with China. Upgrading the Kazakh part of the system is estimated by KazTransGas to cost US$ 2 billion, while Uzbekistan estimates investments needed for refurbishment of its lines at about the same price. The Turkmenistan government estimates the cost of refurbishment at less than US$ 1 billion. All of these numbers are considered by the IEA to be underestimates.

Without clarity on Turkmenistan's gas reserves and the sources and scheme of financing the upgrade of the CAC pipeline network, relying on gas supplies from Turkmenistan for both Russia and Europe (and possibly Asia) might prove a risky strategy. In this light, Gazprom's tactics of binding the financial support to the expansion of the CAC to the event of release of an independent audit of Turkmenistan's gas reserves can be considered as another proof that even Russia is unconvinced that Turkmenistan can deliver more than 50 bcm/y in the reasonable future.

It is important to note that Gazprom will continue to remain quite heavily dependent on Turkmen gas supplies, at least in the coming decade. Due to increasing European gas demand, as well as continuing production decline on Gazprom's largest gas fields, primarily Urengoy and Yamburg, the actual deficit of gas supplies is currently estimated as 10–15 bcm/y minimum.[4] Table 5.2 suggests that, as a matter of fact, Gazprom's mature gas fields have lost around 100 bcm of annual gas production in 2005 compared to 2000 levels.

If Russian gas production continues to lose 100 bcm every five years, that would put dramatic pressure on Gazprom to meet its supply commitments. In fact, as can be seen from the table below, forecast for gas production at two major gas fields, Urengoy and Yamburg, currently delivering more than 50 per cent of

Table 5.2 Gazprom's gas production: matured fields and new additions, 2000–5

	2000	*2001*	*2002*	*2003*	*2004*	*2005*
Total Gazprom gas production, bcm	523	512	522	541	546	548
Zapolyarnoe gas field	0	7	37	64	88	100
New takeovers of independent gas producers	0	0	8	16	16	20
Production from other fields net of Zapolyarnoye and new takeovers of independent gas producers	523	505	477	461	442	428

Source: Institute of Energy Policy estimates based on Gazprom data

Table 5.3 Gas production at Gazprom's matured fields, Urengoy and Yamburg, 1996–2020 (bcm)

	1996	*2000*	*2004*	*2010**	*2015**	*2020**
Urengoyskoye	242.2	193.3	140.6	96	70	52
Yamburgskoye	176.5	168.0	146.3	104	77	54

Source: Jonathan Stern

* Projection

Gazprom's gas production, shows that by 2015 these fields alone (not to mention other declining matured fields) could lose nearly 50 per cent of the annual production as compared to 2004, or around 140 bcm.

Considering the necessary time lag of five to seven years between the initiation of a major development and the start of production at large remote fields like Yamal Peninsula fields (Bovanenkovskoye, etc.), no other super-giant gas field can be brought on-stream before 2010, which suggests that the annual gas supply deficit could reach as much as 80–100 bcm/y in 2010. (Taking into account that in 2001–2005 depleted Western Siberian fields have lost about 100 bcm of gas production, this does not appear to be particularly unrealistic.)

As a result, it should be stated that, for Gazprom, maintaining control over gas supplies from Turkmenistan is not only a matter of commercial interest, but is critical to ensuring the security of gas supplies to its Russian and European customers. Failure to receive the volumes of gas from Turkmenistan envisaged by the Russian–Turkmen 25-year gas supply agreement of 2003 puts the reliability of those supplies at severe risk.

Turkmenistan's alternatives

Since the 1990s, Turkmenistan has been actively considering alternative pipeline solutions that would divert gas exports away from Russia and help the country escape the pressures of Russia's transit monopoly. The discussed and developed alternative pipeline ideas (bypassing Russia) involved:

- 20 bcm/y pipeline Centgas (Central Asia Gas) from Daulatabad (Turkmenistan) via Herat (Afghanistan) to Multan (Pakistan), with possible extension to India (the Trans-Afghani pipeline);
- 16 bcm/y (with possible extension to 30 bcm/y) Trans-Caspian gas pipeline (TCGP) from Turkmenbashi (Turkmenistan) via Baku and Tbilisi to Erzurum, linking the pipeline with the Turkish natural gas pipeline system;
- 28 bcm/y pipeline to Xinjiang (China);
- 8–10 bcm/y (with possible expansion to 13 bcm/y) pipeline from Korpezhe (Turkmenistan) to Kurt-Kui (Iran).

Although these pipelines could have together diverted Turkmen gas supplies away from Russia, most of these plans have not advanced in the past 15 years. The only exception is the Korpezhe–Kurt-Kui (Turkmenistan–Iran) gas pipeline, which has been operational since 1997, and which was extended last year. During the first 10 months of 2006, according to RusEnergy, Turkmenistan had exported 4 bcm of gas to Iran through that pipeline, with the total annual volume of supply expected to reach 6–8 bcm by the end of 2006, while in 2007 Iran plans to import around 14 bcm of gas via the Korpezhe–Kurt-Kui gas pipeline.

The development of other routes has faced substantial difficulties. The most difficult appears to be the Trans-Afghani pipeline, which has to pass through an area of permanent political turbulence and violence. However, it appears that Turkmenistan has put the largest efforts into promoting this particular project. It currently appears as hopeless as the Trans-Caspian gas pipeline which is plagued by unresolved disputes over the division of the Caspian seabed and complicated relations between Turkmenistan and Azerbaijan (see Chapter 4).

However, in regard to the last project – the Turkmenistan–China gas pipeline – a breakthrough has apparently been achieved in 2006. The economic potential of the project, which a preliminary feasibility study done by ExxonMobil made seem uneconomic under the global oil prices in the 1990s, have dramatically changed with the recent high oil prices. During a visit to China on 2–7 April 2006, Saparmurat Niyazov signed a ‘general agreement’ with his Chinese counterpart Hu Jintao on Turkmen gas deliveries to China and the construction of a pipeline to link the two countries. Under the terms of the agreement, China would purchase 30 bcm of Turkmen gas annually during a 30-year period, starting in 2009. To provide the gas, the two countries will jointly explore and develop ‘all deposits’ on the right bank of the Amu-Darya River (eastern Turkmenistan) on the basis of a production sharing agreement to be negotiated. Should additional volumes of gas be required to meet that annual commitment, Turkmenistan ‘may guarantee’ supplies from other gas deposits in the country. In his presentation on Turkmen television following his return to Ashgabat, Niyazov twice named the target date for commissioning the pipeline as 2008, and projected the annual volume of supplies to China at 50 bcm annually from 2010.

While the general agreement does not mention a route for the pipeline, Niyazov announced that it would go through Uzbekistan and Kazakhstan to Urumqi in western China. No estimate of investments in field development or pipeline construction is found in the general agreement or elsewhere, which has been considered by many as an indication that an agreement is still far too preliminary and the proposed timetable is highly doubtful. It should be considered that the possible length of a pipeline is estimated around 2,000 km through Uzbekistan and Kazakhstan, just to reach the Chinese border. Also, the possibility of achieving positive agreements on building the gas transit pipeline network with Kazakhstan and, particularly, with Uzbekistan (whose relations with Turkmenistan had not been perfect, to say the least) remains an open question. However, it appears that the agreement has raised severe concerns in Moscow with regard to the sustainability of the long-term Turkmen commitment under a 25-year agreement signed in 2003, which is the cornerstone in Gazprom's strategy to sustain gas supplies in the period of rapidly declining production at the matured gas fields and delayed development of new deposits in Russia.

It should be also said that, apart from the previous agreements on various projects, the proposed Turkmenistan–China gas pipeline agreement features at least one party – China – which appears capable of providing the project with both the necessary financing and the settlement of complicated transit issues with Kazakhstan and Uzbekistan, with which China has been strengthening relations in the recent period. (The recent groundbreaking of the Atasu–Alashankou oil pipeline from Kazakhstan to China has demonstrated that the successful practical implementation of complicated infrastructure projects in Central Asia with Chinese participation and diverting energy resources away from Russia, is quite possible.) This is one more reason for policy-makers in Moscow to be concerned about the project. If implemented, it may not only challenge Russia's ability to sustain gas supplies through purchases of Turkmen gas, but may also put in serious doubt the success of the Russia–China gas supply agreement that was signed in March 2006 in Beijing (the agreement also envisages the possibility of exporting 30 bcm of Russian gas to China annually from 2010 by building a new gas pipeline from Russia to China, but is also signed rather in the form of a protocol of intent, without clarity on prices of gas supply and legally binding provisions). According to most of the forecasts (for instance, the IEA's World Energy Outlook), Chinese net gas imports in 2015 will be hardly more than 20–25 bcm/y, and are expected to reach about 40–45 bcm/y only in 2030. This suggests that, as far as China is concerned, only one 30 bcm/y gas supply contract – either with Russia or with Turkmenistan – could cover nearly all of the country's projected net imports in the fairly long term. China's being in a position to make that choice and the first-ever appearance of serious competition between Russia and Turkmenistan in gas supplies (to China), has apparently been a point of dramatic concern in Moscow.

It appears that, together with the Korpezhe–Kurt-Kui gas pipeline to Iran, the success of the potential construction of the Turkmenistan–China gas pipeline by 2010 may create a situation where Turkmenistan will acquire gas export options alternative to Russian transit amounting to around 45 bcm/y, or roughly the current level of the country's net gas exports. In practice, that would give Turkmenistan significant opportunity for market manoeuvre in gas supplies and genuinely devalue the Russian gas transit monopoly factor. Given the previous experiences of unpredictable shifts in behaviour of Turkmenistan and its poor record on meeting the country's commitments, this means that there is one more factor that allows the Russian–Turkmen 25-year gas supply agreement of 2003 to be considered unsustainable. This fact alone can have an important impact on the shift in Russian policies on a broad range of issues, including the ratification of the Energy Charter Treaty and the Transit Protocol, and a general change of its 'one-way street' approach to Turkmenistan, when the Turkmenistan–China gas pipeline is finally built.

The major remaining question is Turkmenistan's ability to actually produce and deliver additional volumes of gas. If the country is not be able to do this, Russia will definitely find itself in trouble filling the gap between declining gas production in western Siberia and growing domestic and European gas demand at least over the coming decade.

Commercial relations and the role of intermediaries

The only parties that appear to have benefited from the deteriorating Russian–Turkmen relations seem to be the commercial structures that apparently are backed by quite influential hidden beneficiaries, including possibly the top policy-makers in Russia, Ukraine and Turkmenistan. The history of such intermediaries being established (at least in known history) begins in November 1995, when a joint venture (JV) was founded to handle all exports of Turkmen gas: Turkmenrosgaz, a joint venture of Turkmenneftegaz (51 per cent), Gazprom (45 per cent) and Itera (4 per cent). The mysterious intermediary Itera started in the US as a basic commodities trading company. In 1994, after a sale of sugar to the government of Turkmenistan, Itera obtained rights to export Turkmen gas as payment for its sugar transaction – and since then its business has been successfully expanding. In 1999, after the near-complete shutdown of Turkmen gas exports, Itera was the first company to sign a deal with Turkmenbashi on restarting gas exports to Ukraine, and since then and until 2003 it has been the dominating gas trader on the post-Soviet space with the exception of Gazprom[5]. The latter has been increasingly giving away to Itera not only the opportunities to re-sell Turkmen gas to international markets (which, as some experts point out, Gazprom could have successfully done on its own, or at least outsourced to a subsidiary or a direct affiliate, in order to save some commercial benefits), but also important gas assets in Russia under controversial low prices (such as the rights to develop Gubkinskoye and Beregovoye gas fields in Western Siberia).

An important observation from Itera's participation in the post-Soviet gas market is the fact that the relations in the gas supply area between Russia, Turkmenistan, Ukraine and other post-Soviet countries had been dominated for more than a decade by mysterious intermediaries playing key roles. The involvement of these intermediaries was often justified by the need to involve commercial structures to handle the barter transactions, or the need to have a de-politicized third party able to handle complicated relations between governments.

But, it should be said that such an environment is the ideal background to serve the individual commercial interests of the high-ranked policy-makers in all of these countries, none of which has ever featured particularly low corruption. It is also important that such intermediaries were chosen without any publicity, on an exclusive basis, that they have not disclosed their formal ownership structure for quite a long time, and even after disclosing some ownership information, have raised concerns that the disclosed parties were nominal owners, not the actual beneficiaries.

Further developments in 2002–6 only fuelled increased concern in this respect. First, despite restoring direct control in 2002–6 over Russian gas production assets previously given away to Itera, Gazprom did not try establishing direct control over gas sales of Turkmen gas to Ukraine, from 2003 giving this opportunity to a mysterious intermediary named EuralTransGas, with an unknown structure of beneficiaries. In 2003–4, EuralTransGas' share of gas supplies to Ukraine was around 52–57 per cent of the total Ukrainian gas imports. In 2005, EuralTransGas was replaced with another intermediary, Swiss-registered RosUkrEnergo, which is a 50/50 owned company, where 50 per cent indirectly belongs to Gazprom through a third company Arosgas (Austria), and the other 50 per cent, through an Austrian company named Centragas, to a Raiffeisen Investment AG with an undisclosed structure of beneficiaries.

In January 2006, RosUkrEnergo – under a new agreement signed by Russia and Ukraine after a dramatic three-day gas supply disruption crisis which resonated all over Europe and the rest of the world – became the central intermediary in gas relations between Russia, Turkmenistan, Kazakhstan, Uzbekistan and Ukraine, apparently becoming the single net winner from the crisis. The Russian government and Gazprom suffered severe damage to their reputations during the crisis; Ukraine was forced to pay higher prices for gas, while RosUkrEnergo, on the contrary, was given the status of exclusive gas exporter to Ukraine, very advantageous opportunities for re-export of Kazakh and Uzbek gas (previously captured by Gazprom), very advantageous conditions of use of gas storage in Ukrainian underground gas storage facilities and purchase of Gazprom's gas from these storage facilities for further re-exports, and access to Ukrainian retail gas market through a JV with NaftoGaz Ukrainy, UkrGazEnergo.

A direct illustration of RosUkrEnergo's fundamental non-transparency is the following fact. According to the consolidated financial statements of Gazprombank for the year ended 31 December 2005 (the latest available in November 2006), RosUkrEnergo AG was included in the list of 'associates accounted for at cost', which, as stated in the document, have not been consolidated with the results of the Group nor accounted for under the equity method, as either the Group does not execute control or significant influence over some of the subsidiaries and associates, or the effect would not materially alter the financial position of the group[6].

It is very strange that Gazprombank 'does not execute control or significant influence' over the company where it controls 50 per cent of the shares and where one of the co-directors is the member of Gazprom's executive board. Gazprombank, therefore, only disclosed the cost of interest in RosUkrEnergo, which, according to the financial statement, is US$ 39,000. This only illustrates that, even under Gazprombank's direct co-ownership, Russian authorities are opposed to disclosing the financial results of intermediaries involved in trade of Turkmen gas.

It is also not clear who is behind the remaining 50 per cent of the equity capital of RosUkrEnergo (although, as illustrated before, the 'Russian 50 per cent' is, in fact, not particularly transparent). Although in 2006 there have been some statements that such a beneficiary is a Ukrainian businessman, Dmitri Firtash, still no credible information is available.

The need to further involve the services of such intermediaries is also not clear. Due to rapid monetization of transactions on the post-Soviet space, the need to overcome the barter problem is gone. In this respect, the continuing involvement of intermediaries can only mean one thing: it is a very comfortable environment for top policy-makers of post-Soviet countries to monetize billion dollar cash flows for their individual interests, be it financing of political campaigns or some other cause. It is also important to understand that a transaction which took place between Gazprom, NaftoGaz Ukrainy and RosUkrEnergo on 4 January 2006, could not possibly have happened without direct approval of presidents Vladimir Putin and Victor Yushchenko, and the involvement of RosUkrEnergo in exports of Turkmen gas to Ukraine could not possibly have happened without direct approval of Saparmurat Niyazov. It is, therefore, in the best interest of these leaders to publicly present convincing explanations about the actual need to involve the service of such intermediaries in such an important international transactions; and to disclose the structure of these intermediaries' actual beneficiaries.

Failure to present such an explanation and disclosure will only fuel further suspicions about high-level corruption in the post-Soviet states associated with gas trading and exports.

Conclusion

Turkmenistan, despite relatively moderate gas reserves and production and the landlocked nature of its gas resources, will in the coming years remain a very important 'secondary' player in the Eurasian energy arena, particularly for Russia and Gazprom, taking into account the challenges in maintaining the levels of gas production in Russia sufficient to secure gas supplies to Russian and European consumers.

Political relations between the two countries currently feature a great lack of mutual trust and predictability. A large part of this should be attributed to Turkmenistan's unpredictable politics. Yet Russia, through its 'one-way street' policy approach to Turkmenistan, and its exerting a transit monopoly over Turkmen gas, has contributed a lot to Turkmenistan's seeking of other partnerships and alternative routes for supply of its most important export commodity, natural gas. Recent Russia–Turkmenistan political relations have been showing strong signs of instability and future uncertainty, and can be described not as the developing partnership of the two countries, but rather as their progressive separation, with Turkmenistan definitely drifting further away from Russia's political orbit.

Russia's assertive efforts to increase gas supply prices to the westernmost post-Soviet states (Ukraine, etc.) in 2005–6 have to a certain extent produced a reverse effect, provoking Central Asian gas suppliers (Turkmenistan in the first place) to exert more pressure on Russia in order to increase prices of gas purchased by Russia. A strong further uncertainty about the level of Turkmen gas supply prices appeared after these recent developments. An important factor is the success of Turkmenistan's effort to cut gas supplies to Russia in 2005, forcing Russia to agree to price reconsideration – which gave Turkmenbashi an opportunity to blackmail Russia with the possible repeat of supply disruptions.

Severe concerns exist over the credibility of official gas reserve estimates by the Turkmen Government. Turkmen gas reserves have never been subject to international audit and, despite a promise given in December 2005 by the foreign office of Turkmenistan to disclose the results of an independent audit of its hydrocarbon resources, the results of the reserves' audit have still not been released.

Turkmenistan's ability to deliver larger gas volumes to Russia depends on the existing Central Asia–Centre (CAC) pipeline network's transmission capacity. The lack of maintenance and investment over time has almost halved the operational capacity of the system to about 50 bcm/y, and in order to increase Turkmen exports to 80 bcm/y, major refurbishment and expansion of the CAC system will be necessary. Gazprom has made its financial support of an expansion of the CAC contingent on the release of an independent audit of Turkmenistan's gas reserves. Without clarity on Turkmenistan's gas reserves and the sources and scheme of financing an upgrade of the CAC pipeline network, relying on gas supplies from Turkmenistan for both Russia and Europe might prove a risky strategy.

Recently, Turkmenistan's efforts to develop alternative pipeline solutions that could divert gas exports away from Russia started to bring some results and further expectancies. The Turkmenistan–Iran gas pipeline has been operational since 1997. In 2007 Iran plans to import around 14 bcm of Turkmen gas via this pipeline. As already said, in April 2006, Turkmenistan signed a 'general agreement' with China on Turkmen gas deliveries to China and construction of a pipeline to link the two countries. Although multiple challenges exist with regard to this project (long route, costly construction, controversial reserve base, challenges of gas transit through Kazakhstan and, particularly, Uzbekistan), the proposed Turkmenistan–China gas pipeline agreement features at least one party, China, which appears to be capable of providing the project with both the necessary financing and the settling of complicated transit issues with Kazakhstan and Uzbekistan. This has already raised severe concerns in Moscow with regard to sustainability of the long-term Turkmen commitment under a 25-year Russia–Turkmenistan agreement of 2003, which is the cornerstone of Gazprom's strategy to sustain gas supplies, as well as the threat of serious competition with Turkmenistan in supplying gas to China.

The key conclusion for the international community is that the Russian–Turkmen partnership in gas supplies does not seem to be particularly reliable in terms of volumes of gas delivery, time frame or prices. Further severe conflicts and supply interruptions can be expected. That might somehow improve after certain new gas pipeline infrastructure projects promoting diversity for Turkmen gas exports are begun (e.g. the Turkmenistan–China gas pipeline), which might generate significant change and, in fact, soften Russia's position on the transit and free exports of Turkmen gas to Europe. Moreover, regarding the security of gas supplies from Turkmenistan, a lot is still dependent on the inconsistency and unpredictability of Turkmenistan's leadership. Probably there is little hope for change under the country's present leader, Gurbanguly Berdymukhammedov.

6 Kazakhstan's oil and gas development

Views from Russia and Kazakhstan

Martha Brill Olcott

Introduction

Kazakhstan has made a considerable effort to reach out to all potential international partners in order to become, in President Nursultan Nazarbayev's words, 'a bridge between Europe and Asia'. This notwithstanding, the Kazakh–Russian relationship remains the singlemost important one for Astana, in terms of security cooperation, economic ties, and increasingly in the oil and gas sector as well. The relationship is not one of Moscow dictating to Astana; quite the opposite is true. Kazakhstan is becoming a far more articulate and assertive 'younger brother' than Moscow ever dreamed possible, with Astana oftentimes more than holding its own in bilateral exchanges. Whether this relationship will be transformed by the current global economic crisis remains to be seen, as much will depend upon whether Kazakhstan's economic collapse is greater than that of Russia, and if this is the case whether Russia takes advantage of this to secure economic assets and advantage in Kazakhstan.

The balance in the relationship began to shift quite dramatically when Vladimir Putin replaced Boris Yeltsin. The personal relationship between Yelstin and Nazarbayev was rumoured to be rather strained – largely due to the unelaborated relations between the two in the period between August and December 1991. Yeltsin was in Almaty on the eve of the August 1991 putsch and somehow missed his flight to Moscow, and as a result evaded the cordon sent out to arrest him. But in the months that followed Yeltsin worked to bring down the Union and Nazarbayev strived to recreate it under more equal terms for the 12 remaining republics. This created strains in the relationship between the two men which never fully disappeared after the USSR collapsed. While Nazarbayev never really deferred to Yeltsin, he did recognize the latter's seniority through age and through his position as leader of Russia. However, Nazarbayev never fully relinquished his goal of creating some form of union of post-Soviet states in which all participated as equal players. Nazarbayev viewed Putin's coming to power as a levelling of the playing field between the Kazakh and Russian leaders, with Nazarbayev's diplomatic experience counterbalancing the fact that Putin inherited a more powerful nation. With growing maturity in his post, Putin has taken greater control of the Russian–Kazakh relationship.

Putin has occasionally engaged in public 'one-upmanship' to remind Nazarbayev that Russia's support of Kazakhstan should never be taken for granted, but that the Kremlin's respect must be constantly earned. The current situation of seemingly shared power between Dmitri Medvedev and Vladimir Putin complicates the situation somewhat, as it allows Moscow to simultaneously play the role of good cop and bad cop, and it remains to be seen how long the current division of power between the Kremlin and Moscow's White House will last, and whether Nazarbayev will be able to master this situation and find a way to use it to Kazakhstan's advantage.

In general, though, Kazakhstan's leadership has learned how to better protect its interests vis-à-vis Russia. This is not always done in a very transparent fashion, given that the Kazakh elite are simultaneously seeking to protect Kazakhstan's national interests, the interests of particular Kazakh enterprises, and the personal assets of the Kazakh leaders themselves. The best interests of all three are not always synonymous, so that a blow to one can be absorbed in a way that is to the benefit to another's interest, allowing the Kazakh elite to still come out on top. This complex structure of national, company and personal interest is an important feature in the way that relations are developing between KazMunaiGaz, Kazakhstan's national oil and gas company, and some of its Russian interlocutors, a point that is developed below. The Kazakh leadership has also been able to translate pressure from Russia into something of an international advantage. Because the Kazakhs are able to depict themselves as under relatively constant Russian pressure, there is greater incentive for the US and EU governments to help fund the creation of new transport opportunities for oil and gas as well as sea transport and overland cargo, in addition to various forms of security cooperation between Kazakhstan and the NATO nations, to bolster Kazakh independence against potential Russian encroachments.

This tactic, however, can backfire. Washington pressed Kazakhstan so hard to pledge oil to the Baku–Tbilisi–Ceyhan (BTC) oil pipeline that it became a virtual test of faith for the latter. In fact, President Nazarbayev did not get his much sought-after official visit to the Bush White House until September 2006, just a few months after Kazakhstan formally signed to the project. During this visit, both sides signed a joint statement on strategic cooperation, declaring their 'commitment to a shared vision of stability, prosperity, and democratic reform in Central Asia and the broader region' and 'commend[ing] their energy partnership which has helped move Kazakhstan into the ranks of the world's leading reliable suppliers of hydrocarbon reserves'.[1] Ironically, it was always more important for Washington to secure access to the development and transport of Kazakhstan's oil and gas than it was for Astana to reduce dependency on Russia; for the Kazakhs were both wary of annoying the Russians, and of committing oil that was not yet out of the ground, while Russia's primary interest was in ensuring Moscow preserved the monopoly on the transport of Kazakhstan's fossil fuel.

At the same time, for the Kazakh leadership, it is at least as important to show support for a US and European supported initiative than to find oil and gas pipeline routes that bypass Russia. To be sure, securing alternative export is one of Kazakhstan's priorities, but their first choice would certainly be a north–south route across Iran rather than the Trans-Caspian route through Georgia, Azerbaijan and Turkey.

Overall, the complexity of the relationship the Kazakhs have developed with both the US and the EU has contributed to Kazakhstan's rise as a regional actor. Kazakhstan's rise in Central Asia however also stems from the increasing size of the country's economy. Kazakh citizens (who include Kazakhs, Russians and Uzbeks) have export capital to finance projects in neighbouring countries, and Kazakhstan's oil and gas projects have the capacity to piggyback on those of neighbouring countries, increasing the market reach of otherwise competing producers. In addition, much of the region's oil and gas infrastructure passes through Kazakhstan, as Figure 6.1 makes clear.

Kazakhstan's economy has been a vital factor both in its domestic and foreign policy successes. Yet the grave ramifications of the economic crisis that began in 2008 challenge Kazakhstan's internal stability and economic security. Even by early 2009 the position of Kazakhstan as regional leader was threatened, not by the rise of an alternative power centre in the region, but by the weight of Kazakhstan's own economic crisis. Kazakhstan has effectively already sent home hundreds of thousands of migrant workers from neighbouring states, only rarely by deporting them: most left simply because their jobs in the construction sector evaporated. This produced a strong reaction in Uzbekistan in particular. Tashkent's decision to leave EurAsEC (the Eurasian Economic Community) seems partly in response to this. Once Uzbekistan left EurAsEC, Kazakhstan's businesses lost any preferential status in the country and were effectively forced to leave.

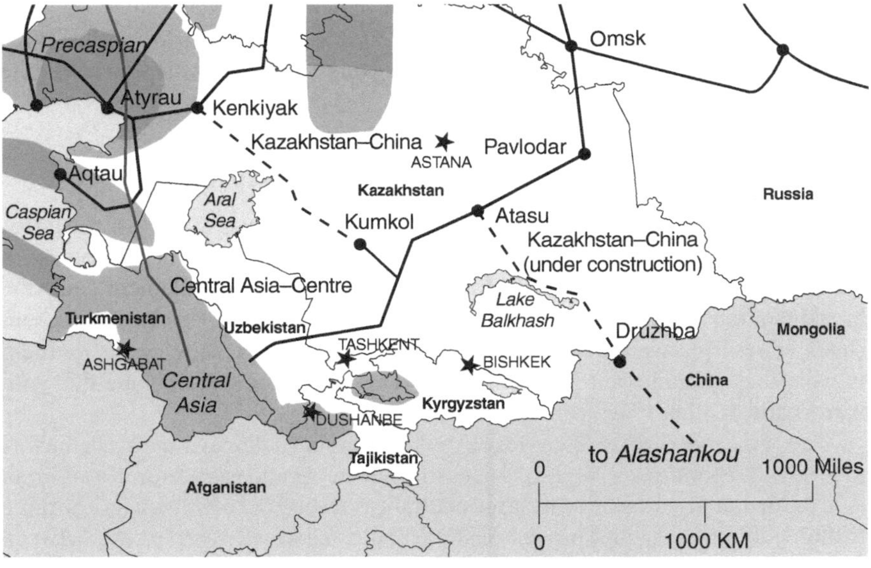

Figure 6.1 Map of Kazakhstan's major oil and gas pipelines

Source: Energy Information Administration

Kazakhstan's banking crisis, which was closely linked to the collapse of the subprime market in the US, seriously damaged its economic position in Kyrgyzstan, and generated a lot of bad feeling between these two countries. More seriously, the banking crisis in Kazakhstan led the country's government to break into its National Fund, which at the beginning of the crisis was over $45 billion, to provide billions of dollars of bail-out money to help meet cash calls from their foreign-held bonds and loans. This money was supposed to support long-term projects designed to achieve economic diversification. Instead, in less than a year, roughly a third of this money was spent: on the banks, on construction of housing, in projects where developers had gone bankrupt after collecting money from those who planned to occupy the units, and for increasing pensions, other social service payments and salaries of government employees. How long the government will be able to make use of this fund is an open question, as the National Fund's rapid growth was the result of high prices for oil and gas, as well as other strategic resources. How long these prices will remain low is quite unclear.

Another real unknown is how the crisis will affect Russian–Kazakh relations. Much will depend upon which country loses its capacity to defend its bank and strategic resources first, assuming either does. Should Russia be capable of doing so, it is likely to increase its presence in the region by buying up shares of Kazakhstan's businesses. This strategy relies on the financial reserves of Russia's companies rather than the liquidity of the government itself. One can already perceive the first signs of Russia's effort to achieve an enhanced economic presence in the region; Russia's Sberbank has been negotiating to buy Kazakhstan's beleaguered TuranAlemBank, which is in state receivership at the time of writing.

Although it is difficult to predict what the outcome of Kazakhstan's situation will be, it is clear that the way in which Kazakhstan emerges from this crisis will not only shape the country's future for years to come but will also affect the region's power balance.

Kazakhstan's oil and gas sector: the Russian perspective

The Russian leadership has always had a difficult time dealing with the idea that the republics of the former Soviet Union have become independent and sovereign states. Therefore, as a means of influence, each Kremlin government since the breakup of the USSR has tried to convince the leaders of these newly independent states, states Moscow termed its 'near abroad', that what Moscow wants for them is what they should want for themselves. They have sometimes done this with carrots and at other times with sticks.

One of the most critical areas for exercising Russian national interests in the former Soviet republics was and continues to be in the oil and gas sector. Kazakhstan is of particular importance as its large natural gas reserves are estimated to be three trillion cubic metres[2] and proven onshore and offshore oil reserves are between 250 million and 1.5 billion cubic metres.[3] When Kazakhstan's major new projects reach full production (not before 2015), the country hopes to produce at least 3 million barrels of oil a day, which would make it a larger producer than Norway, just behind Mexico and Iran. Even today Kazakhstan is an oil producer of consequence, producing 1.45 million barrels a day.[4]

Despite this potential, with the Russian government experiencing a period of relative weakness at the outset of the post-Soviet era, Russia could not gain a foothold in Kazakh energy resources as it lacked the necessary capital to become majority investors in Kazakhstan's world-class development projects. Thus, even though the Russian government never wanted to lose control of Kazakhstan's oil and gas reserves, for nearly a decade after the collapse of the Soviet Union the Kremlin had little choice but to watch the Kazakhs sell exploitation rights to foreign consortia.

The Russians did have some leverage with the Tengiz field, however[5] – a project which originated in Soviet times but was renegotiated by Chevron and the Kazakh government after independence – as Transneft had a share in what was then the only export route for hydrocarbons, the Caspian Pipeline Consortium (CPC), from Tengiz. The rights to develop Tengiz, but not the right to transport gas, were given to Chevron. The transport rights were awarded to the CPC in 1992, which then consisted of Transneft, the Kazakh Government and an Omani group. Transneft stalled on the construction of a pipeline to the Tengiz field in order to protect other Russian ventures, not beginning construction until 1993. As a result, the CPC did not begin transporting Tengiz oil until 2001, almost a year after it was originally projected to become operational. Russia, almost immediately after the consortium was opened, blocked Kazakh exports allegedly because of customs disputes.[6] Transneft does not, however, have the right to operate this project,[7] making it Russia's only oil pipeline outside of the direct control of this state monopoly.[8] Nonetheless, the difficulties in reaching terms with Transneft held up the Tengiz project for a number of years, and as is described below, negotiations over expanding the capacity of the pipeline have proved no less difficult.

By the time Vladimir Putin came to power at the end of 1999, none of the major participants or key bystanders were satisfied with the state of Russian–Central Asian relations. The US and the Europeans were making very slow progress – at best – to getting direct access to Caspian oil and gas reserves, and at that point the BTC (Baku–Tbilisi–Ceyhan) project was still in its earliest stages, and there was still no prospect for Western companies to move gas reserves to help fill the planned BTE (Baku–Tbilisi–Erzurum gas pipeline, or South Caucasus Pipeline, SCP) from the eastern side of the Caspian south to Azerbaijan, and there had been no movement at all in developing a Trans-Afghanistan pipeline (TAP), which would originate in Turkmenistan and end either in Pakistan or, better yet, in India. Russia, meanwhile, had a sense that it was soon going to be a question of now or never to maintain control over the transport of the Caspian reserves. The Caspian states were drawing closer to the EU, NATO and the US; the process was a very slow one, but unacceptable from Moscow's viewpoint of their own strategic interests regardless of its speed.

Adding to Russia's worry of losing the region was China's increasing engagement with Central Asian countries, which Russia was fully aware was being fuelled by the need of China's growing economy for additional energy resources.

Evidence clearly appeared in Kazakhstan when the China National Petroleum Corporation (CNPC) secured assets in the Aktobe oil fields and then when China appeared prepared to support the costly pipeline necessary to move this oil eastward. Nevertheless, while Moscow understood that the export of Azerbaijani crude oil and gas was effectively tied to the BTC and SCP, it also realized that the Central Asian export market had not yet been completely lost even though Russia's long-term access to it was far from secured.

Putin's motivation to control the export, and if possible the production chain, of these assets was even stronger than that of Yeltsin and his team. For Putin, natural resources served as a means of funding Russia's own economic revival as well as its international resurgence. Central Asian reserves then provided Moscow with the cushion it needed to maintain its market position while Russia's energy industry was first reorganized and then investments were made to develop those assets that would secure Russia's position in the global energy market for the first half of the twenty-first century.

Putin's priority was with securing Kremlin control over strategic decision-making in Russia's own oil and gas sector, which he recognized as an enormous challenge and a project for which there was no ready formula to ensure success. By contrast though, the Kremlin thought the challenge of securing long-term access to Central Asian reserves was less daunting, as it did not involve threatening the loss of licences and trying to cajole or force private owners to turn over blocks of stock to state-owned oil and gas companies.

Here, at least initially, Putin tried carrots, travelling to national capitals and generally striving to show respect to the leaders of Central Asia. With the exception of Imamali Rahmon of Tajikistan, all of the Central Asian leaders were his elders and all had been in positions of responsibility decades longer than he had. Many of the Central Asian leaders took something of a personal liking to Putin, and all supported his take-charge attitude with regard to putting Russia's own house in order.

Although 'harmony' on the energy front did not in fact develop, things did improve from Moscow's standpoint. In Kazakhstan, Russia gained a foothold with the creation of a new marketing organization, KazRosGaz. This Kazakh and Russian joint venture was formed in order to move Karachaganak gas to market but it also created the potential for securing a source of supply for Gazprom's Orenburg refinery. The Russian government then further influenced the development of this project in 2006, when Putin and Nazarbayev signed an agreement which guarantees the shipment of a minimum of 15 bcm of Karachaganak gas to the Orenburg gas processing plant, just across the border.[9] But the global financial crisis, which led to a slowdown in the planned Stage–3 of Karachaganak, has kept that project from moving forward with plans to modernize and expand the Orenburg facility.

The Kremlin began strongly encouraging Russian firms to take equity stakes in projects in the Caspian states, with the implicit assumption that if these were not too commercially beneficial they could expect to receive compensations elsewhere. In response, Rosneft increased its activity in the Caspian Shelf section, entering a joint venture with KazMunaiGaz to develop the Kurmangazy field in 2002.[10] In 2003, Gazprom began to press more deeply into Uzbekistan, to help Tashkent ascertain the value of its unexplored assets and better utilize its aging fields. It also improved the terms of trade for transit of Turkmen gas across Uzbekistan.

Not all of Russia's ventures into Central Asia have been met with success, however. For instance, the Russian government has had less ability to affect the sale of rights to the giant offshore Kashagan oil and gas deposit. Additionally, Kazakh–Russian energy relations were particularly mired over the issues of the expansion of the CPC pipeline. As a result, and partly in threat, the Kazakh oil and gas industry began moving towards closer integration with its Western oil and gas partners at an even more rapid rate. It finally signed the agreement with Azerbaijan to participate in the BTC pipeline project in June 2006.[11]

Kazakhstan's increased cooperation with both Azerbaijan and Georgia, in addition to the increased Western interest it received following the Ukrainian gas crisis in January 2006, has made the Kremlin eager to bring Kazakhstan back into the Russian fold. Thus in the spring of the following year, Putin travelled to Kazakhstan (and then to other countries in the region in a one-week Central Asia visit). In May 2007, Putin, Nursultan Nazarbayev, and the newly elected Turkmen leader, Gurbanguly Berdymukhammedov, met in a head-of-state summit in Turkmenbashi, located in Turkmenistan on the Caspian Sea. The three leaders announced at the conference an agreement that the three countries would build a new pipeline that would run along the Caspian coast, as well as refurbish the small Soviet-era pipeline that lies alongside it but is no longer in use. While the project is potentially quite cheap, the Central Asian states have been reluctant to go through with it because of the lack of guarantees that they will be offered commercially attractive prices from Russia long enough in order to receive payback, let alone good profit.

The Russia–Kazakh–Turkmen pipeline additionally poses a direct challenge to the US-sponsored TransCaspian pipeline by diverting Kazakh and Turkmen gas from the SCP. The TransCaspian pipeline is a proposed seabed pipeline that is supposed to circumvent both Russia and Iran, and that Russia opposes. However, Kazakhstan's position had long been that it would likely join a pipeline such as this that would originate in the Turkmen portion of the Caspian, and then join up Kazakh gas to the TransCaspian pipeline by land. The Turkmen leader has been inching closer to this proposal, maintaining that no options are foreclosed, and that the improved Turkmen–Azerbaijani relationship is making the prospect of creating a de facto undersea pipeline linking a number of Caspian Shelf projects to one another (which could carry roughly 7–8 bcm per year, or half the volume of the planned TransCaspian pipeline) a likely development in the near but not immediate future.

At the time of the Turkmenbashi meeting, the rapprochement between Turkmenistan and Azerbaijan was at its earliest stages, so the announcement of the new coastal pipeline – one which the Kazakhs seem to have worked hard to try to sell to the Turkmen as well – was viewed as at least a temporary blow to the US and EU plans to supply not just SCP, but the Nabucco pipeline project as well, as the latter plan had been gaining momentum in Europe following the 2006 Ukrainian gas crisis.

On the eve of this meeting the Kazakh President, Nursultan Nazarbayev, agreed to Russia's Bosphorous pipeline bypass plans, which Moscow had set as a precondition for agreeing to the CPC expansion. The Kazakh leader thus walked away from the Turkmenbashi summit thinking that he was now home free with this critical project, which until Kashagan comes onstream is the single largest revenue producer for Kazakhstan's oil sector.

However, shortly after the meeting, the Kremlin made it quite clear that this was not the case, raising anew questions regarding future tariffs and project debt. The Kremlin then added a tax claim against the CPC.[12] The end result was that the Kazakhs, while remaining committed to the Caspian shore natural gas pipeline project, backed away from fast-tracking it. The Kazakhs have moved steadily forward with plans to substantially increase the freight capacity of Aktau, a goal articulated in Kazmortransflot's (Kazakhstan Maritime Transit Fleet) October 2008 publication of its 2008–2012 development plans, which call for the commissioning of five new 12,000-ton-plus oil freighters during 2009–2010.[13] Chevron, as a major investor on both sides of the Caspian, is also working with Azerbaijani authorities to improve port facilities on the Azeri side and then onward through BTC. Kazakhstan also began sending signals that they would not rule out the construction of a portion of a TransCaspian pipeline in the Kazakh portion of the Caspian.

The ball was then effectively left in Russia's court, for now there was a risk that Moscow would lose the transit fees that CPC expansion would eventually produce. Convenient for the Russians, the Omanis put their seven per cent stake in the CPC pipeline up for sale in October 2008. This Moscow purchased after having elbowed out the Kazakhs from a competing bid,[14] which left Russia the largest single shareholder in the project. In December 2008 Russia dropped its objections to CPC expansion, and finally expansion began moving forward in early 2009 with a target date of 2011 for completion,[15] shortly after the CPC appointed a new board of directors.[16] Had the Russians not held the CPC expansion project up, the BTC might have had trouble getting enough volume to justify its own expansion, but now the project has commitments from both TengizChevroil and Kashagan that at least part of Kazakhstan's new production will be exported along this route.

Russia, though, has not been willing to take the same chances with its gas trade as it has with its oil. Indeed, Russia's gas trade to Europe represents a critical income stream for the Kremlin, while there is no evidence that Russia will be able to meet European demand with the production from Russia's own fields in the near future. As a result, for the past few years the Kremlin has seen little alternative but to increase the income stream from the Caspian producers. Gazprom believes that by offering commercially attractive prices, it will be able to hold the Kazakh gas trade as well as continue the flow of Turkmenistan's gas through Russia. Moreover, Moscow holds out hope that if the price is right, they may also get part of the Azerbaijani gas trade.

The Kremlin continues to fear that a direct sale of Caspian gas to Europe will cause Russia to lose its position in the European market; however, this long-term threat is not Moscow's primary concern. Rather, Moscow's major apprehension today concerns Chinese competition. China has already offered long-term purchase agreements for Turkmen gas with a commercial formula that was initially pegged at just under $200 per thousand cubic metres (tcm) plus a transit allowance, and obtained an agreement to have this flow in a pipeline whose construction is now in process.

It is thus the 'Chinese factor' that caused Russians to offer the Kazakhs, Turkmen and Uzbeks a new pricing formula in March 2008 which promised European prices for 2009 but will use a complicated averaging of Russian, CIS and European pricing, supposedly to compensate for transport costs. Furthermore, while Kazakhstan and Uzbekistan are considering selling their gas to China, the Chinese government already made new moves to penetrate deeper into the Central Asian energy sector by lending $3 billion to Turkmenistan to develop the South Iolotan field in 2009 and $10 billion to Kazakhstan to pay for future oil supplies.[17]

Originally the three Central Asian countries expected to receive upwards of $350 to $400 per tcm given what were then soaring European gas prices. But regardless, given the terms of the pricing formula, which Moscow continues to honour, their payments for the first half of 2009 will be much higher than the current wholesale purchase price of gas in some of the consuming countries. The purchase price for gas is certain to drop in the second half of 2009 because of the averaging formula that Russia is using, as this price will need to reflect the drop in gas prices in Europe. Yet until the price of oil and gas increases, Moscow will face some difficult choices regarding how much to pay the Central Asians to ensure supply, while preserving enough profit for Gazprom to meet its own corporate and national demands.

Kazakhstan's oil and gas sector: the Kazakh perspective

As mentioned previously, in the mid-1990s Kazakhstan's oil and gas industry was the subject of intense international interest. During that time, the Kashagan and Karachaganak projects were turned over almost completely to foreign firms for development. However, since 2002, the government of Kazakhstan has taken a tougher stance with existing and prospective foreign oil and gas developers. New regulations introduced since 2002 require that 50 per cent of any strategic oil and gas project must be owned by Kazakhstan, either in the form of partnership with a private Kazakh firm or with the Kazakh state oil company KazMunaiGas (KMG).[18] The government has also increased pressure on consortium members in both the TengizChevroil and Kashagan projects to renegotiate loopholes in their contracts at the very time when Kazakhstan's three largest projects were increasing or at least gearing up for production. The biggest concessions on the part of consortium partners have come in the Kashagan project, which has taken a delay of roughly five years, and the loss of projected income to the Kazakh government that this will produce created a strong platform for Astana to push for renegotiating the consortium agreement. The Kazakh government was able to buy a full share in the consortium at below market value, and to secure a role for KazMunaiGas in the management of the project, which will give the Kazakh national oil company valuable experience to use in the running of the various onshore projects for which they are the lead company.

Despite their new toughness, Kazakh officials maintain that the government has become more investor-friendly. They point to the fact that in February 2002 Kazakhstan's oil industry was reorganized: the state oil company Kazakhoil (Kazakhmunigaz) was joined with the state oil transport company, Kaztransneftgaz (Kazakh Oil and Gas) to form KMG, and this company has been steadily consolidating its hold over private oil sector in Kazakhstan. In March 2004, the Kazakh government set about the task of transforming KMG into a joint stock company, creating possibilities for broadening ownership. An initial public offering (IPO) for KMG was made available on the London Stock Exchange in 2006. KMG represents the fusion of production and transport units. The company was divided into seven principal units: KMG Exploration and Production, KazTransOil (oil transportation), KazTransGas (gas transportation), Atyrau refinery (oil refinery), Kazmortansflot (tanker shipping), Atyrau International Airport/Eurasia-Air Helicopter Company and KazTransCom (telecommunications).[19] The current head of KMG is Kairgel'dy Kabyldin. His appointment was designed to minimize the influence of Timur Kulibayev, Nazarbayev's second son-in-law, formerly second-in-command in KMG, and was a signal that the company's management was being turned over to professionals.

It is the intention of the Kazakh government that KazMunaiGas take the lead in the production and marketing of Kazakh oil, as well as its considerable natural gas assets. KMG currently controls less than a fifth of Kazakhstan's output, but this figure is rather deceptive as KMG is a stakeholder in virtually every Kazakh oil or gas project of significance. Meanwhile, Kazakhstan's skewed legal environment enables KMG to be an increasingly prominent player in any future oil and gas development. It is the only fully state-owned vertically integrated oil company in Kazakhstan, and has been busily acquiring other assets, or helping oversee the dismantlement of competitors. These included PetroKazakhstan (formerly Hurricane Hydrocarbons), a Calgary-based firm with Kazakh and Canadian stockholders that had $4 billion worth of assets in Kazakhstan, its sole country of operation, which was purchased by CNPC. Nelson Resources was also sold: this was another technically international firm with considerable holdings located only in Kazakhstan. Nelson Resources was a freely traded open-stock company, which was rumoured to include the President of Kazakhstan and his family as large shareholders. It was sold to Lukoil, which further consolidated Russian holdings in Kazakhstan.[20]

KazMunaiGas is also an active partner in the joint venture with the China National Oil Development Corporation to build the Atasu–Alashankou oil pipeline – the second stage of a multi-stage comprehensive pipeline project – and with Russia's Gazprom in the joint venture KazRosGas, which is intended to engage in gas purchasing and marketing, including the gas from the Karachaganak field.

KMG has high hopes for the role that the Atyrau refinery will play in the development of a greatly enhanced domestic refinery industry. The plant is being modernized based on a December 2001 project agreed to by KazakhOil and the Marubeni Corporation, with loans provided by the Japan Bank for International Cooperation (approx $200 million) as well as Marubeni and others. The Kazakhs also gained a controlling interest in the Shymkent refinery from CNPC when they purchased PetraKazakhstan.

For a long time the Kazakhs saw Russia as a less problematic competitor than Azerbaijan. Now Kazakhstan and Azerbaijan no longer view each other as competitors for leadership among the Caspian states. This is a position which Kazakhstan has largely secured, and the leadership in Astana now views the opportunities to ship its oil to market through Baku solely in pragmatic terms. Astana is eager to ship oil along this return, because of the stable access to market that it offers, but it is also wary of antagonizing Moscow too much, as Russia is still the most important route for Kazakh oil and gas to reach markets. The Kazakhs also understand that their access to the Russian pipeline system pushes out Turkmen gas, another source of competition, and that Turkmen gas could push out Kazakh gas from being sent out via Russia in the future.

On another note, it is likely that Kazakhstan will become an important strategic partner of both the Russian and the Chinese national oil companies. It supports the development of a strong consultative arrangement for Chinese and Russian national oil companies. But the challenge for Kazakhstan is to balance Chinese and Russian demands for a growing role in its oil and gas sector without seriously diluting their role relative to Western oil and gas firms.

Russia wants an important voice, if not a near veto, on legal questions concerning the development of offshore Caspian oil and gas reserves, Russian dominance in Central Asia's gas industry, and control of Central Asian hydroelectric power. The legal status of the Caspian Sea is still being worked out by the five littoral nations (Azerbaijan, Turkmenistan, Iran, Kazakhstan and Russia),[21] although it is already divided into de facto national sectors. The Kazakhs and Russians have delineated their national zones, with wide areas of common development, and several joint projects in the Kazakh offshore sector that are already being developed by Lukoil,[22] Rosneft and KazMunaiGas. These projects include the valuable Kurmangazy field[23] with reserves of between 700 million and 1 billion tons of crude oil, which requires a total investment of $10 billion.[24] The Russians may not have acquired their shares under the most competitive of terms, but legislation passed in January 2004 makes them liable for the same increased taxes as other foreign companies, something that has led to Russian complaints.[25] The project is also proving more problematic than the partners originally imagined, with slower output and thus financial recovery. KMG cited technological risks as the cause of two to three years of delays in development. Meanwhile, the consortium charged with developing the Caspian field estimated an increase in overall development costs.[26]

The Kazakhs are forced to cooperate quite closely with the Russians in the transport of oil and gas, and Kazakhstan must balance its interest with those of competing providers, primarily Azerbaijan and Turkmenistan. Yet KMG has managed to become Russia's partner in some transit and refining transactions, allowing them to take advantage of transfer pricing arrangements that Western partners do not receive. For example, in 2006 KMG secured a 50 per cent stake in the Orenburg refinery,[27] somewhat easing the anger over years of bitter negotiation between the Russians and the Kazakhs over a long-term pricing agreement for the supply of gas from the giant field at Karachaganak to that refinery.[28] Of course, some Western firms are also able to benefit from arrangements that are less favourable to KMG, such as ENI, Total, ConocoPhillips and Inpex, who are large shareholders in AIOC and will benefit from oil shipped out through the BTC routing.[29]

Nonetheless, Astana will work very hard to see that it does not advance either its regional or its bilateral relationships with Russia in a way that damages US–Kazakh or EU–Kazakh relations. Similarly, the Kazakhs will continue to ship to China no matter what the US, EU or Russia think. China is increasing its role in Kazakhstan's energy market with Astana's support. Nonetheless, and although this support is not offered begrudgingly, it is offered with some real limitations: Astana shows no signs of giving Chinese national oil companies the same access to its energy assets that it gives the Russians.

Conclusion

The question going forward is whether or not Russia will continue to use a commercial carrot to get Caspian oil or, in a world where Moscow is not afraid to show force when it believes its vital interests are at stake, will they resort to the stick in the future? Russia's attack on Georgia in August 2008 certainly adds a new complexity to the situation, since it demonstrates that Russia will use force against neighbouring states when Moscow believes that it is justified in doing so. What is more, the weight of international opinion seems to have played virtually no role in Moscow's calculations.

All things considered, however, Russia's invasion of South Ossetia and Georgia changes very little with regard to the situation in Central Asia. By comparison to the states in the South Caucasus, the Central Asian states have generally been very willing to follow Moscow's direction in the area of security relations. Only Uzbekistan was willing to flaunt its independence in this sector, though after the Andijan uprising in 2005 they rejoined the Collective Security Treaty Organization, and further deepened the relationship between Gazprom and Uzbekistan's own national oil and gas company. So while relations between Tashkent and Moscow tend to wax and wane, Gazprom is too important a source of income for Karimov to do much flaunting.

In general, Moscow would have very little motivation to use the stick in its energy relations in Central Asia. Unlike Georgia or even Azerbaijan, all of these states share a security agenda with Russia. The problems they have are usually commercial, and are caused by the fact that Russia needs to sell Central Asian gas to satisfy its commitments in the European market. Approaching 20 years of the independence of these states, Russian companies own very little of these countries' current output, and efforts by Russian firms to increase their equity holdings will not get them much more guaranteed supply for several more years. Even then theirs will be a minority interest in projects, and it is highly unlikely that they will be able to acquire any larger ownership stake in Kazakhstan's oil and gas industry, even if they may be able to enhance their position in other economic sectors. So Russia can only secure this supply through its purchase and transit.

Moreover Russia needs this gas badly enough that trying to acquire it through bullying is potentially quite a risky strategy. It backfired on Russia in Turkmenistan in the early 1990s, and once burned is twice learned: Moscow would be foolhardy to try this again. Russians have traditionally negotiated very tough terms of trade for transit fees, but the January 2009 agreement between Russia and Ukraine holds promise that there will be more transparency in that area throughout the CIS. Russia was willing to pay a high price when gas prices were high; they are going to have to make their offers to the Central Asians even more attractive when energy prices are low, for this makes it far too costly to develop many of the permafrost area gas projects that are projected to move into the exploitation phase in Russia proper. This would be true even if Gazprom and Rosneft had a surplus of credit, which neither has due to the current financial crisis.

The good news, from Russia's point of view, is that while energy prices are low and Europe and the US in the midst of major economic crises of their own, the Central Asians may have little choice but to sell their gas to the Russians. The US and EU may keep talking about new pipelines, but for the moment at least no one seems ready to allocate funds to build them (witness how the funds for exploratory technical work on Nabucco have already been spent on more pressing needs). Even the Chinese, with the pipeline from Turkmenistan already partly constructed, could decide to slow down these high-cost projects if Chinese internal demand drops as a result of the global crisis. While most of the Central Asian world-class oil and gas projects are cheaper than those in Russia, they are not cheap, and a prolonged economic crisis, with oil in the $20–40 range, will certainly delay many of these projects, and leave the Russians in a better competitive position to buy Central Asian gas than they have been for several years.

There is also strong evidence that the global financial crisis is causing Kazakhstan to rethink its plans to become one of the 50 top economies in the world, a goal long enunciated by President Nazarbayev. This crisis appears to be reminding them of how difficult it is to become a competitive economy. They also appear to have concluded that this crisis has showed Russia its lack of competitiveness, pushing Russia to value all the more a partnership with Kazakhstan and other close neighbours. Once again, in spring 2009 Kazakhstan and Russia both began talking about a common currency, although Kazakhstan wants a new currency, the 'eurazia', and Russia of course wants the ruble to become an international standard.

It is too soon to tell whether the global crisis will bring Russia and Kazakhstan substantially closer, and whether it will lead to the effective integration of their economies. And if it does, it is impossible to predict whether it will be on terms of a real partnership rather than a relationship in which a strong state dominates a weaker one. Nazarbayev may still be at the top of his powers, but a global economic crisis can prove to be far from a level playing field. If Russia's leadership – be it embodied by Putin, Medvedev, or some combination – comes out of the crisis stronger, with its economy largely intact, then the Kazakh–Russian relationship could evolve in Russia's favour. If Russia emerges from the crisis much weaker, then Kazakhstan might face new economic challenges because of the existing degree of integration between the economies of these two countries, but it will also have new opportunities to carve out its own economic and regional role.

7 Uzbekistan

Central Asian key

Andrew Monaghan

Introduction

Uzbekistan plays a key role in Russia's attempts to restore its power. A glance at a map illustrates Uzbekistan's importance in terms of transit. Furthermore, Uzbekistan is a major regional producer and one of the largest regional consumers of energy. Recognizing these roles, Russia became increasingly active in building relations with Uzbekistan during Vladimir Putin's presidency, particularly in his second term in office. This was illustrated by high profile visits to Tashkent made by senior Russian officials. During one visit on 2 November 2007, then Prime Minister Zubkov discussed Russia's proposal to establish an energy alliance within the Shanghai Cooperation Organization (SCO) to improve energy and economic ties. The visit highlighted the burgeoning relationship between the two states: officially, Russia accounts for over 30 per cent of Uzbekistan's foreign trade and trade turnover amounted to more than US$ 3 billion.[1]

Yet an assessment of Uzbekistan's role is also a timely reminder of the difficulties and complexities faced by all investors, including Russia, in Central Asian states. The region is dominated by autocratic regimes with uncertain futures, poor human rights records, opaque decision-making structures and non-transparency and uncertainty regarding both energy reserves and investment possibilities – and thus prospects. Moreover, relations between the states in the region are complex and unstable. Thus as one analyst has lucidly described it, the energy trade in Central Asia is a 'dangerous game'.[2] Uzbekistan is an important element of this.

In examining Moscow's attempts to restore Russia's power in the region, specifically Uzbekistan, this chapter first focuses on Uzbekistan itself, looking briefly at the domestic and foreign policy context before assessing its energy reserves and domestic energy policy. It then turns to Uzbekistan's role in the Central Asian region. Finally, the chapter examines Uzbekistan's relations with Russia, establishing the evolution of the relationship, briefly assessing Russian aims and the results of the relationship so far.

The key points to emerge from the paper are that Uzbekistan features prominently in Russian regional thinking – not only is it a key transit state, but without its energy production contribution, the region's and indeed Russia's gas supplies would be considerably constricted. Furthermore, Russia and Uzbekistan have a growing partnership. This has been developing for some years. Uzbekistan is now a major partner for Russia, behind only Kazakhstan in importance. This partnership benefits both Moscow and the current Uzbek leadership. Yet this remains an essentially fluid and evolving situation – Uzbekistan's immediate political future is uncertain and other potentially attractive partners for Uzbekistan are emerging, particularly in the Far East. Russia also faces risks in the further development of relations: Moscow does not want to become too bound to Uzbekistan and thus be obliged to take on the full burden of bailing Uzbekistan out of economic or political chaos.

Two caveats should be noted at the outset. First, Uzbekistan's energy sector is highly opaque, even compared to the broader context of Central Asia – on the one hand, decision-making remains opaque, and on the other, figures for reserves and production are disputed. Official statistics are widely considered to be inflated and highly unreliable. As one European official noted, forecasts often coincide with political visits. Thus, 'it is difficult to know what to believe' and there is no reliable forecasting of the picture beyond the short term, this official observed, even four to five years ahead.[3] Furthermore, political tensions mean very few experts feel comfortable speaking on record.

Second, this chapter assesses a range of issues which form a dense and complex network, not all of which can be sufficiently examined here. The chapter focuses on Uzbekistan's oil and particularly gas reserves (but not electricity and coal production), its transit capabilities, and the evolving energy relationship with Russia and, to a limited extent, other partners such as China where this affects the Russian–Uzbek relationship. Other issues, including the domestic political situation and Uzbekistan's political and security relations with its neighbours and Russia, though important, are only lightly touched upon.

Uzbekistan

Domestic and foreign policy issues provide an important context for understanding Uzbekistan's energy sector and foreign energy relationships. Uzbekistan is the most populous state in the Central Asian region, with a population of some 26 million.[4] Its capital, Tashkent, was the administrative, military and industrial hub of the region with both a relatively developed manufacturing sector and transport infrastructure. These factors contributed to a comparatively high level of energy consumption, particularly of gas. Though Uzbekistan was one of the largest gas producers in the USSR, it also profited from having a favourable resource endowment not limited to energy or mineral abundance. It thus differed from other Newly Independent States following the collapse of the USSR because its exports were more based on cotton and gold, which benefited from more flexible transport systems and could be readily diverted to world markets.[5]

Uzbekistan ranks as one of the most autocratic regimes in a region not known for high levels of democratic reform. The centralized political structure places exclusive power in the hands of the executive, particularly the office of the President, Islam Karimov. The judiciary and administrative framework function largely as a tool to implement the president's policies and rule of law continues to be arbitrary. Uzbekistan's government is influenced by clan ties and is riddled with nepotism and inherent corruption; the president, his family and other senior officials are immune from scrutiny.

This political and legal atmosphere undermines the ability to manage public policy effectively, particularly economic and energy sector development. Business disputes cannot be resolved efficiently without an independent judiciary; burdensome administrative procedures complicate the granting of authorization and permission. This is particularly relevant given inconsistencies and non-transparency in the energy sector, which generate a climate of uncertainty about the applicability of legislation.[6]

The clannish nature of the Uzbek regime destabilizes its foreign relationships and constricts them to ones which provide *personal* security and economic benefit rather than reflecting a broader *state* policy. The revenue from natural resources plays an important role in maintaining influence in the clan structure. In the past, President Karimov has used free, or very cheap gas as a popular benefit (though this has largely ended),[7] and distributes the wealth generated from them to secure the loyalty of key elites, including the National Security Service, its rival the Ministry of Internal Affairs, which controls the police and is thought to control Uzbekneftegaz, and the Prosecutor's Office, which is thought to control Uztransgas. This provides the majority of the income of the security services and they in return implement political 'stability' through suppression.[8]

This political tension has an important link to energy policy and foreign relations. The Uzbek government faces a number of challenges, particularly the rise of Islamic radicalism. Though this paper does not seek to address these issues directly, they are relevant here for two contextual reasons. The first concerns the government's response, including robust suppression and reprisals.[9] This contributes to broader regional instability, causes migration and provides grounds for accusations by the Uzbek government that neighbouring states such as Kazakhstan and Kyrgyzstan are harbouring anti-Uzbek militants on their territory, and threatened action to remove the threat. This is an important reason for the tense atmosphere between the region's states.

Second, in seeking to address such challenges, the Uzbek leadership has sought international allies. Sustained Western condemnation following the Andijan massacre in May 2005 also contributed to a broad refocus of foreign policy in a search for allies that would not seek to interfere in internal politics but instead support the status quo, including Russia and China. On 14 November 2005, Presidents Putin and Karimov signed a Treaty on Allied Relations, notable because Uzbekistan had rejected military cooperation previously and Tashkent had been suspicious of Russia's military presence in Tajikistan. Subsequently, Uzbekistan withdrew from GUUAM, joined EurAsEC (January 2006)[10] and began to extend further opportunities to Russian energy companies.

Though there is little detailed information about Chinese investments and agreements with Uzbekistan, these seem to be growing significantly. In 2004, Chinese premier Hu Jintao spent five days in Uzbekistan, coming to 10 separate agreements with President Karimov. He reportedly promised US$ 1.25 billion to Uzbekistan. China's National Petroleum Corporation (CNPC) also agreed a cooperation pact with Uzbekneftegaz to conduct exploration work at 23 smaller fields in the Bukhara–Khiva region (these fields are near a main gas pipeline, so it appears that associated gas may also be available).[11] Beijing's trade turnover with Uzbekistan reportedly exceeded $1 billion in 2007.[12] This does not mean that the Uzbek leadership seeks to allow Russia or China to dominate it: to the contrary, as noted below, Tashkent appears to encourage competition between Russia and China to enhance its own influence and room for manoeuvre.

Uzbekistan's energy sector

Uzbekistan was the only state of the FSU which did not suffer a serious decline in oil and gas production after 1991, largely because of strong political support for the oil sector. Indeed, during the first half of the 1990s, Uzbekistan became both the largest Central Asian gas producer and largely energy self-sufficient, ceasing to be a net oil importer in 1995. This was important given Uzbekistan's relatively high energy consumption. Moreover, growth in domestic energy demand provided a favourable context for the expansion of new fields.[13]

Uzbekistan's hydrocarbon reserves, located in five main oil and gas producing regions (Ustiurt, Bukhara–Khiva, South-west Gissar, Sukhandarya and the Ferghana Valley) are significant, though the oil is of low quality and the gas high in sulphur.[14] Oil reserves are estimated to be some 600 million barrels. But since achieving self-sufficiency in 1995, production has fallen from a peak of some 191,000 barrels per day (bpd) in 1999 to 114,000 bpd in 2007. Consumption has fluctuated during this period, but Uzbekistan again became a net oil importer in 2005, buying from Russia and Kazakhstan, and in 2007 was consuming approximately 119,000 bpd.[15]

Uzbekistan's gas reserves are estimated to be approximately 1.87 trillion cubic metres. Production rose in the late 1990s, then remained at approximately 53 billion cubic metres (bcm) from 2000 to 2003. The following year saw an increase to 55 bcm.[16] Over the last three years, 78–80 per cent of this production was consumed domestically. Production mostly (i.e. 95 per cent) comes from the Amu Darya basin and Mubarek area in south-west Uzbekistan. A recent report suggests that the Ustiurt plateau is considered the most promising area for exploration and development, though its reserves are thinly spread and not very economic.[17]

Uzbekistan faces a number of problems in its domestic energy sector, however, including low investment, high levels of corruption and mismanagement. Thus the Energy Charter Organization recently stated that 'years of low cost recovery… have left the domestic energy infrastructure in urgent need of rehabilitation'.[18] Another report notes that the gas sector is 'mismanaged and deteriorating', resulting in considerable losses in capacity: the intention was to maintain production at 60–65 bcm until 2010. Domestic consumption seems to have been reduced to planned levels (43 bcm) but, as noted above, according to some analysts there is evidence of gas being withheld from domestic consumers to facilitate export.[19]

One Uzbek analyst asserts that to obtain even a four per cent growth in GDP, some US$ 7.6 billion must be invested in Uzbekistan's fuel and energy complex. 'It is quite obvious,' he argues, 'that the required amount of investments to maintain sustainable development of the fuel and energy complex cannot be provided by... the country's resources alone.' Particularly important, he notes, is the continuing need to address current consumption efficiency in all sectors of the economy, with increased supervision and tracking of consumption levels, increased efficiency in competition on the domestic market (and with foreign companies active on the Uzbek market), including the break-up of Uzbekenergo to establish a real energy market and the establishment of two economically independent vertically integrated companies. Gas supply enterprises should also be independent, he argues.[20]

Tashkent has sought to encourage foreign investment – to the tune of US$ 400 million through production sharing agreements (PSAs). It has offered 80 fields, of which 78 are in 16 exploration blocks. Eight individual fields with significant reserves have been opened to foreign participation, including four in the Gissar Basin and four in the Amu Darya region. One analyst has noted that though there are risks for foreign investment, there are advantages, not just in the relatively large gas reserves, but also cheap labour and lower overhead costs.[21]

The government's search for foreign investment began in 1998 with a programme to attract investment to develop reserves in the Ustiurt plateau in south-west Uzbekistan. On 28 February 2000, the government adopted the Oil and Gas Investments Decree, granting companies the possibility of conducting exploration in the Ustiurt region and newly discovered oil and gas deposits for 25 years on a concession basis. Foreign companies would be exempted from all types of taxes, payments and related customs duties in force in Uzbekistan during the period of prospecting and exploration. The limited impact of this decree led to the passing of a full PSA Act at the end of 2001, granting exclusive rights to explore and extract mineral deposits in a specified area of subsoil.[22]

The government has attempted to hone the PSA Act, and enacted some important measures to improve efficiency and financial performance in the energy sector. Barter payment is now banned. In the gas sector, tariffs have been increased significantly and a gas metering programme implemented, resulting in increasing payment collection.[23] These measures have generated some rising international interest, as described below, predominantly from Russia and Far Eastern partners. Western companies, however, have shown little interest.

Yet a number of problems exist for investors, which limit Uzbekistan's attractiveness, including for Russian companies. The legislative atmosphere remains uncertain, partly because of the incongruence of the many investment laws (particularly relevant given the nature of the administrative system noted above). Furthermore, the most attractive fields are kept under Uzbek control, the riskiest offered to foreigners. Indeed, the approach of the government to privatization is considered ambiguous and the energy sector is kept under heavy state control.

Political instability, both domestic and regional, is often cited as a problem. Compounding this, Uzbekistan's geography is inhospitable – it is landlocked and surrounded by desert and mountain ranges, making raw material transit expensive. While infrastructure was sufficient in the 1990s,[24] it has not been sufficiently overhauled and consequently is now unsuited to increased production and consumption requirements.[25]

Uzbekistan's energy relations

Uzbekistan plays a double role in the region, supported by Russia. First, it is a key transit state. Two main pipelines, the Central Asia–Centre and Bukhara–Ural lines, were key routes for Turkmen and Uzbek gas in the USSR and remain important today: Turkmen gas reaches Russia through Uzbekistan. With Russian assistance, these lines are to be developed and expanded.[26] Tashkent plans to expand its pipelines by a few hundred kilometres, reconstruct and expand gas distribution plants and compressor stations and build additional branch lines to increase exports to some 16 bcm by 2014.[27]

Second, Uzbekistan plays an important role as a producer. Its energy production capacity is largely taken up by domestic consumption – without this capacity to meet the gas demands, the gas would have to come from elsewhere, constricting other available supplies. But Uzbekistan also conducts a limited export trade in gas and the intention is to increase the proportion of gas exported to 20 bcm.[28]

Uzbekistan is an important gas supplier to Russia, of which more in the next section, and the wider Central Asian region: Uzbekistan supplied Kazakhstan with 1.8 bcm in 2006;[29] Kyrgyzstan with 0.51 bcm; and Tajikistan with 0.57 bcm. Yet, as noted above, Uzbekistan has tense relations with its neighbours. Negotiations between the parties about energy are frequently difficult, and Kyrgyzstan's and Tajikistan's lack of hard currency to pay for gas has led to frequent cut-offs, especially during winter.[30]

Uzbekistan–Russia relations and the energy dimension

An evolution has taken place in Uzbekistan's relations with Russia, from Tashkent asserting its independence during the 1990s towards building a partnership in the 2000s. This, of course, has not been a black and white changeover. Even during the 1990s, Uzbekistan maintained some economic and foreign and security links with Russia, prompted by instability in Tajikistan and the rise of Islamist opposition which both compromised political stability in Uzbekistan and illustrated the need for Russian assistance.

Nevertheless, a clear acceleration has taken place since 2000. This is partly because Russian economic recovery made it a more attractive partner for Uzbekistan, and partly because Vladimir Putin pursued a more active policy toward the region, investing Russian policy with more coherence than had been the case during the 1990s.[31] Subsequent to security agreements in 1999 and 2000, Uzbekistan and Russia signed the Treaty on Strategic Partnership (June 2004) and agreed a treaty on alliance relations (November 2005). A range of economic and business links has grown up in this context.[32] Russia represents 30 per cent of Uzbekistan's foreign trade (up from 19 per cent in 2005[33]). There are over 450 joint ventures with Russian participation working in Uzbekistan and a further 285 Uzbek–Russian joint ventures in Russia.[34] This, in theory, provides a solid context for the development of energy relations.

The relationship is important to Russia: Russia faces gas shortages – Gazprom seeks to import significant quantities of Central Asian gas to maintain its own energy balance. Moreover, Moscow perceives the construction of alternative pipelines to be a potential threat to its economic interests. Thus Moscow has sought to engage partners to ensure that it is not bypassed. These two points have contributed to greater Russian activity in the region: to find gas and ensure that it is not left without partners and cut out of the profitable pipeline networks.

A number of leading Russian energy companies are investing in Uzbekistan. Analyst Sergei Luzyanin eloquently describes the December 2002 agreement between Gazprom and Uzbekneftegaz as simply the 'first swallow' of serious energy cooperation. This was succeeded by the 'first systematic steps' taken during a working visit of then President Putin to Samarkand in August 2003 during which President Karimov signed documents about deepening cooperation with Russia in the oil and gas sectors. In 2003, Gazprom acquired control over the main gas pipelines in Uzbekistan. Luzyanin notes that engagement was for 'economic reasons', but that, at this stage in 2003, it was only the first step and was still a long way from 'serious relations'.[35]

These have begun to grow, particularly in the gas sector. During Putin's visit to Uzbekistan in June 2004, LUKoil signed a PSA to develop the Kandym–Khauzak–Shady project for the next 35 years. Kandym, the largest of these fields, contains some 150 bcm. LUKoil thus anticipates investing US$ 1 billion and that its annual production will rise to 9 bcm.[36] In July 2004, Soyuzneftegaz bought a controlling block of shares in Uzpek, which has licences to two major gas fields. Uzpek plans to sell to Russia, Kazakhstan and China.[37] In April 2006 Gazprom signed a PSA for a 15-year project for developing the Shakhpakhty gas condensate field; in August that year, a consortium including LUKoil, CNPC, Petronas and the Korean National Oil Company signed an agreement with Uzbekneftegaz to explore and develop resources in Ustiurt plateau. This means US$ 100 million initial investment.[38] As part of this agreement the same consortium agreed to develop projects in the Aral Sea. Russian companies PetroAlliance and GeoFizika will carry out the exploration.[39] In 2007, Zarubezhneft began drilling in the Ustiurt region, and Stroytransgaz signed a PSA with Uzbekneftegaz for a five-year exploration and 36-year joint development of Central Ustiurt and South Western Guissar regions, with Stroytransgas pledging $466 million.[40] Stroytransgaz is also investing in an Uzbek liquefied natural gas plant venture.

These planned Russian investments, taken with Russian involvement in developing pipeline infrastructure and oil sales to Uzbekistan, encourage an impression that Russian interests and influence begin to dominate; a view further encouraged by the attempts to establish greater cooperation between Russia, Turkmenistan, Kazakhstan and Uzbekistan in the shape of a gas pipeline network development plan, one of the elements of which, as noted above, envisages rehabilitating and expanding the Central Asia-Center pipeline from Turkmenistan and Uzbekistan to Russia to 90 bcm.[41]

Yet some Russian, Western and Uzbek analysts remain critical of the achievements of Russian policies in the region, noting that Russian leaders still lack a clear set of goals for the future of the region, Uzbekistan included.[42] There is also a distinction between plans, contracts and agreements on one hand and real achievements on the other. First, many of the agreements are based on *exploration*: there is no guarantee that Russian investors will find anything. And, though offering PSAs, the Uzbek government has kept a hold on the most profitable deposits.

Second, the relationship appears to lack real substance. The specifics of Uzbek gas exports to Russia remain unclear. Some suggest that in 2003, 2.4 bcm was exported to Russia,[43] others note that exports did not start until an agreement in 2004 which envisaged rapidly increasing imports, starting with 7.7 bcm that year. Moreover, while some suggest that this had risen to 8 bcm in 2006,[44] others note that recent agreements only cater for 3 bcm and that the Russian company Itera was only able to obtain 1.5 bcm recently.[45] Other reports suggest that Gazprom bought 9 bcm from Uzbekistan in 2006 and persuaded Tashkent to sell it 13 bcm in 2007.[46]

Furthermore, Russian commitments to invest are not always fulfilled – one report notes that in early 2007, Gazprom had still only invested a tenth of what it was supposed to have done by then, just US$ 30 million instead of US$ 300 million.[47] This has led to friction between Russia and Uzbekistan, with the latter threatening to revoke Gazprom's exploration licence. Gazprom blamed the delay in investment on the failure of Uzbek officials to provide the appropriate permits. Gazprom has failed to persuade Tashkent to increase the amount of gas available to it to 17–18 bcm.[48]

The agreements between Uzbekistan and Russia benefit the Uzbek leadership by binding Russia closer to it or its designated successor. As one analyst notes, the Uzbek regime considers foreign powers to be pawns in its own game. The agreements with Russia are perhaps the current regime's best chance of prolonging its rule.[49]

Finally, the lack of real substance leaves the relationship dependent on personal ties and open to problems. Some analysts note that an 'unmistakable chill' descended on the relationship towards the end of Putin's second presidential term and in the early stages of President Medvedev's presidency.[50] This is illustrated by the significant competition Russia faces, particularly from China – though Chinese companies also face difficulties in investing in the Uzbek energy market. As noted above, China has invested heavily in Uzbekistan during the same period as Russia. In April this year, Tashkent announced a plan to build a 530-km pipeline to China with a capacity of 30 bcm in cooperation with CNPC.[51] Such agreements, by creating alternative markets, maintain Uzbekistan's room for political manoeuvre and give it negotiating weight to increase prices paid by Russia. Indeed, one analyst has remarked that Hu Jintao had developed relations with Uzbekistan 'perhaps even more effectively' than President Putin and

'Chinese policy toward Central Asia is more coherent than that of Russia', though the central government does not appear to control fully China's oil giants which have a shorter-term horizon and are more profit-oriented than the government.' Central Asia has welcomed Chinese investment and trade, he argued, not least because it allowed the states to apply a competitive tension between Russia, the USA and China.[52] Moreover, companies from Korea and Malaysia have also begun to invest heavily, becoming involved in areas where Russian companies have also been active.[53]

The influence of other powers is therefore increasing. Indeed, another analyst has noted a change in foreign policy style emerging in Uzbekistan, as Tashkent shows signs of moving more towards a multi-vector foreign policy, signs which do not rule out a return to building relations with Western states and organizations and China.[54] It is by no means certain that the regime which succeeds Islam Karimov will be so pro-Russian. While becoming part of a summit agreement with Russia, Turkmenistan and Kazakhstan to develop pipelines that would increase Russian influence, Tashkent has agreed separate pipeline deals with China. Moreover, Tashkent is attempting to oblige Gazprom to pay more for Uzbek gas: prices have risen significantly over the last couple of years to some $160 per thousand cubic metres. Discussions in 2008 have focused on bringing prices for Uzbek gas into line with 'European' prices. This will put a strain on Gazprom and appears to be the result of Tashkent playing Russia and China (and other markets) against each other.[55]

Conclusions

Uzbekistan is a key energy player in what is a complex and diverse region. Without Uzbekistan's gas production, there would be a serious shortfall in regional gas supply with all the consequent ramifications for spreading local political instability and knock-on shortages to other states in the region and beyond. As a transit state, Uzbekistan occupies a key role both in supplying Russia with Turkmen gas and contributing to efforts to develop East-bound infrastructure.

In Uzbekistan's energy sector there are signs of reform. Clearly, however, Uzbekistan remains a difficult environment in which to work, a position exacerbated by brewing political uncertainty. This conclusion is doubtless shared by the Russian companies investing there, which are exploring inhospitable deposits with mixed success and which have failed to persuade the Uzbek government to increase its gas production or sales significantly and have found their licences under threat from international competition. Predictability, such an important asset for energy sector investment and development, is plainly lacking. And, of course, analysts note that any handover of power in Uzbekistan is unlikely to be as smooth as has so far proven the case in Turkmenistan, which, given the powers vested in the presidency, is important.[56]

Nevertheless, Uzbekistan's relationship with Russia is growing and is an important feature of the region's political landscape. Energy cooperation clearly stands alongside security as a priority for both parties: the links between the two assist

Gazprom in balancing supply and demand in Russia, strengthen Russian influence in the region and assist the Uzbek government. This is *mutually* beneficial to a considerable degree: the two states share a number of interests, particularly with regard to the threat both perceive to be posed by Islamic radicalism. Russia has interests in the region and considers the development of economic relations to be an important asset in maintaining political and economic stability. The Uzbek leadership, of course, benefits economically from such interaction: selling gas to Russia is not necessarily a bad deal, since it is effectively trading bad-paying regional customers (Kyrgyzstan and Tajikistan) for a paying Russian market even if the prices are lower than those of Europe.[57]

Yet Russia continues to face problems in restoring its power and influence in Uzbekistan, which particularly in the shape of the Karimov regime, continues to protect its own interests and does not shy from refusing Russian demands, and even criticizing and threatening Russia companies. Furthermore, Uzbek foreign policy is not solely focused toward Russia: China, particularly, has played a very active role in the region and Uzbekistan more specifically.

The stakes in Uzbekistan are high for Russia: Russia needs stability on its southern flank and gas for its own energy balance. The balance is a fine one: Russia seeks to counteract the influence of other actors in the region, particularly Western, Turkish and Chinese, and to ensure Russian participation in the region while not increasing its responsibility too significantly or being drawn too far into the region's political complexities whereby it might have to take sides, which would undermine its broader aims. All this has to be achieved with essentially limited resources. Despite its influence, Russia's power, in Uzbekistan at least, is not fully restored.

Recommendations

While the US has put considerable effort and resources into Central Asia, Europe lags far behind. Though perhaps not high on the EU's agenda, the 'details of Central Asia' are important to understand, particularly with regard to the priority the EU accords the Nabucco pipeline project. Given the current large but finite production and export capacity of Turkmenistan and its burgeoning export contracts, questions must be raised about whether it will be a serious provider for the Nabucco project. Consideration should be given to a potential role for Uzbekistan. Nevertheless, any policy recommendations must start from the very basics, particularly with regard to Uzbekistan.

Considerable effort and resources should be dedicated to learning more about the Central Asian region. Like the South Caucasus, Central Asia is not a monolithic entity: the interests of the constituent states are many and diverse. It is not enough to have briefings on 'Central Asia' – as one of the few experts on the region frequently complains: 'I am usually given 10 minutes to brief policy-makers on "Central Asia": that's two minutes per state!'[58] Furthermore, while Russia is an important regional actor, the region should be understood on its own terms, not simply through the prism of Russia/USSR. The same applies to Uzbekistan more specifically: Uzbekistan's energy role is perhaps one of the least known of any major producer. This gap must be filled.

EU sanctions on Uzbekistan should be reconsidered – are they effective? It seems that Tashkent largely shrugs off the limitations. The net effect of the 'isolation' of Uzbekistan by Europe may simply be to heighten our own lack of awareness of evolving events there. The direct consequence of this will be that Europe will be unprepared to take advantage of any opportunities that may emerge.

Consideration should be given to EU expertise and experience – where lie its advantages over potential competitors in the region? Where can it contribute? In this respect, the EU could particularly contribute to improving industrial efficiency in gas consumption, thus releasing gas for export.

8 The uncertain trajectory of Russia–Azerbaijan relations in the multiple-pipeline era

Pavel K. Baev

Introduction

A favourite political maxim since the mid-1990s in Azerbaijan has been: 'Happiness is multiple pipelines.' In the past decade, four 'strategic' oil and gas pipelines have been built in the Caspian/Black Sea region: Tengiz–Novorossiisk in mid-2001 (Caspian Pipeline Consortium, CPC), and Baku–Tbilisi–Ceyhan (BTC) in mid-2005, for oil; Blue Stream (Beregovaya–Samsun–Ankara) in early 2003, and Baku–Tbilisi–Erzurum (or South Caucasus Pipeline, SCP) in mid-2007, for natural gas. 'Happiness', however, has remained elusive.[1] The political meaning of this axiom has always been clear: broadly, several pipelines would guarantee the country a greater capacity to assert its independence; and specifically, they would diminish its dependency on the omnipotent northern neighbour, Russia.

Azerbaijan has indeed achieved some freedom of manoeuvre since two of these four new pipelines connect it directly to Western and Turkish consumers, while the old Baku–Novorossiisk (oil) and Baku–Mozdok (gas) pipelines remain in use. The long-awaited and happily consumed increase in revenue has sharply contracted since autumn 2008 and has also brought new risks, while the absence of reliable allies has left Baku exposed to potentially fatal security challenges. Russia, in the meanwhile, has remained uncharacteristically sanguine vis-à-vis Azerbaijan, while subjecting Georgia to economic pressure and military 'punishment'. The Russian leadership does not currently feel compelled to block Azerbaijan's Westward drift, and is instead comfortable waiting for the moment when the consolidation of an autocratic regime would propel this newly emerged petro-state back into its 'big-brotherly' embrace.

This chapter will examine the key dimensions of Russia–Azerbaijan relations in order to evaluate their stability and assess the probability of a break in trends. It begins by analysing the tensions in the energy sector, which are latent for the most part but still generate an irreducible conflict of economic interests. Political dialogue and security estrangement are examined first in general terms, and then in three particular directions, involving Iran, Turkey and Georgia. The conclusion attempts to identify the most probable breaking points in this trajectory in the short- to medium-term perspective.

Cracks in the 'Caspian Energy Security' construct

There is certainly far more to the economic ties between Russia and Azerbaijan than energy-related trade and transit, but it is the oil and gas business that has been the most politicized in Russia, to the degree that exporting hydrocarbons has become not only the main instrument for foreign policy but also its main content. While Moscow and Baku are both fixated on energy matters, their direct interactions in this sphere were reduced to a surprisingly minimal level.

At the beginning of the last decade, the Kremlin decided to discontinue Russian participation in Azerbaijan's oil and gas development projects and not to take part in the BTC's construction.[2] At the same time, Moscow stopped trying to sabotage the BTC project by erecting political hurdles, apparently thinking that it was letting the project collapse under the burden of its own economic inefficiency. Moscow assumed that the high costs of construction and the limited amount of oil available for export in the Azeri sector of the Caspian Sea (given the impossibility of settling its eastern and southern borders) would ensure the pipeline never made a profit.[3] As an extra precaution, Russia insisted that Azerbaijan fulfil its contractual obligation to export some of its oil through the Baku–Novorossiisk pipeline, which was quickly re-routed in 2002 so that it no longer transited through Chechnya.[4]

Despite delays, the opening of the BTC in mid-2006 seriously but not radically altered the landscape in the oil sector.[5] A major factor was certainly the breathtaking rise in oil prices, to approximately $150 per barrel in July 2008, which helped to recover high investment costs for the BP-led consortium and guaranteed an astounding increase in revenues for Azerbaijan's national budget before the oil price's dramatic fall in late 2008. It is impossible to get an accurate estimate of oil reserves even for the Azeri–Chiraq–Guneshli oilfields currently being developed, but it is nevertheless clear that even without additional oil from Kazakhstan, the BTC project will be profitable on the current price-plateau of $75–90 per barrel.[6] Moscow has clearly miscalculated and the only thing it can now do is prevent this 'hostile' pipeline from being connected with the oilfields in Kazakhstan's sector of the Caspian, where production is set to double in the second half of this decade.

The first step toward accomplishing this aim was the agreement to build the Burgas–Alexandroupolis pipeline bypassing the Turkish Straits signed in March 2007 by President Putin and the Prime Ministers of Bulgaria and Greece.[7] The second step was the agreement with President Nazarbaev about exporting Kazakhstan's oil through Russia, reached during Putin's visit to Astana in May 2007.[8] The necessary third step was investing to double the capacity of the Tengiz–Novorossiisk pipeline from 32–35 to 65–70 million tons a year, but on this dossier the Kremlin stumbled: Transneft, the state-owned oil-transit monopoly, has only a minority stake in the Caspian Pipeline Consortium (CPC) that operates the Tengiz–Novorossiisk pipeline and so it was only by the end of 2008 that the compromise on its debt restructuring and increased tariffs was hammered out.[9] That delay has created for Kazakhstan, which very carefully controls Russia's penetration in its energy sector, some space for 'multi-vector' manoeuvring and exporting symbolically significant quantities of oil through the BTC pipeline.[10] Putin and Medvedev tried to entice Nazarbaev to give priority to the Novorossiisk route by offering Kazakhstan a stake in the Russia–Bulgaria–Greece pipeline, but the financial foundation of the bypass pipeline softened as oil prices collapsed in autumn 2008 and its political integrity eroded as Bulgaria developed second thoughts after the change of government in mid-2009.[11]

The controversies related to natural gas are no less intense, even if they are less visible on the surface of Russia–Azerbaijan political relations. Moscow has shown no concern whatsoever about the South Caucasus pipeline that started delivering gas from the Shah Deniz field to Erzurum in eastern Turkey in December 2006. The opening of this significant new supply source has nevertheless changed the Caucasian gas equation. Smarting after the 'gas war' with Ukraine, Gazprom demanded in 2007 the doubling of prices from both Azerbaijan and Georgia to the 'market' level of $235 per thousand cubic metres.[12] In the face of such arrogance, President Aliyev Jr saw an opportunity, and announced that Azerbaijan would stop both importing gas from Russia and exporting oil via the Baku–Novorossiisk pipeline. Energy ties between the two states have thus been reduced to a rather insignificant trade in electricity since the beginning of 2007. Moscow was somewhat surprised by this disengagement, but nevertheless concluded that covering domestic demand and supplying most of Georgia's needs would leave Azerbaijan with only a limited amount of gas for export to Turkey – and next to nothing for export to Europe.[13]

The South Caucasus Pipeline, therefore, could only become a 'strategic challenge' for Russia if it connected Turkmenistan's vast reserves of natural gas to European markets via the planned Nabucco pipeline and projected TransCaspian pipeline. President Aliyev has lobbied hard for these projects in Brussels and received unambiguous even if ineffectual support from Washington.[14] Moscow, nevertheless, remains confident that it has sufficient leverage over President Gurbanguly Berdymuhammedov to secure privileged, if not exclusive, rights on importing gas from Turkmenistan.[15]

In mid-2008, Gazprom's CEO Aleksei Miller made an unexpected proposal to Baku: to buy any amount of Azeri gas at 'European' prices. President Aliyev was not ready to commit to such a deal, under the assumption that Azerbaijan's oil and gas industry was becoming increasingly competitive with that of Russia, so in 2010, a symbolic amount of 1 bcm was exported, and in 2011, it can perhaps be doubled.[16]

Overall, at the start of the new decade, the intensity of securitization of the energy business in the wider Black/Caspian Seas area has significantly diminished and there are good reasons to expect that the slow economic recovery, not to mention a possible new downturn, would keep political passions depressed into the second half of the decade, when the oil production in Azerbaijan would begin to decline.

Defying the Kremlin's political field of gravity

Ever since becoming president, Putin has wanted to give a new impetus to Russia's relations with Azerbaijan, leaving behind the anxiety and animosity of the 1990s and emphasizing the fundamental compatibility of the political regimes crystallizing in the two countries. Visiting Baku in January 2001, Putin sought to impress President Heydar Aliyev not with the accompanying gunboats of the Caspian flotilla but with insightful understanding of the delicate problem of transferring power to his son Ilham. Putin's plan to make the elder leader dependent on Moscow's support was built on the premise that the West would not accept such a dynastic succession. He obviously underestimated Aliyev Sr, who managed to 'sell' his family plan to the US administration despite President George W. Bush's disappointment at the failure of the Key West talks.[17] Aliyev Jr was duly installed as president in October 2003 and the objections of Western institutions, including the OSCE, were dismissed as irrelevant.[18]

Putin could not fail to see that the old *Parteigenosse* had outsmarted him but still made every effort to build a cordial rapport with his scion. According to many observers, personal chemistry between the two 'new generation' leaders was very good, despite the significant difference in their upbringing.[19] The issue that could have provided crucial content to this budding 'beautiful friendship' was the threat of 'colour revolutions' that engulfed the post-Soviet space in 2004, bringing the quasi-democratic rulers close to panic. Indeed, the violent unrest that accompanied Ilham Aliyev's accession to power in autumn 2003 could very well have escalated during the parliamentary elections in November 2005 and become a mass uprising. Russian mainstream media was in fact full of speculations that the 'irreconcilable' opposition relied on US support.[20] In reality the protests in Baku were swiftly and efficiently suppressed, and in just half a year Ilham Aliyev established impeccable credentials by securing an invitation to visit Washington, even if such conservative voices as *The Economist* (4 May 2006) ventured advice to the Bush administration to 'use a long spoon'.

As the tide of the 'colour revolutions' subsided, it has become increasingly clear that excellent personal relations between these leaders cannot compensate for a crucial deficiency in Russia's policy toward individual countries; in the case of Azerbaijan, the inability to resolve the Nagorno-Karabakh conflict. Putin and Medvedev may never have developed personal chemistry with presidents Robert Kocharyan and Serzh Sargsyan, but Armenia nevertheless remains a key security ally that welcomes Russian military presence – and this fact ties Moscow's hands in moving anywhere from the problematic status quo in the conflict that it helped to 'freeze' back in May 1994. Every time Putin and Medvedev tried to bring the conflicting parties to the negotiation table, for instance at the tri-lateral summit in Moscow in June 2009 or at the OSCE Astana summit in December 2010, the initiatives fell flat; the recycled proposal to send Russian peacekeepers has proven to be a non-starter.[21]

Unlike some of his lieutenants, Putin remained remarkably calm with regard to the expanding US–Azerbaijan military ties and Azerbaijan's professed enthusiasm for entertaining security relations with NATO.[22] As a self-proclaimed pragmatist,

Putin can posit that symbolism heavily outweighs substance in this 'mature partnership', while tensions between Azerbaijan and Armenia will continue without fail to complicate their relations with NATO. By the end of his mandate, Putin had become confident that Russia's ability to dominate the Caspian area militarily was beyond any 'Caspian Guard' challenge, particularly since the intensity of combat operations in Chechnya had greatly diminished. Orchestrating the transfer of presidential authority to Medvedev, Putin had to assume that the 'personal chemistry factor' might weaken, but what was far more important was the trend toward the consolidation of Aliyev Jr's authoritarian regime, confirmed during the presidential elections of October 2008.

The Kremlin now expects that the EU will recognize the futility of its ambition to play a meaningful role in the South Caucasus and will in particular find it increasingly difficult to embrace the corrupt petro-khanate (particularly as oil prices stay reasonable). It was taken by surprise by the half-baked initiative of the Obama administration to facilitate a Turkish–Armenian rapprochement (more on that later) but assumes that the setback would bring home the plain fact that the US energy interests in Azerbaijan are minuscule.[23] The sustained decline of Western engagement guarantees that Moscow's field of gravity will become irresistible.

Whether this assumption holds or not, Moscow clearly recognizes the need to increase the content of its political dialogue with Baku, particularly since the energy connection remains thin. A promising avenue has been to restore industrial ties dating back to the Soviet Gosplan economy as well as establishing new joint ventures that would serve the purpose of diversifying the economies of both Russia and Azerbaijan, which have become heavily tilted toward the energy sector. The flagship project is the aluminum plant to be built by Rusal in Azerbaijan in order to use the advantage of cheap electricity production, but there is a large number of smaller enterprises that aim to reconnect Azerbaijan's agricultural sector with the Russian market, or to secure contracts in the massive programme for building infrastructure such as bridges and airports in Azerbaijan for Russian construction companies.[24] There is a certain economic rationale behind these projects but the economic crisis of 2008–9 delivered a devastating blow to Russia's non-energy sector well before such projects had a chance to come out of the proverbial pipeline. Medvedev's attempted strategic shift towards 'modernization' has resulted in a de-prioritization of economic ties with Azerbaijan, which is also facing with a challenge of diversifying its petro-economy set on the track of diminishing growth.[25]

The emergency measures adopted in Moscow for alleviating the unexpected economic recession might worsen Russian–Azeri relations, already significantly damaged by several ill-conceived campaigns against illegal migrants in Russia. The campaigns' first targets were Georgians: not only were their remittances frozen, but several hundred people who had been working in Russia without the necessary paperwork were also expelled. After that, Moscow has seen several sweeps of *Gastarbeiters* that rounded up all 'Caucasian suspects', but perhaps more important was the legislation aimed at 'normalizing' trade in street markets and bazaars adopted in 2007–8. As a result, a large number of Azeri citizens (up to 2 million by some estimates) who made their living in Russia by engaging in, and in some cases controlling this semi-legal trade encountered serious problems and had to adjust to the loss of income.[26] The much trumpeted Year of Russia in Azerbaijan ended in January 2007 on a sour note, as the authorities in Baku switched two mainstream Russian TV channels off the air.

Overall, Putin and his courtiers charting a rapprochement with Azerbaijan (Medvedev playing a supporting role at best) have quite possibly underestimated both the leadership qualities of Ilham Aliyev and his determination to anchor his fiefdom to the Western security 'platform'.

The Iranian troubles, the Islamic factor and Gabala

Iran constitutes an important variable in the complex equation of Russia–Azerbaijan relations, though not in the same way as it shapes Russia–US relations, since the issue of the Iranian nuclear programme, which is absolutely pivotal in the latter, is not directly relevant to the former. At the risk of over-simplifying, Moscow's official position is that Iran's controversial nuclear programme does not constitute a direct security threat, so it could be effectively bracketed out of the framework for Caspian interactions, particularly since Azerbaijan as well as Central Asian states prefer to stay clear of the high stakes involved in the game of non-proliferation.

One conflict involving the interests of Azerbaijan, Iran and Russia is focused on the deadlocked question of dividing the Caspian seabed and delimitating the maritime borders of all five littoral states. At the beginning of his first presidential mandate, Putin tried to resolve this question and initiated a Caspian summit in Ashgabat, Turkmenistan, which failed to produce any real breakthrough.[27] While this embarrassing outcome was widely interpreted as a setback for Russian diplomacy, in hindsight it is possible to assume that Putin was quite content with the disagreement prevailing. He subsequently ordered the staging of large-scale military exercises on the Caspian, presumably to intimidate the riparian states (summer 2002), which indeed were alarmed, so such an exercise has not been repeated since.[28]

In the following years, Russia settled its maritime border issues with Kazakhstan and Azerbaijan, so that only in the southern part of the Caspian Sea do disputes continue to hamper exploration work. Such a deadlock also represents a significant obstacle for building Trans-Caspian pipelines that could have connected Baku with Aqtau, Kazakhstan or Turkmenbashi, Turkmenistan as Medvedev emphasised at the November 2010 Caspian summit in Baku. So long as border issues remain unresolved, proposals for demilitarizing the Caspian basin remain stillborn, and Russian naval forces in the Caspian remain relevant and useful political instruments.[29] For these reasons, Moscow has never applied any pressure on Baku or Ashgabat, or for that matter Tehran, to hammer out a compromise, even while it continues to endeavour in high-profile diplomatic activities. At the October 2007 Caspian Summit in Tehran, for instance, Putin reminded Azerbaijan about 'the impossibility of offering one's own territory to other countries in the event of aggression or any military action against one of the Caspian states'.[30]

Another delicate and potentially explosive issue that connects Russian and Azeri interests and involves Iran concerns the growth of Islamic extremism and the spread of its underground networks. Indeed, Moscow has been so worried about the activities of so-called *jamaats* (Islamic communities) in the North Caucasus that Putin asserted, during a brief visit to Grozny, that Russia has always been 'the most reliable, trustworthy, and consistent protector of the interests of the Islamic world', and lashed out against the 'distorted interpretation of the Koran' by radicals.[31] The claim was hardly convincing, but by the end of 2006 the scale of Islamist activity across the North Caucasus had noticeably diminished (the death of arch-terrorist Shamil Basaev constituted a significant blow to this network); however, since mid-2009, a new rise in terrorist activity has been gaining momentum and religious extremism is a key factor in this trend.[32]

It should be noted that Moscow has never perceived Iran as a significant source of Islamic proselytizing, funding or any other subversive influence in the North Caucasus; Baku, on the other hand, sees things from a very different angle. The Azeri leadership keeps a wary eye on Iranian support for Shia mosques and *medresse* (theological schools), primarily in refugee camps and among the ethnic groups in the Lenkoran region.[33] Presiding in the Organization of Islamic Conference in 2007, Azerbaijan sought to embrace moderate trends in Islam, while emphasizing the secular nature of its state.[34]

It is possible to argue that the potential for growth of Islamic radicalism in Azerbaijan, with its painful war traumas, hundreds of thousands of 'internally displaced persons', widespread corruption and fast growing social inequality, is vast – yet the real scale of this phenomenon has remained quite limited. Various explanations could be advanced to explain such a counter-intuitive fact, from the weak Islamic tradition in the Azeri communities to the strong influence of Turkey, which includes the spread of Sunni mosques competing with Shia institutions. Nevertheless, Azeri authorities have decided to leave nothing to chance and have identified the monitoring of Islamic networks as one of the key tasks for the generously financed special services. Meanwhile, Moscow has been carefully exploiting these concerns and phobias relying mostly on expanded 'anti-terrorist' cooperation between the FSB (Russia's Federal Security Service) and Azerbaijan's Ministry of National Security, targeting primarily the Chechen diaspora in Azerbaijan, accused of maintaining ties with Al Qaeda.[35]

Using various channels, Russian authorities have been spreading the subtle message that the prevalent attitude among the numerous Azeri community in Northern Iran makes it prudent to stay clear of any possible conflict between the US and Iran.[36] There is also a stream of far less subtle warnings about the risks of cultivating military cooperation with the US that could, allegedly at least, make Azerbaijan automatically a party to military confrontation.[37] Import of Russian S-300 surface-to-air missiles is presented as a safer alternative.[38] Baku indeed has many reasons to feel worried about the consequences of escalating tensions between the US and Iran, since any violent destabilization of the latter would inevitably affect its security. This precarious posture was further complicated by the loudly trumpeted initiative on joint use of the Gabala early warning station by Russia and US that was delivered by President Putin at the G8 summit in Heiligendamm in June 2007 without asking Aliyev's opinion.

The Gabala issue has been an irritant in Russia–Azerbaijan relations for a long time; the agreement on leasing this Soviet-built radar (operational since 1985) was signed in January 2002 for a 10-year period, but issues concerning the reduction of environmental damage and increasing rent come up at every high-level meeting. President Aliyev had little choice but to support the initiative, but many experts in Baku accused Putin of taking too much liberty with assets that Russia had on time-limited lease.[39] As the practicalities of the possible joint use and necessary modernization of the Gabala radar came under discussion, it became clear that the initiative was designed as a PR trick aimed primarily at blocking the plan for deploying elements of a US strategic defence system in Poland and the Czech Republic. When US President Barack Obama cancelled that plan in mid-2009, the Gabala initiative acquired new currency, despite Moscow's decision not request an extension of the lease on that station after 2012, since new early warning radars were built in Russia.[40]

The Prague Treaty (April 2010) on reducing strategic arsenals became President Medvedev's main foreign policy achievement but he found himself in a tight corner in the follow-up discussions on missile defence. He had to demonstrate to Obama that Russia was indeed starting to share the United States' threat assessment of the Iranian nuclear programme and so supported new UN sanctions.[41] At the same time, he wanted to convey to Tehran that despite certain disappointment, there was no shift in Russia's policy toward Iran, so unilateral sanctions imposed by the US and the EU would not be 'legitimized' by the UN. Gabala was at the centre of this tricky manoeuvring, since Medvedev sought to keep the cooperative momentum in strategic arms control but could not afford to commit Russia to any practical project that would bring an escalation of tensions with Iran.[42]

Russia's ambivalent stance generally answers Azerbaijan's own preferences but as risks of a military clash and massive destabilization increase at the start of the 2010s, Baku comes to the understanding that it can neither stay out of this Iraq–Iran–Afghanistan zone of conflict, nor count on any meaningful support from Moscow.

The Turkish connection and the Nagorno-Karabakh stalemate

At the beginning of Putin's mandate, traditional prejudices and suspicions concerning Turkey's ambitions in the Caucasus, rooted in 225 years of war from the Azov campaign of Peter the Great to World War I, still prevailed in the Kremlin. Azerbaijan, for that matter, was perceived as the main conduit for the Turkish 'penetration', which was seen as inherently hostile to Russia's interests. It was Turkey's very cautious political line toward the second Chechen war that helped change these perceptions, so that gradually the thinking in Moscow – driven by both energy and security considerations – became more constructive.[43] In the security realm, the key turning point was the start of the Iraq War, when Turkey opted not to participate in the US invasion; Moscow had good reasons to assume that the turmoil in northern Iraq would attract most of Ankara's attention, thereby reducing its appetite for expansion in the Caucasus. With regard to energy, Moscow decided that the BTC pipeline would not constitute a vital link between Azerbaijan and Turkey since that oil would merely be transported to the European market, while Russia would always be able to offer Turkey significantly better deals in natural gas.[44]

The breakthrough occurred in December 2004, when Putin went on an official visit to Turkey, aimed at paving the way for a new energy partnership and security cooperation. Turkish Prime Minister Recep Tayyip Erdogan made a reciprocal visit to Moscow in January 2005. In June of that same year, Erdogan met with Putin in the Russian presidential retreat near Sochi. In November, Putin travelled again to Ankara. Such unprecedented intensity of top-level bilateral dialogue between Russia and Turkey was underpinned by the rushed development of gas-related projects spun from the Blue Stream pipeline. This expensive subsea project was completed back in 2003 (Putin's ceremonial opening in November 2005 was somewhat belated) but had remained a 'white elephant', as the two sides continued to quarrel about volumes and prices.[45] The new plan was to add another underwater pipe in order to double Blue Stream's capacity (from 16 bcm to 32 bcm a year) and to export the extra volume to Italy via Gazprom's networks in southern Europe and a new pipeline (hence the presence of Prime Minister Silvio Berlusconi at the ceremony).[46] Feasibility was not in doubt, but the geo-economic rationale seemed nevertheless questionable, since the intention was clearly to overtake the slow-moving EU proposal for a 'Caspian corridor' and to block the access to the European market for the gas from Azerbaijan.

Putin could well have succeeded, but his 'gas war' with Ukraine in the first days of 2006 sent a massive shockwave through European consumers, so much so that the EU swiftly breathed new life into the Nabucco pipeline project and embraced President Ilham Aliyev, inviting him to Brussels in November 2006.[47] Moscow expected that Turkey, frustrated by the procrastinations and underlying hostile attitudes to its long-standing application for accession to the EU, would turn to Russia, who showed readiness to form an alliance built on its new energy power.[48] Ankara, however, opted for a more flexible approach that sought to concentrate as many energy flows as possible – yet such a 'gas hub' was not an option for Gazprom, which was firm set on preserving its exclusive rights. From the start of 2006 and until mid-2008, high-level political dialogue between Russia and Turkey was effectively discontinued, and in 2007–8, several energy initiatives in the Black Sea area that bypassed Turkey were put forward by Moscow, starting with the Burgas–Alexandroupolis oil pipeline and including the South Stream gas pipeline towards Varna, Bulgaria, aimed at the Balkan and Italian markets.[49]

The pause in high-level dialogue ended in early 2009, when President Abdullah Gul visited Moscow; then Putin went to Ankara in August; Erdogan came to Moscow in January 2010; Medvedev made an official visit to Turkey in May 2010; and Putin held talks in Istanbul in June.[50] It is possible to speculate that it was the second Russian–Ukrainian 'gas war' of January 2009 that delivered a strong impulse for reviving the dormant 'beautiful friendship' as Moscow abandoned the intention of cutting Turkey out of the energy strategizing and opted instead for maximizing gas flows in order to reduce the transit through Ukraine.[51] The planned capacity of the South Stream was doubled (and the route of this pipeline across the Turkish exclusive economic zone was approved), the deal on expanding the Blue Stream was re-confirmed, and an oil pipeline Samsun–Ceyhan was agreed upon, quite possibly instead of the delayed Burgas–Alexandroupolis project.

Focusing like a laser beam on energy issues, the Kremlin has tended to underestimate the impact of the Nagorno-Karabakh conflict, assuming that the status quo is essentially acceptable for all parties. Hence the ambivalent reaction to President Obama's initiative on Turkish–Armenian rapprochement centred on lifting the transport blockade of Armenia; and hence also the barely hidden relief when the initiative collapsed.[52] Moscow made a half-hearted attempt at reasserting its leadership in managing the negotiation process, but generally sticks to the proposition that Azerbaijan could be gradually pulled into Russia's political sphere of influence without putting forward any concession on the deadlocked Nagorno-Karabakh non-settlement. Moscow is not particularly concerned with the modernization of Azerbaijan's armed forces according to its new military doctrine and is inclined to dismiss statements from Aliyev and his courtiers concerning their willingness to resort to force in order to restore control over lost territories, perceiving them as empty rhetoric.[53] There is indeed every reason to assume that Georgia's crushing defeat in August 2008 will dissuade Baku from experimenting with the use of military force to regain the enclave.

Russia will continue to treat Turkey as its main partner in the Black Sea region, though it would still prefer to contain that country's influence in Azerbaijan and quietly sabotage its initiative on establishing a Platform for Stability in the Caucasus.[54] Russia is also interested in limiting the scope of the South Caucasian 'energy corridor'. Moscow expects that, for Turkey, relations with Azerbaijan will be far less important in the short term than developments in northern Iraq, and in the medium term might only shrink further, due to a decline in Azeri oil production.

The Georgian war makes a difference

The five-day war between Russia and Georgia in August 2008 might at first appear to have had surprisingly little impact on Russia–Azerbaijan relations, in spite of the significant resonance the conflict had on the international scene. That resonance has by now disappeared without much trace; Baku, however, has had to make serious adjustments to the strategic aims and tactical goals in its foreign policy, in order to respond both to the precedent of Moscow's massive use of military force and its unilateral recognition of Abkhazia and South Ossetia as independent states. Aliyev has also taken into account the very real risk of a new escalation of this conflict.

The causes and ramifications of the Russia–Georgia war are too convoluted to be examined in this chapter.[55] Nevertheless, there are several elements to this often emotional drama that pertain directly to Russia–Azerbaijan relations. At the risk of oversimplifying, it can be argued that Moscow counted on the trend that political tensions between the 'revolutionary' regime of Mikheil Saakashvili, who portrays himself as a champion of democratic reforms, and the authoritarian regime of Ilham Aliyev, who has inherited a mature system of patrimonial corruption, would drive the two states apart. A crucial test for that assumption came during Aliyev's visit to Moscow in November 2006, when Putin tried to secure his support, even if implicit, in escalating the confrontation with Georgia. Not even a symbolic nod of approval did Putin receive from his presumed soulmate; instead, paying a visit to Tbilisi in early February 2007, Aliyev in fact expressed solidarity with Georgia and offered tangible energy gifts.[56]

The Kremlin had to swallow its disappointment but still assumed that the situation whereby Azerbaijan became Georgia's solo energy supplier, which in turn controlled the transit of Azerbaijan's Western exports, would generate numerous tensions between these neighbours, since Tbilisi had to cover its energy needs as cheaply as possible while Baku could afford to be charitable only to a point. That point, as it happened, has never been reached, perhaps because the vacillation of oil prices since the August war has distorted the combination of the Caspian energy equations. Aliyev took a very cautious position during the long-envisaged yet unexpected outbreak of hostilities, refraining from condemning Russia's behaviour even when it appeared safe to follow the EU lead in criticizing 'excessive' use of force.[57] Despite US Vice-President Dick Cheney's emergency visit to Baku, Aliyev could not avoid the conclusion that, in a crisis, the West could only issue ineffectual protestations while Russia could, after such an easy and complete victory, in fact resort to force.

Moving its tanks at will across Georgia's territory, Moscow took great care not to inflict any damage to the strategic pair of pipelines carrying oil and gas from Azerbaijan to Turkey. The 'energy corridor' was nevertheless rendered useless during the conflict, since BP decided to shut down the South Caucasus pipeline and KazMunaiGas closed down the Batumi oil terminal (the BTC was damaged earlier by an explosion on Turkish territory). The route's high vulnerability was thus demonstrated even without Russia actually targeting the pipelines, leaving Azerbaijan to consider anew the rationale for greater diversification of its export channels.[58] Gazprom's offer to buy any amount of Azeri gas was subsequently reiterated with added emphasis, and today Aliyev is finding it increasingly difficult to refuse. Aliyev must also acknowledge that while feverish talk in Brussels and Washington about Nabucco and the Trans-Caspian gas pipelines' indispensable role continues non-stop, the projects' feasibility remains problematic.[59]

Beyond these energy issues, Azerbaijan's leadership must perceive that Russia's recognition of Abkhazia and South Ossetia as independent states, combined with the recognition of Kosovo's independence by the US (and a few dozen other states), greatly weakens its case for re-establishing control over Nagorno-Karabakh.[60] While Medvedev has been careful to maintain a diplomatic balance in Russia's relations with Armenia and Azerbaijan, he also works on reasserting Russia's control over negotiations between them. This effectively implies that in order to achieve minimal progress, Baku will need to make serious concessions. Russian military bases in Armenia played no role in the war with Georgia, but together with the new military bases in Abkhazia and South Ossetia they complement the recently expanded system of bases in the North Caucasus, so that Russia's military dominance over the whole region has strengthened. Turkey, meanwhile, has shown no inclination to challenge this dominance, and the issue of NATO expansion has been reduced to irrelevance.[61]

This unchecked military prevalence underpins Russia's power and fuels its political influence in the region; Azerbaijan has good reasons to expect that the main target for pressure and provocations would be a defiant Georgia – and would much prefer to stay clear of this confrontation. At the same time, Baku cannot ignore the risks involved in potential domestic unrest in Georgia fostered by the trauma of humiliating defeat and fuelled by the desperation caused by the economic crisis. Yet a new 'colour revolution' in Tbilisi might very well be more violent than the one that took place in late 2003, as Saakashvili could resort to force against his opponents, and it could profoundly affect regional stability, possibly triggering a new wave of secessions. At the same time, civil unrest in Georgia could at any moment break Azerbaijan's vital connection with the West, which requires Baku to stay close to Moscow.

Conclusions

The cordial façade of Russia–Azerbaijan relations in fact stands on a rather shaky economic foundation and camouflages their fundamentally unbalanced structure, which produces oscillating political and security tensions. In 2008, both states held presidential elections, but if the closely orchestrated re-election of Ilham Alieyev hardly attracted any attention, the reconfiguration of the Russian system of power has brought much uncertainty and generally weakened Moscow's networking efforts in the Caucasus and Central Asia, since Medvedev – who now has to personify these efforts – is no Putin. This weakness can still be corrected if Putin returns to the Kremlin in 2012, as many in Russia expect, but this self-restoration would hardly rescue the integrity of the corruption-corroded political system. In the meantime Russia's standing in the Caucasus, and particularly vis-à-vis Azerbaijan, has been strengthened by the punishing military intervention deep into Georgia in August 2008.

The war has revealed, at least in the near term, the fundamental and irreducible weakness of Western geopolitical engagement in the Caucasus, at least in part caused by the US's overemphasis on Iraq and NATO's overstretching itself in Afghanistan. Washington, as well as key European capitals, has also grown weary of Mikheil Saakashvili, who until recently was cheered as the champion of democratic reforms. This could potentially translate into declining interest for the region, and Baku cannot expect that energy matters will, in a period that could very well witness shrinking demand and low prices, secure enough Western attention to provide the country with reasonable space for political manoeuvre. The readjustment of the balance of power after the Russia–Georgia war makes it imperative for Azerbaijan to accommodate some Russian demands, from granting Gazprom symbolic volumes of gas export to accepting an indefinite prolongation of the deadlock over Nagorno-Karabakh.

The sudden arrival of a devastating economic crisis would inevitably affect relations between Russia and Azerbaijan since their 'petro-economies' are extremely vulnerable to the sharp decline in energy revenues, while power-sharing among the elites and political apathy in society were shaped by – and conditional on – expectations of lasting prosperity. These expectations have since been shattered. While in Azerbaijan an increase in domestic instability (including the spread of Islamic extremism) can be expected, in Russia the consequences could also include a shift from the present-day 'best behaviour' back to a more aggressive foreign policy. The relative stability in the North Caucasus cannot be sustained by new massive transfers of federal funds and there are clear signs that isolated terrorist attacks signify a development of a complex region-wide political crisis. Moscow's responses to this security challenge could very well be disproportional and 'asymmetric', and aimed at shifting their epicentre outside of Russia's territory. While Georgia would most probably remain the target of choice in such a scenario, Azerbaijan cannot consider itself safe. The geopolitics of Caspian hydrocarbon production and transit, in which Baku is a focal point, might go through a period of declining interest – and stakes – should oil prices settle on a plateau twice lower than their June 2008 peak. Regional security, meanwhile, might deteriorate as two key players, Turkey and Russia, are hit by economic hardship and political turmoil, while Iran defies international sanctions and persists in advancing its nuclear programme.

Part III

Russian export policy and relations with transit countries

9 Russia, Gazprom and the CAC

Interests and relations

Leonid Grigoriev

Introduction

Russian foreign and energy policy was formed along several tracks during the 1990s and early 2000s. First, the dissolution of the USSR brought about the unbundling of the Soviet integrated pipeline system transiting through all newly independent states (NIS). Second, the period was marked by low demand and therefore low gas prices on the international markets. It was changes in global demand and thus prices after 2003 that provided a new stimulus to Russia, spurring the search for a new strategy. Third, income from Gazprom was used by the Russian government as a source of domestic social and economic spending until at least 2004: low domestic gas tariffs imposed by the government after the 1998 crisis further dragged down the country's investment in its resources. Lastly, during this period the Kremlin imposed politically motivated low gas prices for some CIS consumer countries, leaving Gazprom hostage to Russian foreign policy.

Increased world demand for oil and gas during the global economic boom of 2003–8 changed the situation completely, notably by pushing up prices. In this new context, Caspian oil and gas became one of the most promising sources of energy outside the Middle East for future growth potential. As new sources of gas were being developed in Kazakhstan and expectations were growing with regard to Turkmenistan's export capacity, Central Asia's gas pipeline network, the Central Asia–Centre (CAC), became a focus of international politics.

In the natural gas market, Russia enjoys multiple roles as a producer, consumer and exporter as well as a transit state, moving huge amounts of gas every year. As outsiders (especially the US and EU) focused on this region, they tried to separate oil and gas issues from the local economic and political life (including in Russia), leading to limited understanding of various actors' interests. This was the wrong approach: gas and transit issues cannot alone satisfy Russia's or other players' interests. Several other issues are thus critical in the NIS, and are considered important by numerous governments, including in Russia, Central Asia, the Caucasus and Eastern Europe, for instance, migration, trade, anti-drug cooperation and energy. Natural gas and transit issues are just one aspect of Russian policy, though evidently they are more critical for Gazprom.

There is no obvious or explicitly stated strategy or policy on the part of either Russia or Gazprom toward the CAC specifically. One may nevertheless identify fairly recurrent attempts to attain certain objectives: Russian energy security (transit interests, flexibility of supplies) and Gazprom commercial interests (transit fees and flexibility of supplies). Russia's pipeline network, the Unified Gas Supply System (UGSS), is one of the few remaining infrastructure systems which still binds the post-Soviet space together. The Russian government is clearly keen on supporting some form of compatibility across the NIS economies and maintaining low transaction costs for business, trade and labour. Such a goal makes the CAC and the UGSS a valuable resource for all countries and companies involved in the NIS.

While oil and gas issues are central to the region's politics and economy, the countries' key interests remain domestic political stability and economic development as well as reliable demand for oil and gas for the maximization of profits. Some countries, such as Uzbekistan, Russia and Kazakhstan, play the dual role of producer and transiter. In this the interests of outside actors and local producers differ; this situation has only been made more complicated following the significant fluctuation of energy prices between 2007 and 2009. As a major supply route of gas from Central Asia to Ukraine onward, the CAC has become a critical factor in European energy security. As a result, upstream production of gas in Central Asia and the Caspian region (CACR) and the pipelines linking them to Europe have become the focus of numerous serious studies.[1]

The CAC: Soviet legacy

The pipeline system inherited from the Soviet Union came as a free gift to many newly independent countries: it was available for use for post-Soviet countries at zero investment cost, with no corporate debt but with ready engineering and human resources as well as long-term planning into exploration and development already made. It is hard to imagine such a vast pipeline network being built after independence, with all the associated commercial and economic risks at a time of significant political and economic transitional crises in the region.[2] What is more, the CAC system survived the transitional crisis of 1990s largely due to Russian demand for over 17 years. Since then, Russia has continued to build various extensions to its UGSS (Blue Stream, etc.), diversifying its capacity to deliver gas on a long-term basis for its EU consumers.

The whole and its parts

The FSU's pipeline network, as a legacy of the planned economy, is sometimes considered by Western observers as inconvenient – if not outright detrimental – to the interests of the oil and gas actors in the region today. In all fairness, however, it is better conceived as an historical achievement, in terms of scope and technology, by a planned economy bent on gathering gas from fields far dispersed to consumers across the Eurasian continent. It is a complex system which depends on the proper functioning of all parts for the whole. The planning system operates on the basis of securing balanced demand and supply as well as their respective investments.

By 1990, the UGSS totalled 220,000 km of transit pipelines (as distinguished from distribution pipelines), linking together over 200 gas fields, six gas processing plants, 46 underground gas storages, 4400 switch stations, and thousands of compressor stations with an overall capacity of 50 million kV. This required huge investments, made in the 1970s and 1980s. The system still moves over 700 billion cubic metres (bcm) every year. Gazprom has managed to avoid major decline in the UGSS's gas output within Russia in the 1990s while all other industries in Russia were severely hit (witness the 40 per cent drop in oil output, for example). Outside Russia, the UGSS became a free gift from the Soviet planners to a few neighbour countries of the post-Soviet space. While the network has been legally divided between countries, the UGSS's management requires the system's integrity. Until 2004, most NIS producers certainly could not finance the system's maintenance and repair, let alone its development, outside its most immediate needs.

Upon Soviet dismemberment, the massive UGSS and all the investment it represents was divided between newly created countries. In Tajikistan the ownership of pipelines was given to the state enterprise Tajikgas. In Kyrgyzstan Minpromenergo assigns ownership of the gas transportation system to government-owned Kyrgyzgas. The Uzbek section also fell under control of a single corporation, Uztransgas, part of the country's Uzbekneftegas holding. State-owned Turkmengas owns Turkmenistan's pipelines, which are managed by subsidiary Turkmentransgas. Somewhat more complex is the Kazakh system: KazTransGas, which operates the Kazakh section of the CAC, is wholly owned by KazMunaiGaz, while it controls, either directly or indirectly (through Intergas–Central Asia), all transit pipelines (CAC and Bukhara – Ural). Despite this, the entire system linking the various national pipelines is a near-single entity, regulated by international contracts and agreements. This bond is especially strong between Kazakhstan and Russia due to their intertwining supply sources in border regions.

Maintenance and repair

After nearly 20 years of separate existence, all actors in the gas world now seem to have a stake in the state and condition of the CAC pipelines. Yet the Ukrainian and CAC parts of the UGSS are estimated by industry specialists to have benefited the most from general repairs during these two decades. Though there is some reserve capacity in the system in the event of a disruption of supply, notably in the case of accidents, the system essentially works at full capacity. The CAC's economic benefits are divided between countries according to the geographic distribution of the main routes.

At the beginning of the transition period, in 1991, the gas production from three countries, Kazakhstan, Turkmenistan and Azerbaijan, totalled 134 bcm.

Table 9.1 Capacity of Central Asia–Centre pipeline system, bcm

	Capacity, bcm	*Length, thousand km*	*Construction start date*
Bukhara–Ural	21	4.5	1963
Central Asia–Centre	68	3.4	1968

Source: Gazprom

Export for these countries totalled 71 bcm, and declined to 10 bcm by 1998 (while output dropped to 76 bcm). Today, exports have been restored to 77 bcm (of which 6 bcm goes to Iran). Industry experts have long been sceptical about the working conditions of these old pipelines. Nevertheless, by the 2000s the main pipelines' capacity had been restored, though it was admittedly not sufficient to handle future exports from the region.

Since then, projections for future output of gas in the region have been widely debated. As usual, actual implementation of big infrastructure projects is delayed by years and decades. By the 2000s the main pipelines' capacity was mostly restored; however, it is already clear that this old network will not be sufficient to handle the region's future exports. The more technical aspects of the CAC's conditions have been questioned for a long time: what has been the quality of maintenance and repair (some pipes may be affected by corrosion, etc.); what more is needed; what may be losses of capacity?

Most CAC pipelines are operational but at reduced levels from their original design capacity. Today, this means transit is roughly 50 bcm per year – down from the maximum of 70 bcm transit capacity to Europe (see Table 9.3). Experts also debate the result the 1990s' crisis had on the system's technical capacity. As was stated by the head of Giprorechtrans, Vladimir Rudometkin, in 2008, 'for years there was no full-fledged study of pipelines and their maintenance.'[3] Pavel Klimov, First Deputy of Intergas Central Asia, has moreover stressed that securing the reliability and effectiveness of the main pipelines by modernization is his company's top priority. Wear and tear on the CAC does indeed appear rather serious, due to the network's already long lifespan. This problem can be solved, however, in one or two years, given the appropriate investment.[4] This evaluation means that the CAC's present seems sustainable. Adding new pipelines, as is projected, would give an additional incentive for further maintenance and repair.

Russia and the CAC

Gazprom is planning to buy 70–80 bcm and possibly up to 90 bcm per year from the CACR via the CAC, securing these amounts through formal agreements. Russia's advantage, in this race, is that its pipeline system can already deliver this Central Asian gas to Ukraine and onward. In 2006, the Russian network carried

56 bcm of CACR gas: 39 bcm from Turkmenistan (this should reach 50 bcm in 2007), 9 bcm from Uzbekistan and 7.5 bcm from Kazakhstan.

Given 'market prices', the best route for new transit capacity in the CACR is obviously through Russia along the old CAC. This includes the enlargement of CAC4 from 4 to 10 bcm and building another line with a capacity of approximately 20 bcm (these numbers are subject to frequent changes by the media). The idea for this project dates back to 2003. At that time it was a Ukrainian–Turkmen project to circumvent the Caspian Sea by the north (1,745 km, 30 bcm and a $1 billion investment). KazMunaiGaz is also planning to enlarge the capacity of its CAC section, bringing it from 58 to 100 bcm. These additional routes – the PriCaspiy, in Russian – will be sufficient to absorb the immediate gas supply growth in both Turkmenistan and Kazakhstan. According to this plan, Gazprom should have enough gas for delivery to Ukraine and beyond.

In the long run, however, the key issue will not be transit capacity but the CACR's actual supply into the system. Most experts, for example, still do not know precisely how much gas there is in Turkmenistan. Turkmenistan's official numbers for future gas output go far beyond what is actually feasible, given present public knowledge. For instance, Turkmenistan's gross gas output for 2007 is estimated to have been approximately 78 bcm, but is projected to reach 160 bcm in 2015 and 250 bcm in 2030. Domestic consumption will meanwhile also grow, from 18 bcm presently to 50 bcm, expanding export capacity from 58 presently to 125 bcm in 2015 and 200 bcm in 2030.[5] Naturally, should these projections be valid, Turkmenistan's pipeline system alone ought to be enlarged by more than 100 bcm a year.

Such plans are supported by the price setting changes to the 'market formula' for Uzbek and Turkmen gas – a mechanism essentially very close to the net-back approach. Such a radical change makes it much easier for gas producers to receive adequate export payments through any pipeline corridor. Differences between routes for producers now come down to direct cost differentials. Gas transit from countries in this region now comes under the general regulation of the Energy Charter Treaty.[6] Russia signed the treaty in 1994, though like Norway it has not yet ratified it.

Rival consumers

There is also more than one party interested in Caspian gas, China, Iran and the EU being prime examples. This author believes one can rank the political feasibility of each project in a geographically clockwise order, starting from the north. After Russia, China will receive its long-term supplies by pipelines, beginning with 30 bcm annually in 2009 via a new 7,000 km pipeline. Turkmenistan's future output now appears as though it will be absorbed by China. New Delhi is also looking for supplies via continental lines. The next increment will be for its neighbour, Iran. Subsequent increments will be the subject of competition between India and the EU. Meanwhile, the legal division of the Caspian Sea is still pending, blocking the TransCaspian route (Iran still objects to any pipeline crossing

Table 9.2 Net export of energy products, $bln, 1996–2007

	1996	*1997*	*1998*	*1999*	*2000*	*2001*	*2002*	*2003*	*2004*	*2005*	*2006*	*2007*
Russia	37.0	38.0	27.0	30.0	51.0	52.0	55.0	70.0	89.0	117.0	188.0	214.0
Ukraine	–7.8	–7.2	–5.4	–4.5	–5.2	–5.1	–5.0	–5.2	–6.4	–7.4	–10.0	–13.4
Belarus	n/a	n/a	–1.5	–1.0	–1.1	–0.9	–0.7	–0.9	–0.9	0.0	0.2	–1.6
Azerbaijan	0.4	0.4	0.4	0.7	1.4	1.9	1.6	1.9	2.6	2.8	4.8	4.8
Turkmenistan	n/a	0.5	0.3	0.7	2.0	n/a	n/a	n/a	n/a	n/a	n/a	n/a
Kazakhstan	1.1	1.6	1.3	2.2	4.0	4.0	4.9	7.0	11.2	17.4	23.2	27.9
Memo												
OPEC oil and gas export	171.0	167.0	110.0	158.0	256.0	214.0	210.0	256.0	361.0	533.0	650.0	731.1

Source: Comtrade, OPEC, national statistics, IEF estimates

the Caspian), and instability in Afghanistan makes a trans-Afghan pipeline a very difficult endeavour. In the long run, all the promised gas will be delivered, but the TransCaspian route to the SCP (South Caucasus Pipeline) may well be further delayed.

Two sets of competition are taking place for the CACR's hydrocarbon resources. The first concerns competition between actors for determining the final destination of the region's incremental growth via continent pipelines. This pits China, India and the EU against each other. Indeed, the region's new energy output helps support economic growth, but its final destination – delivered via pipelines – will be locked for years. Given geographical and political considerations, the markets for CACR are to a large extent inevitable – the issue concerns both the timing of the new pipelines and the volumes that will be locked up. The EU will therefore have to wait a little bit longer for additional supplies from the CACR. The EU's advantage lies in the higher prices its market commands – but in natural gas, it is still a seller's market, and the EU has no choice but to negotiate.

There is also a second competition revolving around *how* these new resources can be brought to market, especially in the case of bringing gas to Europe. Such a competition basically boils down to whether there will be Russian participation (and probably transit via Russian territory) or whether Russia should be avoided entirely. Particularly since the January 2006 Russian–Ukrainian gas conflict, most of the rhetoric in recent years has been devoted to bypassing Russia, this chiefly for political reasons. The old Nabucco project (3,300 km, 31 bcm capacity, €4.6 billon, projected to open in 2012) has thus re-emerged as a counterweight to Russia. This project nevertheless appears to be overtly political and devoid of real economic rationale. In other words, Nabucco's importance is exaggerated. There are rather alarming forecasts for the EU's growing gas supply gap: by 2015, it may reach 100–200 bcm, maybe even more (though recent data in fact suggests less). Nabucco, with its 30 bcm potential capacity, could alone supply 15–30 per cent of this gap – an important step, to be sure, but not exactly a panacea. In order to fill this gap, the EU will have to cooperate with all suppliers.

There is a stamp of strategic importance on the EU's project. Some suggest giving the project a special status and financing. Others speak of excluding Gazprom from participating in the project, this for strictly political reasons. Meanwhile, several Russians have voiced their scepticism concerning the project. Normally, pipelines are built for a particular market and for a particular source of gas. As far as Nabucco is concerned, however, nothing is yet clear: the purchasing market, the actual source of gas, the financing mechanisms and source.[7] Perhaps realizing this, representatives of the Nabucco Gas Pipeline International Company have in fact invited Gazprom into the project. Of Nabucco's 30 bcm capacity, 15 bcm would be reserved for companies participating in the construction, while the rest would remain available for auction.

Gazprom counter-attacks

Gazprom has responded to the Nabucco project with the more commercially viable South Stream project, linking Russia to Burgas in Bulgaria via the Black Sea, and onward to Balkan and central European markets. South Stream, contrary to Nabucco, already has a partner (ENI) for the marketing, financing, technology (Blue Stream was built with Italian support) as well as throughput from Russian and Central Asian fields. The Nabucco–South Stream commercial competition includes important political implications, however. Gazprom wishes to build pipes like South Stream (such as Nord Stream) for the supply of the EU market, but has no interest in projects blatantly aimed at avoiding Russia.

Gazprom's head, Alexei Miller, has not completely eliminated the possibility of Gazprom taking part in Nabucco, however. What is more, if Caspian states actually had the additional gas supply in the range of 200 bcm that is needed by 2030, Nabucco would become commercially interesting for big companies. The crucial issue is timing – all this while Gazprom secures its position on transit routes and strengthens its own supply elasticity. A global recession may yet change the intensity of development, however, due to reduced demand in the EU and the humbled financing capacity of the main actors involved.

Thus, while the CAC, as an old gift from an extinct economic system, remains a functional network, supplying gas to far-flung consumers, it remains a contentious subject. Yet the network, as is, works, even if it needs more repair. It can be and will be enlarged, but it is fundamentally viable from engineering and economic points of view. Future gas deliveries from the CACR are thus a geopolitical issue, and the secret to unlocking this lies in both local and global politics.

Actors and interests

Beyond being a critical component for gas deliveries as well as for supplying cash to producers, the CAC also constitutes an important link between different countries in the region. One cannot study international pipeline networks without taking into account the various interests of the countries involved: their economic development, the political stability of their elites, etc.

The legacy of transition

The post-Soviet space has only just emerged from the long and exhausting transitional crisis of the 1990s. Then, all countries went through serious industrial, budgetary and unemployment crises. By overlooking Russia's 10-year crisis, during which its GDP declined by 43 per cent, one may also overlook the economic and sociological reasons why Russia supports the few national companies that successfully survived this period and made money for the country. The post-Soviet space, in turn, had few transition models to follow and adapt to: some restructuring and focusing on services, some trying to save a semblance of industrial capacity, some dependent on remittances or transit fees, while others relied

Table 9.3 GDP and oil and gas production, Ukraine and Central Asia, 1999 and 2007

	Unit	*Kazakhstan*		*Turkmenistan*		*Ukraine*		*Uzbekistan*	
		1999	*2007*	*1999*	*2007*	*1999*	*2007*	*1999*	*2007*
GDP	*$bn*	*17.0*	*104.9*	*3.9*	*26.2*	*31.3*	*141.6*	*17.0*	*22.3*
GDP per capita	$1000	1.1	6.7	0.8	5.1	0.6	3.1	0.7	0.8
Natural gas production	bcm	9.9	29.6	21.3	67.4	18.1	21.1	55.6	58.5
Natural gas export	bcm	4.0	15.2	10.0	43.3	1.0	0.0	6.0	14.7
Natural gas import	bcm	3.0	10.0	0.0	0.0	60.3	50.6	2.0	0.0
Oil production	mln ton	30.1	68.7	7.1	9.8	3.8	4.5	8.1	4.9
Oil import	mln ton	0.5	7.7	0.0	0.0	9.4	9.8	0.0	0.0
Oil export	mln ton	25.2	60.8	1.5	1.8	1.5	0.0	0.6	0.0

Source: UNSD, IEA, Euromonitor

on their natural resources, mostly fuel, for their development.[8] Naturally, Russia tried to save some of its manufacturing capacity, but only succeeded in a few sectors, notably in mining and hydrocarbons, metals and chemicals, as well as a few natural resources-based industries. There is no doubt that the hydrocarbon industry is favoured in Russia and enjoys state support.

In such an economic environment, the political elite of Central Asian countries and Ukraine/Belarus were genuinely concerned about their regimes' stability, fearing social and economic problems during most of the transition period. Independence brought numerous social problems which required the active support of a fragile equilibrium of political forces: GDP decreased by half, encouraging the respective governments to focus on the stability of the export and transit incomes.

In many cases, new projects in oil and gas upstream with foreign investments supplied these governments with jobs, investments, good PR and at least a limited flow of funds for the budget. In spite of low oil and gas prices during most of the transition period, countries with natural resources could at least generate export incomes even while their manufacturing and industrial sectors remained in deep crisis. During the transition period, very little domestic financial resources were available for economic development or new infrastructure projects. Higher oil prices came rather late for containing the transition's economic and social crisis. Only recently have the region's budgets become mostly balanced and investment patterns changed. The Azeri and Kazakh examples are representative of a certain

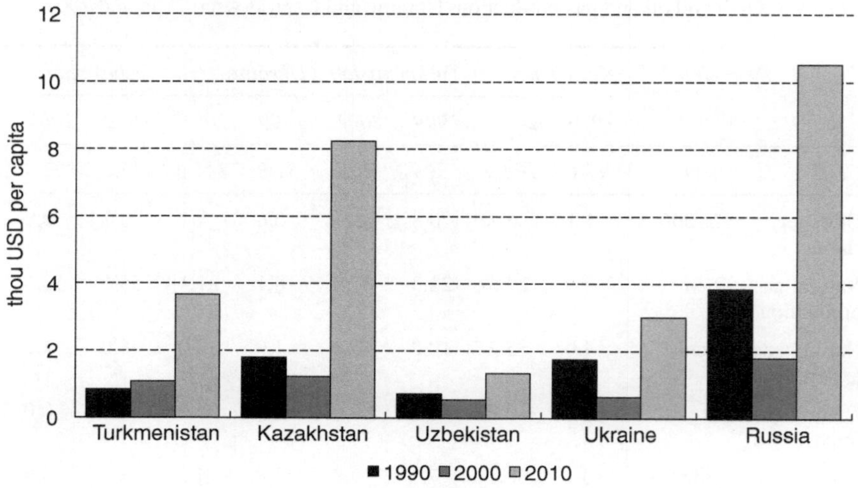

Figure 9.1 Russia and countries in the region, GDP per capita, $1000s, 1990–2010

Source: UNSD, IMF

change in development goals in the past five years. Financing huge long-term investments in the gas industry as these two countries have done, however, is a complicated task, especially with price volatility. Prices, moreover, are bound to stay low during the period of recession, undermining such financing. It is worth noting, as an aside, that since 2004 liberalized markets (primarily Anglo-Saxon ones, i.e. the US and UK) have paid higher prices for gas than those in continental Europe (see Figure 9.2).

The CACR's growing importance

Russia and the CACR have recently become the focus of politicians and experts in global energy and politics alike. Tough negotiations on gas transit, and the repeated Russia–Ukraine gas conflicts in particular, suggest, quite correctly, the substantial political interests of all parties involved in the patterns of energy production and export routes. This became even more political in the autumn of 2007 after the International Energy Agency (IEA) finally admitted the serious supply problems facing the EU in its flagship *World Energy Outlook*. Taking into account China and India's growing manufacturing might and the increased energy consumption and import from a new global middle class, supply problems could occur as early as 2015. These facts led the IEA to note that the 'global energy system is on an *increasingly* unsustainable path.'[9] According to the IEA, the shortage could for example reach up to 12 million barrels a day (mbd) in the case of oil.

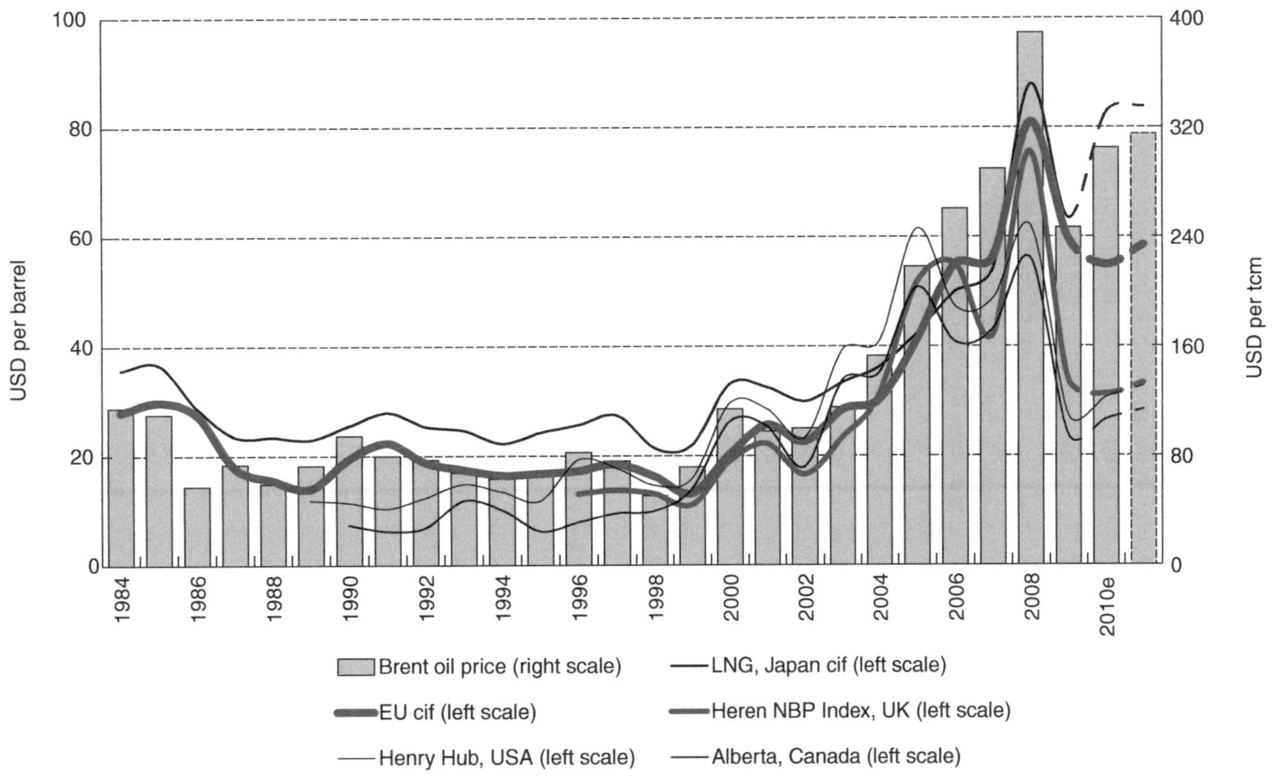

Figure 9.2 Natural gas prices on main markets, \$/tcm, 1984–2010

Source: UNSD, IMF

This essentially means that the IEA has changed from displaying a balanced picture of the future global energy market to an alarmist stance. This new recognition has elevated the political importance of Caspian supplies and has shortened the timeframe in which these supplies play the critical role of balancing the global market in the future. Obviously, without Iranian supply the region is limited in its capacity – and still very much landlocked – for any drastic changes to occur with regard to the EU market. This author would paint a more balanced picture, however, stemming from a projected further decrease in the energy-per-GDP ratio.

In such a context, the Caspian Sea region clearly appears as a convenient source of oil and gas supply in the coming decades, being partially able to fill the gap after Qatar wraps up its LNG projects, by 2010. Inevitably for the EU, other actors and interests are at play (China, India, the US…). Thus, CACR supplies as well as upstream and transport energy projects have increasingly been the subject of more global geopolitical manoeuvring. Realizing this, countries are less trusting of global market mechanisms for their hydrocarbon supply and are looking for some assurance of delivery through special relations with governments, companies' access to fields and assuring the transit ways to export markets. Groups of 'interested consumers' from the US, China, India and the EU are involved in bringing about their respective energy policy objectives in the region, and therefore in its politics. There is, by contrast, no such thing as a 'pure' Russian energy policy or Russian foreign policy vis-à-vis its regional neighbours that can be isolated from myriad other issues. Instead the country has a complex mix of conflicting objectives and policies, in which energy supplies may very well be sidestepped in order to reach other goals.

From a buyer's to a seller's market and back again

For the purposes of this analysis, two time periods may be considered: before and after the oil price increase (roughly around 2003–4). These two periods are fundamentally different in terms of the gas policies of all countries involved. It is important to remember that oil prices in the summer of 1998 dropped as low as $8 a barrel and caused (among other factors) a financial crisis in Russia. Demand for Turkmen and Uzbek gas sank before the turning point of 2004. Today, in the face of a global recession, new changes are to be expected that are again unfavourable to producers.

Global economic growth generated high energy prices and changed the strategies of the region's actors, countries and companies alike. Who could imagine, only nine years ago, the kind of long-term planning Turkmenistan or Gazprom can presently afford in the region? Only 10 years ago Ukraine was paying for Turkmenistan's gas through barter trade.

The combined effects of the economic upturn in developed markets (particularly in the EU and the US) and high energy prices fundamentally changed the situation in the CACR. Suddenly, the political elites in the hydrocarbon-rich countries discovered the virtues of new Production Sharing Agreements (PSA) and old Soviet-era projects, including transit projects. Seeking and retaining these transit opportunities has become a common trend in the region.

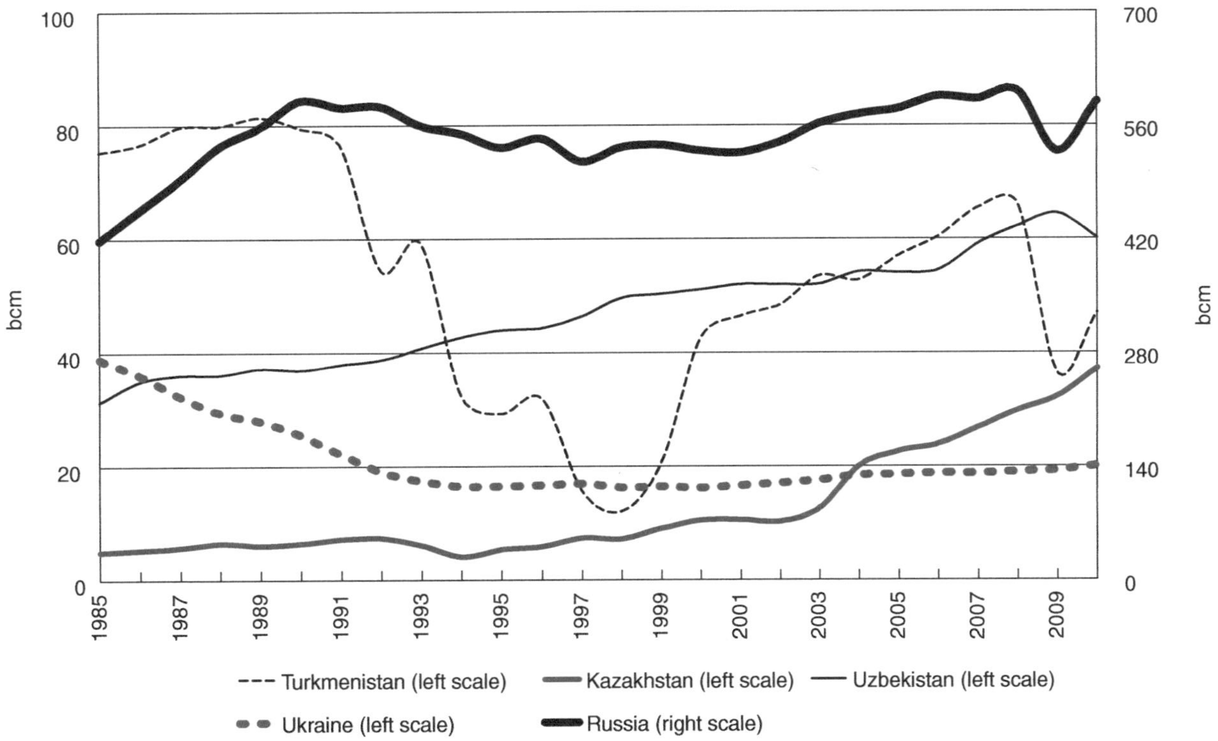

Figure 9.3 Natural gas production, bcm, 1980–2009

Source: National Statistics Committees, BP

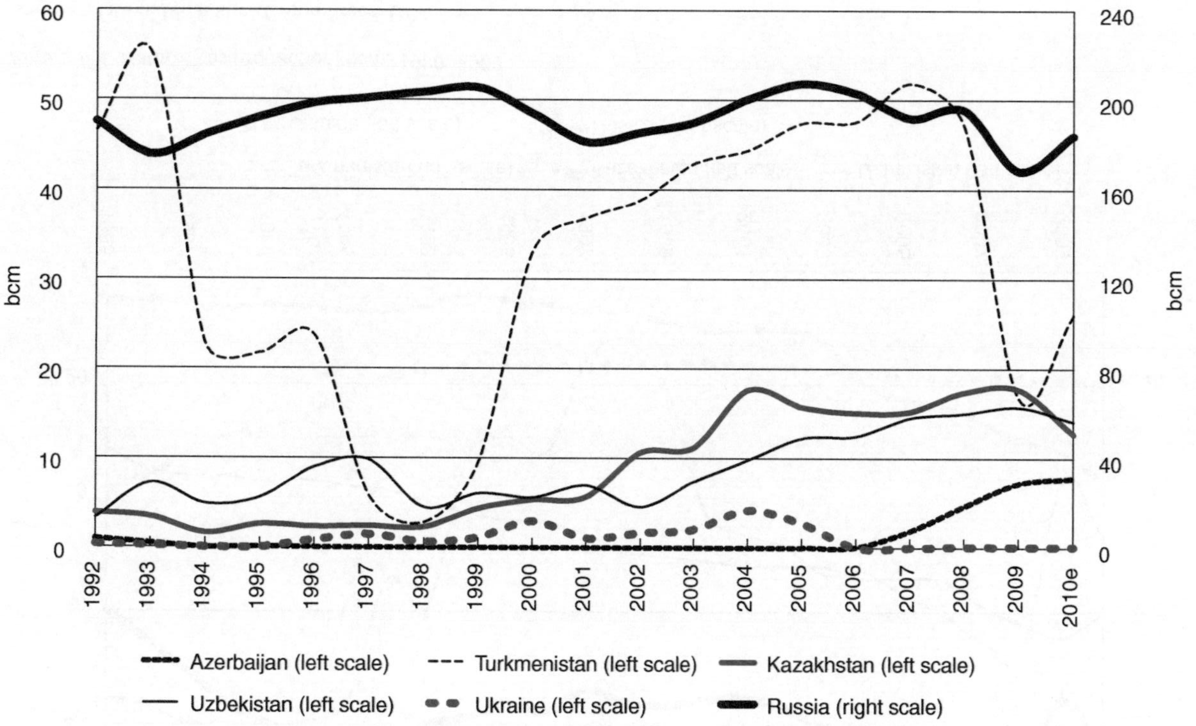

Figure 9.4 Natural gas export, bcm, 2001–10

Source: National Statistics Committees, BP, IEA, IEF estimates

Their previous cocktail of domestic problems began to dissipate by 2006–7: no longer did they suffer budgetary crises, no longer were they ignored by world powers (and pressed on issues of democracy). Instead, CACR countries found new mechanisms for social engineering, new capacity for funding prestigious projects, development projects, or simply their military. At the same time, high energy prices generated increased disparity between exporters and importers of hydrocarbons. Human relations between NIS citizens and business relations in the region extending from Central Asia to the western borders of the FSU have nevertheless remained fairly 'traditional'. Given their overall prosperity, conflicting issues were mostly set aside, seemingly resolved, which further helped to stabilize local regimes. Now, after the 2008 crash in oil prices, this relatively short period of windfall energy income is over and old frictions may re-emerge, casting a cloud on regional stability.

Consumer priorities

Regional energy projects in Eurasia have always been intertwined with international politics. Some players pursue complex political objectives that go far beyond the simple commerce of oil and gas. It is fairly easy to identify the priorities and interests of the region's major outside players (i.e. large consumers):

1 Increasing flows of oil and gas from the region to global markets;
2 Setting up legal grounds for investments by international oil companies;
3 Securing access to investment and transit routes so as to create competition between suppliers (and hence price reduction);
4 Creating additional transit routes to avoid Russian territory.

In this list, only the two first objectives can be defined as commercially driven, and can be considered typical of global energy markets – companies and governments trying to set up favourable conditions for investment and thus output. The third objective (if it really exists) is distinct from the previous two in one simple regard: such a goal goes against the interests of regional suppliers. The creation of the Gas Exporting Countries Forum (GECF) is an example of this basic conflict between producers and consumers; blind competition between producers, so dear to consumers, now seems unlikely to occur even in gas markets.[10] The last objective, though entirely unrelated to commercial concerns, nevertheless carries significant economic implications. From the Russian point of view, there is of course no reason to insist so heavily (especially in the media) on avoiding Russian transit; in Russia, therefore, such a priority is seen as fatally flawed, and met with incomprehension. The two Ukrainian gas conflicts may have worried European consumers, yet Gazprom has never intended to break its contractual obligations. On the contrary, Gazprom is extremely keen to deliver a steady supply of gas, and thus continue to be regarded as a reliable supplier. These conflicts, however, may in fact make Moscow more intent on changing the pattern of its supply routes by injecting political considerations and non-commercial financing in future projects.

Producer priorities

Countries with oil and gas resources and transit capacity have had a rather complicated set of objectives in recent years, as various combinations of resources lead to different combinations of practical objectives for local exporting countries:

1 Income for the budget and political stability (at least survival);
2 Generating resources for enhancing economic development;
3 Better export choices for maximizing companies' proceeds through diverse and reliable access to infrastructure;
4 Diversifying export routes to maximize upstream production and counterbalance transit problems.

The first priority in the early part of the transition crisis for local governments was obviously to mitigate the pressing budgetary crisis. Oil and gas export and transit incomes were therefore used for immediate social and socio-economic needs. Only later did increased incomes from PSAs and other projects begin to deliver the financial resources required for further development. In such an unstable environment, however, oil and gas price fluctuations were particularly problematic, given that always-expensive price of transiting hydrocarbons to market. Such price fluctuations thus presented all exporting countries of the region with a challenge.

Gazprom and Russian priorities

It is vital to note that the interests of the Russian authorities and those of Gazprom are extremely complex and not necessarily the same in all points. For instance, Gazprom is primarily interested in:

- Producing enough gas for domestic obligations (decided by the government) and for foreign long-term contracts;
- Getting domestic gas prices raised – from $50 currently to $100 by 2010 (two-thirds of Gazprom sales are domestic), but preferably to EU net-back levels, i.e. 'market prices';
- Vertical and horizontal diversification within and outside Russia – in electricity generation and downstream chemicals;
- Becoming (according to official objectives) the largest global energy company;
- Achieving non-discriminatory regimes for access to foreign markets, particularly the EU;
- The predictability of export demand (for investment decisions) and export prices;
- Transit incomes from the use of Russian territory and the creation of 'mirror' routes;
- Securing commercial (i.e. non-political) conditions for transiting its gas to markets, including the diversification of export channels (e.g. LNG);
- Particularly securing access to more profitable segments of foreign gas markets (i.e. retail).[11]

By contrast, the Russian government's economic interests in the gas segment of the economy are harder to describe in full. What is certain is that there is a natural difference between the interests of Gazprom on the one hand and those of gas consumers on the other: industries (metals, chemicals, fertilizers, etc.) as well as people (individual consumers, i.e., the broader population).

While the Russian government is allowing the rapid increase in domestic gas prices, up to 20–22 per cent annually, Gazprom would clearly want the adjustment to be faster. Gazprom was the government's cash cow for a long time (especially under Premier Minister Victor Chernomyrdin); until the early 2000s the company therefore had limited capability to finance its activity before the recent energy price hike. Moreover, the present recession will probably result in domestic prices being increased somewhat more slowly than was originally designed.

Russia's 2003 'Energy Strategy to 2020' also clearly reflects the period of low prices and more prominent private oil and gas companies in Russia. At the time, the 2003 Strategy was in fact very concerned with securing markets for Russian exports and much less about securing sufficient output upstream to supply distant markets, as is the case today. Yet as energy prices climbed, Russia's energy output increased (see Table 9.4). A new Energy Strategy was to have been developed by the end of 2008, but it has not yet been approved. A global recession may well change its content yet again.

During the transition period, gas output in Turkmenistan for instance declined by a factor of six. In this context, the idea of developing a new transport infrastructure for export (especially one avoiding Russian soil) was everything but logical; it was in reality entirely political. During this time, the Russian energy sector was not experiencing much pressure from domestic or global demand, nor could it count on high prices or government funds to support it in the long run. Gazprom also struggled to secure revenue, given the lack of both domestic and foreign income, but this was the work of the government: low gas prices in the FSU and especially the CIS (notably in Ukraine and Belarus) were essentially a government subsidy, and weighed heavily on the company. In other words, Gazprom long suffered a political burden.

Table 9.4 Russian Energy Strategy (2003): actual data and optimistic scenario targets

	Actual data		*Energy Strategy targets – optimistic scenario*		
	2005	*2008*	*2005*	*2010*	*2020*
Oil (mln t)	470	488	445	490	520
Gas (bcm)	641	664	615	665	730
Coal (mln t)	298	326	280	330	430
Electricity (TWh)	952	1 037	935	1 070	1 365

Source: Federal State Statistics Service, Ministry of Energy

Despite these differences, the interests of Gazprom and the Russian government have lately converged. Lately, Russia's politicized approach to the gas industry largely mirrors Gazprom's own interests. There has been a well-documented contradiction between the various hikes in natural gas prices (in Ukraine and the Caucasus countries) and Russian political influence in those countries. One might better turn the argument on its head, and simply highlight the fact that, for a decade before these hikes, Russia received little in exchange for subsidizing these countries' gas consumption. It is not worth dwelling on this issue, especially since this stance has since changed. Today, Russia's state interests in the gas industry are at least clear and straightforward:

- Allow Gazprom to conduct its business and domestic gas prices to increase;
- Make sure Russian gas output is sufficient for both domestic needs and foreign contractual obligations;
- Help to the companies to develop and become more diversified;
- Counterbalance systemic risks and help its companies access foreign markets or investment opportunities (typical support for national companies);
- To negotiate the non-discriminatory legal regime for Russian export and investments;
- To counterbalance the politically motivated 'avoiding Russian territory' approach of some outside actions.

The situation was more difficult for countries with a role in transiting Russian oil and gas. For a long time this business was considered as a means of barter trade, a way to secure side services and achieve political ends. The business of transiting resources has its specificities: it depends entirely on the throughput; transit fees go directly into the national budgets as income; it is regulated, at least partially, by international law. Some countries, such as Ukraine, Moldova and Georgia, have been seriously dependent on transit income for balancing their national budgets. Yet transit pipelines also require maintenance and repairs.

For the past few years, events in the CACR have seemed surprising to most outside observers. Yet gas agreements, price changes and new pipeline projects are coming to life with 'regional ingenuity'. The price and transit crises in Ukraine in 2006 and 2009, and in Belorus in 2007 were also resolved in a manner that was mostly satisfactory for all countries directly involved.[12] Despite this, outside observers have been left with a feeling of increased Russian control, be it statist or corporate. Russian policy is always under scrutiny for non-commercial objectives. Yet in reality, both Russian companies and political actors are trying to protect their respective interests. All parties were acting in very much the same way as most observers would in the same situation, with the same set of assets, liabilities, obstacles, advantages and disadvantages.

Foreign policy is business

Diplomacy is required to secure the best possible terms for all parties: states, companies and citizens. Russia's approach to regional energy issues may be better understood if one takes seriously all the objectives of outside actors, particularly their attempts to return to a buyer's market and build transit routes to bypass Russia. On both sides the feeling of insecurity is persistent and even strengthened by political events. It sometimes appears as though Russian decision-makers have perused Zbigniew Brzezinski's *The Grand Chessboard* (1997): in this case they might have taken too seriously Brzezinski's prescriptions as a ready list of instructions for the foreign policy of respective countries.[13] Such an approach on the part of Russia has the merit of being clear: let commercial competition go on, so long as critical threats are avoided. For Russia, two pitfalls to this policy loom on the horizon.

The first would be a third crash in oil prices, after those of 1986 and 1998. This means realizing that, behind the rhetoric of competition, one must beware of steps that are purposely taken on the part of consumers for returning to a buyer's market. The collapse of oil and gas prices in the second half of 2008 has given a strong incentive to secure transit incomes. Of course, gas exporters will probably never again be pitted against each other in a blind competition for the sake of creating a consumer's market. Long-term investment decisions in the oil and gas industry are determined by future price expectations and related costs (currently at $60 minimum for 'new oil'). In certain cases government decisions play a strong guiding role – witness Qatar's decision to limit its LNG programme by 2010. To some extent, better understanding of the gas market may be acquired via the GECF. Statements by participants of the December 2008 meeting in Moscow reflect the decision to coordinate activities in four specific areas: relations between countries and gas consumers; exchange of information on forecasts and investments between producers; introducing new technologies; and cooperating in LNG production.

The second threat – responding to politically motivated Russian bypasses – is already amply covered in political circles, the media, and even expert opinions. From the Russian perspective, Gazprom ought to be considered as the most reliable supplier of natural gas for decades. Europe's own industry and business circles fundamentally trust Gazprom as a supplier. It is difficult to imagine Russia's diplomatic stance as ignoring this fact. Russian governmental control over the country's major pipeline system to the EU market should not be understood as a dependence on the Russian state, but as a broadly convenient way of delivering huge supplies of gas. Russia also exports a third of its coal production and two-thirds of its oil and refinery products. If one takes in the entire picture, it becomes clear that Russia is heavily dependent on physical demand and fluctuating prices for these exports (Russia produces about 10.3 per cent of the world's primary energy). Conversely, the global economy largely depends on Russian energy supply. No amount of diplomacy can change this basic equation, though it may well change respective incomes.

Getting behind the rhetoric

Consumers' objectives, such as avoiding Russian territory, came as early as the mid-1990s – long before Russia began its more assertive policy in the CACR. The Baku–Tbilisi–Ceyhan oil pipeline (BTC) was to a large extent a political objective; this pipeline was completed only recently (2006), several years behind schedule. The same happened to the Baku–Tbilisi–Erzurum gas pipeline (South Caucasus Pipeline, or SCP). Both enjoyed huge political support and were portrayed by many politicians, journalists and experts as an 'avoiding Russia' political brand. Not surprisingly, no Russian companies were involved in the construction or supply of oil and gas via these bypasses: since the economic rationale of both projects depended on the region's future output, Russian companies instead focused on building export infrastructure for their own resources, while still trying to build reserve capacity for any forthcoming increase in oil and gas production from Kazakhstan or Turkmenistan. At the same moment, Russian companies were losing markets to other suppliers (though not to a large extent) and were trying to defend their own market access.

The era of low gas prices and increased competition for market access in the early 2000s has left some traces in present-day analyses of the gas industry. A typical comment looks like this:

> Russia – in its turn – is interested to a great extent in securing of transport of energy resources in the preferred direction. For Russia it has not only economic, but political meaning. Russian participation in gas transport may allow it to retain a certain control over local competitors in energy export.[14]

It is interesting to note that, in 2004, the authors of this note were clearly confusing the interests of Russia and those of Gazprom. Since then, a clearer difference and division of labour between the two actors has emerged: Gazprom busies itself with domestic projects while Russia focuses its foreign policy on protecting the national company's interests abroad (export volumes, legal regimes and investment conditions). This is very different from how other countries conduct their own foreign policies.

'The "PriCaspiy" corridor has a good perspective,' said Russian Foreign Minister Sergei Lavrov in an interview for *Kazpravda*:[15]

> It is well-known that many consumers of energy resources including, first and foremost, economically developed countries, would like to develop of multidimensional alternative routes for hydrocarbon supply to the importing countries. That is normal. It is not normal however when energy problems are politicized to the detriment of producer countries' interests, and even common sense. After all, hydrocarbons, like capital, flow in the direction where it is most profitable. If a purely economic problem (even if strategically significant) is politicized, there emerge attempts to consolidate the energy resources consumers on the basis of opposition to 'the Russian energy monopoly'. The

> energy supply corridor 'PriCaspiy' has great potential in connection with the agreement of Russian, Kazakhstan and Turkmenistan on the 'North–South' gas pipeline construction. The quadripartite activity (Kazakhstan, Russia, Turkmenistan and Uzbekistan) on reconstruction of the gas-transport system 'The Central Asia–Center' is being conducted in the same key.

Key agreements for new energy corridors to Europe and the EU have been quick to emerge in the high energy prices context of 2007–8. Some were signed in Ashgabat at the end of November 2007. The enlargement of CAC pipelines circumventing the Caspian Sea by the north has for now ended all hopes of a rival TransCaspian route. This does not however preclude the return of a TransCaspian scenario (with additional gas output), but only after additional lines are built toward Russia and then China. Another agreement was signed in January 2008 between Russia (and Gazprom), Serbia and Bulgaria for the South Stream project. This new corridor will serve as an outlet for CACR gas and a natural commercial viable route for gas to reach the most profitable market for countries of Central Asia for the time being.

The price for Turkmen gas for purchase by Gazprom for the first half of 2008 has been set at $130 and $150 for the second. Going by 2007 volumes, this price would supply Ashgabat with approximately $8.5 billon (this is discounting oil and refinery products). This deal has essentially priced Turkmen gas at close to net-back market prices, seriously threatening the Ukrainian economy which is heavily dependent on Turkmen gas. This in turn creates immediate pressure on the Ukrainian polity. It is also indicative of the trend facing the Caucasus and Black Sea regions. Normally, Gazprom would sell gas coming from the CAC by adding its own transit fee ($30 in 2007), yet in the case of Ukraine the extremely complex nature of economic and political relations clouds rational accounting. Should the TransCaspian pipeline option materialized, the SCP could be easily filled and gas would be transported to the Balkans by Nabucco or another similar line. Yet in this case politics would not cease be a problem for setting volume and price to Ukraine. Political supporters of the TransCaspian pipe overlook such a consequence of completing the SCP. Gas from the Caucasus has already gone up from $100 to $230 per mcm for the simple reason that Azeri gas was westbound to Europe. The same awaits Ukraine if Nabucco is to be completed. Both the TransCaspian pipeline and Nabucco may still come to pass, but these remain distant options.

Conclusion

As we know, it is impossible to have one's cake and eat it too. That wisdom should be a guiding line for EU–Russian energy cooperation. It is hard to understand the EU's logic: how can the EU hope to encourage Russian companies to cooperate in supplying its growing gas needs while simultaneously limiting Russian companies' range of action within the EU? The EU's transit policy, which is almost entirely based on creating alternatives to Russia by avoiding Russian territory, is a particular case in point.

The EU is a global energy consumer and Russia a global energy supplier: the only solution lies in interdependence. The policy of building diversification pipelines that avoid Russia on principle is not one that is based on the reality of resource distribution. It would be foolish to assume that Russia and Gazprom would simply discard the country's natural advantages in terms of gas resources and the convenience its territory represents for transiting gas from the CACR to Europe. This is in fact the founding of Russian and Gazprom foreign energy policy.

Russian foreign energy policy is in fact quite straightforward: retaining both access to the EU market *and* transit incomes from Central Asia. It is only natural that Russia would defend its economic interests in the corridor between the Caspian Sea and central Europe, which the EU is ignoring. There are a number of political interests in the region, yet basic economics are sufficient to explain Russia's approach to foreign policy.

The EU's '20–20–20' energy strategy, as unveiled in the Commission's 2008 Energy Package, may well change the scope and structure of the EU's future energy demand – and therefore its imports. The programme could however be delayed, for technological reasons or as a consequence of the global recession sure to follow 2008. Meanwhile, the EU's gas output is still declining, and energy prices have gone back to acceptable levels after having crashed from their July 2008 heights. As a result, we may be at a turning point in terms of import patterns, whether or not the Energy Package is implemented. The EU's energy policy is, after all, a very complicated affair. One thing seems certain, however: so long as the EU seeks more supplies from Russia while at the same time seeking more 'independence' from Russia, it will be rather hard for the EU to find a solution to its predicament.

As far as the CAC is concerned, its future will depend more on the relative speed of upstream output growth in the CACR than on the existing network's capacity to enlarge. The upgrading of the entire CAC network and its dependencies has only just begun – this for the first time since the USSR collapsed. The new export capacity development may still be influenced by the political interests of outside actors (that is, non-regional powers), but commercial logic still stands for the efficient use of the existing system, especially when taking into account the factor of time. Clearly, some countries' foreign energy policies favour pipeline projects that are based not on commercial logic but on certain geopolitical consideration. Russia's foreign energy policy, on the other hand, seeks to secure the country's long-term commercial and corporate interests while at the same time maintaining stable relations with regional producers and transit countries. Attempts to ignore Russia's long-term energy interests in the region have been and will continue to be met with a reaction based on the country's history with the CACR. Such a history also comes with deep understanding of the local interests and the motives behind the region's governments and elites.

Oil and gas flows from the CACR will reach world markets one way or the other, adding to the physical balance and stability of world supply. At the same time, the reality of geography and the present configuration of the pipeline networks favour Gazprom in retaining its position as a major supplier and carrier of CACR hydrocarbons. What is more, the various interests of the region's elites, who wreak their own long-term political stability, further help Russian foreign policy objectives, namely, minimal costs and consideration of Russian economic interests. The web of conflicting interests in Eurasia (especially with outsiders) was affected by the two Ukrainian gas conflicts of 2006 and 2009 as well as by the conflict between Russia and Georgia in August 2008. The global recession that began in 2008 will have a very significant impact on the transit system linking the CACR with the EU market. The EU's energy-saving programmes (especially the '20–20–20') may reduce long-term demand for gas; a drop in demand will give more time for diversifying this transit system. A recession and low prices reflecting excess capacity will, in turn, reduce investments in both upstream projects and transportation networks. Meanwhile, the CAC system will likely remain the backbone for transiting CACR gas to Ukraine, Turkey and the EU for years to come.

10 After the war

The Southern Corridor

John Roberts

The corridor in context

The development of new pipelines from the Caspian to the Black Sea and the prospect of major trans-Caspian shipments of both oil and gas to European and Mediterranean markets ensure that Azerbaijan and Georgia occupy key positions in regard to European Union initiatives to enhance Europe's energy security. But while these two countries constitute the heart of the Southern Caucasus Energy Corridor linking Caspian producers to European markets, the corridor's utility as a route for further extensive development remains the subject of intense controversy.

On the one hand, the Georgia–Russia war of August 2008 has highlighted the vulnerability of a corridor passing through particularly problematic zones of conflict; on the other, the Russia–Ukraine gas crisis of January 2009 has highlighted European concerns about the reliability of Russia as an energy supplier, spurring particular interest in the prospects for expanding the Southern Caucasus Energy Corridor to carry large volumes of Caspian gas directly to mainstream markets in the EU.

The EU has long argued that what is required is the transformation of the Southern Caucasus into a corridor for both oil and gas. The term 'corridor' – applied in various ways but with all including the concept of a link from Caspian producers to Europe – is used not only by the EU and the Caucasus countries themselves but also by the US as an aspiration for the attainment of energy independence by the Caucasus countries of Azerbaijan and Georgia and, by extension, Turkey.[1] The logic of the corridor is simply that it offers a route that does not involve either Russia or Iran, two producer countries which might be said to be in competition with both Caucasus and Central Asian suppliers seeking to use the Caucasus to gain access to outside markets. When the European Commission issued an action plan to accompany its Second Strategic Energy Review in November 2008, it declared: 'A southern gas corridor must be developed for the supply of gas from Caspian and Middle Eastern sources, which could potentially supply a significant part of the EU's future needs. This is one of the EU's highest energy security priorities.'[2]

In the Caspian context, the EC's Action Plan stressed the need to work 'with partners such as Azerbaijan and Turkmenistan' and in the Middle East context to work with 'Iraq and Mashreq countries, amongst others'. In effect, although Iran was once considered a prime initial source for input into the proposed Nabucco pipeline to central Europe (see below), the current EU attitude is that Iran is not an appropriate partner at present. This essentially relates to the EU's concerns about Iran's nuclear programme rather than to any problems that might arise in connection with Iran's role as a potential rival producer to those in the Caspian. Indeed, the EC's Action Plan specifically states that, along with Uzbekistan, Iran should be considered as a potential supply source 'in the longer term'.

Both Georgia and Azerbaijan hope that the concept of a Southern Caucasus Energy Corridor will boost their political as well as their economic independence by making Europe – and to a certain extent America, through the participation of US companies in various major energy development projects – more dependent on the Caucasus route. They hope that this will in turn lead to increased political support for their countries' independence. Long before the August war of 2008, Georgians, in particular, would question whether Russia was truly reconciled to their independence. For its part, and particularly so long as it operates as if energy pipeline development constitutes a zero-sum game rather than a win–win proposition, Russia will remain highly nervous concerning the growth of energy transit through the South Caucasus.

With the creation of both the Baku–Tbilisi–Ceyhan oil pipeline (BTC) and the South Caucasus gas pipeline (SCP), in practice the Southern Caucasus Energy Corridor has already been achieved. However, although it is expected to grow in importance as oil and gas volumes build up, the corridor remains subject to political constraints. In particular, as the Georgian war demonstrated, it runs the risk of frozen conflicts erupting again. These conflicts include the Nagorno-Karabakh dispute between Azerbaijan and Armenia, Georgia's breakaway regions of South Ossetia and Abkhazia, which Russia has recognized as independent states and, more broadly speaking, the attitude of Russia toward the former Caucasus republics of the Soviet Union.

Before considering the abilities of Azerbaijan and other Caspian producers to contribute to Europe's energy requirements, it is useful to summarize the specific political and overall energy security context within which any Caspian contribution to Europe's energy balance has to be assessed. In practice, this is a context shaped essentially by two events: the Georgian war of August 2008 and the Russia–Ukraine gas crisis of January 2009.

The Georgian war of August 2008

The debate still rages as to how the war began. On 7 August 2008, Georgian troops first launched an artillery bombardment and then a ground assault on Tskhinvali, capital of the breakaway Georgian territory of South Ossetia. Saakashvili said after the war that Georgia's military operations were intended to halt the entry of large numbers of Russian troops into South Ossetia, while there had previously been some bombardment of Georgian military positions from areas controlled by Russian troops stationed as peacekeepers in the breakaway territory. At first sight, the Georgian assault appeared to be a campaign to secure the full return of central government authority to all of South Ossetia. But the overwhelming response that Russia brought to bear in the region, coupled with an extensive build-up of forces in the North Caucasus and positioning of heavy munitions in Abkhazia and South Ossetia before fighting broke out, makes it hard to determine which actions can be said to have triggered open warfare.

What is certain is that Russian forces acted both swiftly and effectively. First, they ousted the Georgian forces from Tskhinvali and then from the rest of South Ossetia, including areas that the Georgian troops had held before 7 August. At the same time, they removed Georgian forces from positions in another breakaway territory, Abkhazia. Russian forces also entered Georgia proper, cutting the main east–west highway, entering the city of Gori and sending reconnaissance and demolition units into the port city of Poti where Georgia's naval facilities were also subjected to Russian naval bombardment.

A ceasefire arranged by French president Nicolas Sarkozy halted the fighting but Russian forces continued for some days to demolish installations they considered to be of prospective use to Georgia's military forces. Moreover, nine days after the ceasefire agreement, Moscow announced it was officially recognizing the independence of both South Ossetia and Abkhazia.

The war was not about energy, but it had profound implications for energy security. First of all it highlighted the Southern Corridor's fragility – and the problems inherent in transiting a country ruled over by such a poor practitioner of the art of brinkmanship as Saakashvili. Second, it highlighted the way in which Russia would, at least to various Western observers, over-react. Third – so long as there is nothing to connect Russia to the still unresolved issue of the mysterious attack on the BTC line two nights before the outbreak of general fighting on 7 August – the war demonstrated that though energy installations were vulnerable, even Russia seemed wary of attacking them directly.

The fragility of the Southern Corridor is the hardest to assess. On the one hand, the war produced several immediate reactions. Azerbaijan began re-evaluating its energy relations with Russia, prompted in part by a Russian decision to halt its relatively small volume of imports of Azerbaijani oil at the start of the crisis in order to effect what it said was necessary maintenance on the Baku–Novorossiysk pipeline.

Before the war, the European Union – and Turkey – had specifically been looking to the Caucasus as the initial stage in developing what Ankara termed a 'Fourth Corridor' or a 'Fourth Artery' to carry Caspian and Middle Eastern gas from the Caspian to Europe, thus matching existing corridors bringing gas from Russia, Norway and Algeria. However, one consequence of the combination of the Georgian war and the financial crisis that began to engulf so much of the world in late 2008 is that it has become even harder for Western companies to raise commercial finance for new gas lines in the Caspian, the South Caucasus and the Black Sea, and for companies to secure the necessary commitments alone, both to provide the gas upstream and to buy it downstream.

Against this, however, is the fact that the world's international oil companies are flush with profits in the wake of the high oil prices of 2007 and the first half of 2008, and with relatively few prospects for new investment. In addition to the question of investment, there is the issue of transit. It is significant that neither Chevron nor the Kazakh government have backed away from continued reliance on the Southern Corridor for the export of Kazakh oil from the giant Tengiz and Kashagan fields. Chevron proceeded to increase, on schedule, its oil flows from Tengiz, both by rail and through the BTC, in the autumn of 2008. At the same time, the Kazakh and Azerbaijani governments re-committed themselves to developing a 500,000 barrels per day (b/d) shipping service across the Caspian Sea to bring Kashagan crude to Baku for input into an expanded BTC.

If oil is the immediate concern, however, the biggest long-term concern for Europe, as it grapples with the consequences of the Georgian war, is its gas policy. Even before the war, government guarantees on volumes were thought to be required; now they are considered absolutely essential. What this means for governments is that they will have to play a far more active role than they have so far.

The European Commission is moving to address this issue in two ways. With regard to Azerbaijan, it is essentially content to exercise political muscle, coupled with a judicious use of EU funds for initial project development, to further the cause of pipeline projects that it supports, notably the planned Nabucco pipeline from Turkey to Austria and the series of interconnections – the ITGI system – that would link Turkey with Italy by way of Greece. Both systems – and a third, the Trans-Adriatic Pipeline (TAP) project that would connect Greece to Italy via Albania – would enable an expanded South Caucasus Pipeline to convey as much as 20–24 billion cubic metres (bcm) of Caspian gas to Turkey and the EU annually.

At the same time, it is contemplating a way of underwriting the long-term purchase of Turkmen gas so that instead of receiving its income in the standard manner – with limited revenues in the early years as gas development is ramped up, then a middle era of relatively high income, followed by a final decline – the EU would commit to a principle of revenue equalization, so that from year one the government could expect to receive close to average annual revenues. This concept, initially seen as a way of encouraging gas development in Azerbaijan, was by 2010 seen as a way of encouraging Turkmenistan to move swiftly to develop its gas resources – and to commit them to export rather than, as so many other oil producers have done, primarily using them to develop indigenous industry and as a substitute for oil. By working to ensure pipeline development, the EU would help to enable both Azerbaijan and its prospective customers in Europe to have a genuine choice as to which export system would deliver the right combination of commercial and practical viability for Azerbaijani gas exports to Europe.

Associated with this approach is the idea of creating a Caspian Development Corporation as an umbrella group for European companies – and perhaps for North

American firms as well – to develop and export the gas resources of countries on the Caspian's eastern shore. Initially, this group would seek to work with Turkmenistan for development of Turkmen gas resources but the concept could apply to Kazakhstan and Uzbekistan as well. Creating a CDC would enable European and other international companies to mount the kind of integrated exploration, production and export project similar to the development by the Chinese National Petroleum Corporation (CNPC) of Turkmenistan's Bagtiyarlik field and the export of Turkmen gas to China by means of a 2,200-km pipeline through Uzbekistan and Kazakhstan to a connection with China's existing West–East gas trunkline.

Here, the attitude of Turkmenistan as a prospective source of gas for Europe by means of a duplicate SCP line and such prospective projects as Nabucco and White Stream is likely to prove critical (see below). Officially, Turkmenistan holds to the view that pipelines across the Caspian are purely a matter for the states at either end of the line; in practice, its immediate response to the Georgian war was a resounding silence and the end of all reference to Trans-Caspian pipelines. This silence lasted until the meeting summoned by the Hungarian government to develop the Nabucco pipeline project in Budapest on 26 January 2009, at which the principal Turkmen official handling foreign energy relations, Yagshygeldy Kakaev, again professed his country's interest in supplying Europe with gas via the Southern Corridor. At present, Turkmenistan quite definitely favours the development of an export system that would carry Turkmen gas to Europe without passing through Russia. When European Commission President José Manuel Barroso held talks with President Gubanguly Berdymukhamedov in Ashgabat in January 2011, both leaders stressed their support for Turkmen gas exports to Europe, and set up committies to assess trans-Caspian gas transport either by pipeline or using tankers carrying compressed natural gas.

Russian interests

The Georgian war demonstrated that Russia is quite prepared to use force in pursuit of its interests. This is a lesson that will not be lost on Azerbaijan and its neighbours. Yet it also demonstrated – as far as we can tell – that Russia was anxious to avoid any outright responsibility for severing the Caspian's energy links with the outside world. Russia stated that it did not target pipelines and, indeed, Russian forces did not take any steps either to control the above-ground Baku–Supsa line – which their forces crossed at least once – or to impede operations at the Kulevi oil terminal on the Black Sea, which at one point was cut off by Russian patrols. However, Russian ordnance was responsible for the destruction of a bridge on the main railway, thus cutting rail traffic, and some Western defence sources have stated that it did look as if one stick of bombs that landed near the underground BTC pipeline might have constituted a deliberate, though misplaced, attack on the line.[3] Others considered that these were simply bombs that fell somewhat short during a Russian attack on a nearby airfield.

During the war, BP shut down the SCP, citing *force majeure*, as well as the Baku–Supsa oil pipeline (150,000 b/d), which passed through territory that was, at least briefly, controlled by Russian forces. Yet there was no physical damage made to either line. The BTC remained closed throughout the conflict, initially as a result of a bomb attack by Kurdish separatists on a valve station in central Turkey on the night of 5–6 August and subsequently pending the return to semi-normality in Georgia.[4] It was up and running again in September, albeit at a reduced flow rate due to upstream problems in the giant Azeri–Chirag–Guneshli field.

The issue of 'frozen conflicts'

One overriding security issue is the prospect of renewed warfare. It is still disputed as to whether President Saakashvili was engaged in a striking act of brinkmanship when he ordered Georgian troops to capture Tskhinvali and so resolve a supposedly frozen conflict. If so, this action proved that an initial intervention could create more problems for Georgia and Azerbaijan's supposed allies in Europe and North America than it does for Russia. This has obvious implications in terms of potential international security guarantees for Georgia, should such guarantees be deemed necessary to ensure energy transit through the South Caucasus. But it also has implications for neighbouring Azerbaijan, which has poured some $2 billion into rebuilding its armed forces in 2007–8 in an effort to pose a credible military challenge to Armenian forces controlling the disputed territory of Mountainous Karabagh (Nagorny Karabagh). Put simply, the entry of Georgian troops into Tskhinvali proved that no one can ever assume a military campaign will go as smoothly as planned, or that its consequences can be predicted.

On the other hand, if the war is viewed as a trap into which Saakashvili fell, with the Georgian army reacting to a Russian move to pour thousands of troops into South Ossetia by means of the Roki Tunnel, then this raises equally serious questions concerning Russia's willingness to resort to force to achieve its goals – and to the nature of the EU's response to such a policy.

Thus Russia's own swift campaign to degrade Georgia's military capabilities may itself lead to consequences unforeseen by Moscow. As for Azerbaijan, with the BTC and BTE pipelines passing within 15 km of the Armenian frontline, the dangers should be self-evident. Nevertheless Baku has still publicly declared that it considers military force an option in its efforts to resolve the Mountainous Karabagh dispute.

Azerbaijan's role in providing oil to world markets

One key problem is that just as the frozen conflicts are heating up, so is the relative importance of Caspian reserves increasing, particularly with regard to Azerbaijan. When the first oil through the BTC pipeline arrived in Ceyhan on 28 May 2006, Azerbaijan's role in global oil markets was generally predicated on two main assumptions. One was that the production plateau at Azerbaijan's giant Azeri–Chirag–Guneshli (ACG) field complex would only last for two or three years, from around 2010 to 2013; the other was that the amount of oil the BTC line could carry was limited to roughly its 1 million barrels a day (1 mb/d) initial design capacity.

Both these assumptions now appear misplaced. Azerbaijan looks set to extend plateau production of 1.2 mb/d at ACG to at least eight years, maybe even 15 years. At the same time, use of drag reduction agents (DRAs) has already increased the effective capacity of BTC to 1.2 mb/d while further investments in additional pumping stations and looping are capable of raising capacity to around as much as 1.6–1.8 mb/d.

It is still early days in terms of actual development, but these concepts are under serious study by the cluster of companies that are developing ACG and which built the BTC.[5]

Data on ACG reserves has grown steadily since the fields were first discovered in the Soviet era. As of 2006, BP considered that ACG held 5.4 billion barrels of proven reserves available for production under the 1994 contract, with the field containing around 900 million barrels in further proven reserves that would be available for development at the end of AIOC's 40-year concession in 2034. In 2007, Azerbaijan Oil Minister Natig Aliyev allerted that subsequent drilling had shown that recoverable reserves were at least 50 per cent higher. Aliyev has, indeed, said that Azerbaijan's proven reserves could eventually total as much as 30 billion barrels.

Azerbaijan's role in providing gas to world markets

Azerbaijan is now also moving to become a major gas producer in its own right. Indeed, the then US Deputy Assistant Secretary of State Matt Bryza argued in 2007 that Azerbaijan could be producing as much as 50 billion cubic metres of gas a year (bcm/y) beginning around 2016.[6] Bryza was almost certainly being overly optimistic concerning the timing, but it is increasingly looking as if Azerbaijan possesses the reserves to support gas production on this scale, even if it takes a few years longer to raise actual production to such a level. In the wake of the gas supply crisis of winter 2006–7, when a failure to reach agreement on pricing led to a halt in all Russian gas supplies to Azerbaijan, Azerbaijan was forced to increase its own production sharply to compensate for its annual routine import of some 4.5–5 bcm of Russian gas. Aided by the start-up of Phase-1 production at the giant Shakh Deniz gas field, Azerbaijani gas output climbed swiftly in 2007, to 10.3 bcm (from 6.3 bcm in 2006). In 2009, with Shakh Deniz moving towards its full Phase-1 capacity of 8.8 bcm, and with other fields yielding around 6–7 bcm, gas production reached 13.3 bcm. Before it became clear that the start of the $22 bn Phase-2 production programme (SD-2) would be delayed until 2016, actual gas operators in Azerbaijan were anticipating the country producing around 20 bcm in 2010, 30 bcm in 2016, as much as 45 bcm in 2018 and stabilizing at just over 50 bcm from 2021 onward.[7] With due allowance for a two- to four-year delay while SD-2 comes on line, this trend is not unreasonable.

These estimates were also prepared before the official publication in November 2007 of the results of the SDX-04 appraisal well at Shakh Deniz, which in turn led to significant upward revisions in the volume of gas output to be expected from the second phase of Shakh Deniz. SDX-04 confirmed the existence of a further major reservoir at the giant field, which BP officials considered would add between 50 and 100 per cent to what was its then publicly stated 849 bcm in proven reserves. In terms of output, this means that while Phase-1 output will hold steady at around 8.8 bcm/y (or perhaps a little higher) Phase-2 will add a further 16 bcm/y, with output starting in 2016 and full production possibly reached in 2018, taking total field production to around 25 bcm/y in 2019 or thereabouts. At the same time, BP is in negotiations for the development of the gas reserves located beneath the giant ACG oilfield complex (and which do not constitute part of the ACG production sharing agreement). These are expected to be able to produce as much as 10 bcm/y.

Azerbaijan's current and prospective gas output is expected to reach European markets by means of several pipelines, notably the SCP and its Turkish-operated extension to Erzurum, which has been carrying Shakh Deniz Phase-1 gas to Georgia since 2006 and to Turkey since late 2007 – with some of the gas sold to Turkey then re-sold to Greece by means of the Turkey–Greece Interconnector. As of early 2011 there are three main contenders for carrying major volumes of Azerbaijani gas, essentially from the SD-2 project, to EU markets. These are an extension of the Turkey–Greece Interconnector by means of an 8 bcm/y line under the Adriatic to Italy, a project known as the Interconnector Turkey–Greece–Italy (ITGI); a 10–20 bcm/y project known as the Trans-Adriatic Pipeline (TAP) to link Greece and Italy via Albania; and Nabucco, a planned pipeline from Turkey to Austria, intended to have an eventual 31 bcm/y capacity.

Getting Nabucco off the ground

Nabucco is the prime focus of a coordinated effort, backed by the European Union, to develop a major direct link between the Caspian and the core EU gas markets of central, northern and Western Europe. First proposed in 2002, Nabucco is a partnership between five founding partners, the principal gas companies of Austria, Hungary, Romania, Bulgaria and Turkey, with Germany's RWE coming in as a sixth partner in 2008. Currently estimated at €8.5 bn – all reported pipeline costs should be taken with a pinch of salt since they reflect so many changeable factors – Nabucco is designed as a system capable of carrying gas from a variety of sources to a variety of customers. This makes it somewhat unusual, since most long-distance gas pipelines are developed by the producers of the gas or by producers in cooperation with consumers. In theory, the line could draw on a variety of suppliers in an arc going from Russia, the Caspian, Iran, Iraq, and even Egypt. In practice, the problem is that while all of these countries or regions might be able to contribute gas in the long run, a combination of political and practical reasons mean that only Azerbaijan – and perhaps Turkmenistan – can be considered as Caspian region suppliers in the short term; it is also the reason why northern Iraq is the only significant short-term Middle East supplier for Nabucco. This is not an absolute distinction, since half of the line's capacity is to be made available to whichever shippers submit the best open bid for space, but it does mean that prospective Russian and Iranian suppliers cannot expect to secure guaranteed long-term space in the line.

As of early 2011, Azerbaijan and Turkmenistan, along with producers in northern Iraq, were being wooed to supply the 50 per cent to be made available on a long-term basis.

Nabucco is being planned in stages, with the first stage intended to carry some 8–10 bcm/y beginning in 2014 or 2016, depending on whether it has to wait for SD-2 gas to come on line or whether it can secure Turkmen or northern Iraqi supplies before SD-2 gas starts flowing into the South Caucasus Pipeline. In terms of mainstream throughput, Nabucco would rely on using an expanded SCP to carry Caspian gas to the Georgian–Turkish border. It could then choose either to use existing Turkish infrastructure before entering entirely new pipe between a junction near the Turkish capital of Ankara Turkey and the Austrian gas hub at Baumgarten or to lay new pipe all the way from the Georgian border to Austria. In either case, much of the gas would be used en route by the Nabucco partners themselves. If built in two stages, the second stage, to be completed in or around 2018–19, would take input to 31 bcm/y with the construction of new purpose-built line across Turkey and the delivery of some 20–25 bcm/y to Baumgarten, with the rest offloaded en route.

As of early 2011, getting the initial input was the main problem, since Nabucco has to compete in this arena with both the rival ITGI and TAP proposals. All three projects – Nabucco, the Greece–Italy section of ITGI, and TAP – are looking to the same source, the second phase of Azerbaijan's Shakh Deniz project, to supply the vital initial long-term input for their line. Officially the EU is neutral: it supports the concept of the corridor without taking sides between different projects. But at a meeting in Budapest on 26 January 2009, summoned to stress Nabucco's importance to government leaders, the European Commission committed itself to supporting the project with up to €250 million, while the European Investment Bank declared its readiness to finance up to 25 per cent of Nabucco's costs once the necessary intergovernmental agreement was in place. In effect, the Commission undertook to provide sufficient finance to carry the project through all the preliminary feasibility and detailed engineering study phases to the point at which a clear decision can be taken on whether or not to build the line, while the EIB was saying it is prepared to lend the project close to €2 billion.[8] The EU has also pledged €100 million for development of the Greece–Italy component of the ITGI system.

The crucial element remains input. Short-term input for Nabucco could come, for example, from Azerbaijan and Iraq, with the former committing up to 10 bcm/y and the latter 2–3 bcm/y. These remain possibilities, and rely on Azerbaijan opting for Nabucco over its rivals and on northern Iraq agreeing to supplying input via a line that has yet to be built to connect the Kurdish autonomous region of northern Iraq with Turkey's steadily expanding domestic grid. Eventually, Nabucco – and its rivals – would be looking to secure additional volumes of gas from Azerbaijan, which is why the European Commission is working so hard to secure a commitment from Azerbaijan that it look beyond Shakh Deniz to the development of the deep gas underlying ACG, and the prospective development of the Absheron field, where Chevron found gas in 2002–3 when it was actually looking for oil. In this context, it should be noted that for Azerbaijan to export substantially more than 23–25 bcm/y would either require expansion of the South Caucasus line by laying a second pipe alongside the first or development of an additional export system. A second pipe would certainly be necessary if Turkmenistan were to contribute significant volumes to Nabucco.

Wooing Turkmenistan

OMV, the Austrian gas company which first took the lead in developing Nabucco, has recently been wooing Turkmenistan for prospective input into Nabucco and the necessary infrastructure that would feed Nabucco. In December 2008, OMV and RWE announced they were initiating their own studies on a possible Trans-Caspian gas pipeline. This could take various forms, the most obvious being the laying of full-scale pipelines from either the Turkmen or Kazakh coasts of the Caspian to Azerbaijan. However, there is also the possibility of laying a smaller system intended to serve, at least initially, only Turkmenistan's offshore gas resources, since this could entail a new sub-sea pipeline of barely 100 km to connect existing platforms off the Turkmen coast with their counterparts off the Azerbaijani coast. In this context it should be noted that Turkmenistan's offshore fields are generally reckoned capable of producing between 5 and 10 bcm/y for export, and, since their development has already begun, they could thus make a very useful contribution to getting Nabucco off the ground.

In contrast, Turkmenistan's onshore resources are much more plentiful but would require a much longer development period. For example, the Yoloten–Osman field complex is considered capable of producing as much as 70 bcm/y, but given technical complexities and their own reluctance to commit to onshore production sharing agreements with international companies (the CNPC agreement being the exception that proves the rule), major new onshore production dedicated to European markets could not reasonably be expected to yield actual output for seven to 10 years. Onshore Turkmen field development might thus be harnessed usefully to serve the second phase of Nabucco, or, were Nabucco to have secured full capacity from other sources a decade or so from now, perhaps to launch the White Stream project. White Stream is a proposal for a sub-sea gas pipeline from Georgia to Romania which would carry Caspian gas to the heart of Europe, with a possible end point in or around Trieste. The project's developers are the GU–EU (Georgia–Ukraine–European Union) group, so called because their original plan was for a gas pipeline that would reach the EU via Ukraine. The EU is currently funding a White Stream basic feasibility study.

Nabucco and South Stream

It is Nabucco that is the prime focus at present, however. And here Russia enters both the wider picture and the immediate frame. For the EU, the significance of Nabucco lies in its ability to become a fourth artery for delivery of pipeline gas to Europe, thus both reducing overall European dependence on Russia for gas supplies and also, by creating a further degree of competition to Russian gas, putting Russia (and the other major pipeline gas suppliers, North Africa and Norway) under greater pressure in terms of prices charged to European customers. The paradox, however, is that if there is no significant early supply of Azerbaijani gas for input into Nabucco, then either Nabucco's proponents have to turn to other sources of gas, or the financing of Nabucco becomes problematic.

There are other elements of serious concern both to Nabucco's developers and to its best-placed supplier, Azerbaijan: will Russia seek to use its influence with major gas purchasers, notably Italy, to develop an alternative pipeline system, known as South Stream? Will there be room for both South Stream and Nabucco? Is South Stream in fact designed as a spoiler for Nabucco? The Gazprom proposal to develop South Stream as a pipeline across the Black Sea from Russia to Bulgaria and thence through the southern and western Balkans to Italy, possibly with one spur crossing the Adriatic to southern Italy and another entering northern Italy overland via Croatia and Slovenia, appears intended to serve much the same market as Nabucco. Indeed, some formulations proclaim an intention to deliver gas to Baumgarten.

However, South Stream suffers from two inherent weaknesses. The first is that it is expected to cost some €25 billion to build the line to its full projected 63 bcm/y capacity – and this at a time when Gazprom's debt is high and Russia's financial reserves increasingly strained in the wake of the global financial crisis. Second, South Stream is clearly intended to serve as a new transit system for carrying existing supplies to Europe that currently transit Ukraine and/or Belarus rather than as a line that would serve new sources of Russian gas production. There is also a peculiar double irony in the roles to be performed by Nabucco and South Stream. Nabucco is intended to be a transit line, but its creation would actually serve to bring on-line additional gas supplies for producers that, at present, either cannot access Europe at all or else have to accept the need to allow Russia to profit from its pipeline monopoly at considerable expense to their exchequers. As for South Stream, it is a producer's line that is clearly intended to carry existing, rather than new, gas output.

Trans-Caspian issues

The development of the Southern Corridor is profoundly important for other Caspian countries. The Kazakh and Azerbaijani authorities have already set up a shipping line capable of handling some 500,000 b/d of Kazakh oil for delivery from a new port to be built at Kuryk on Kazakhstan's Caspian coast with input into the BTC system near Baku. Timing the ramp-up of such trade will depend on various factors that remain difficult to assess at present, including the construction of a new class of 63,000-ton Caspian tankers, construction of loading and unloading facilities and, above all, the speed with which new Kazakh production at Kashagan comes on line and the availability of other exporting systems, notably the timing of actual implementation of the long-projected expansion of the CPC pipeline from Atyrau to Novorossiysk.[9] Although the partners in CPC, which include the Kazakh and Russian governments, have formally agreed that the line should be expanded from its current capacity of around 30–32 million tons a year (600,000–640,000 b/d), it looks as if it will still take two or three years for this work to be carried out. Expansion work should have begun in 2004; it now looks likely to start in early 2011. Nor is it clear that the expansion will reach as high as the originally intended 67 million tons a year (mt/y) target, with at least one prominent source suggesting that 50 mt/y might prove to be its next plateau capacity level.

Pipeline security

In developing the East–West corridor the security of the pipelines is obviously a critical factor. Because it is so important it is used by some in Azerbaijan and by many in Georgia as the justification for joining NATO. Although Georgian President Mikheil Saakashvili has set NATO membership as a policy goal, NATO itself has taken a rather different view, stating that it is prepared to cooperate with Georgia and Azerbaijan but that it does not envisage full membership at any time soon. This view was strongly reinforced by what was seen at the time as Saakashvili's action in initiating the Georgian war of August 2008. There would appear to be concerns in NATO – there certainly are in senior European Commission circles – that President Saakashvili's displays of brinkmanship have proven counterproductive in terms of enhancing the security of the east–west corridor.

In the meantime, security for the pipelines takes two forms. One is routine security of the kind designed to prevent illicit tapping into either the Baku–Tbilisi–Ceyhan oil pipeline or the smaller Baku–Supsa oil pipeline for criminal use. This practice was widespread in the Caucasus and indeed was one of the main reasons why Russia re-routed the pipeline that runs from Baku to Novorossiysk in order to bypass Chechnya, through which it used to run.

BTC developers, following the example set by Baku–Supsa, have put a wide range of security measures in place, ranging from posses of local horsemen capable of patrolling large distances to satellite surveillance. But the best security, BP officials have said, is cooperation with the villagers and farmers through whose land the pipelines pass. In terms of major conflicts, of course, the main way of defending the pipelines is diplomatic, however.

Russia and the corridor

From a Russian perspective, the initial development and anticipated expansion of the corridor constitutes a profound challenge to its energy policy, particularly with regard to gas. Oil is impacted in that the more oil that flows through pipelines in competition with Russia, the more Russia comes under pressure to adopt competitive practices – notably with regard to transit cost pricing and open access – for its own pipelines. But because oil is an essentially global, fungible market, and since oil can be transported in a variety of ways, there are limits to Russian ability to profit from the landlocked status of various Caspian producers and the continued reliance of much Caspian oil production, especially from Kazakhstan, on export outlets transiting Russia.

Yet gas is a very different matter. Here Russia is able to use its geographical location athwart the old Soviet-era pipeline routes linking Central Asian gas producers with the hard cash markets of EU and non-EU Europe. This enables Russia to buy gas from Caspian producers at prices much lower than it secures for the gas that it then sells – whether of Russian or Central Asian origin – to hard cash customers.

In December 2008, Russian Prime Minister Vladimir Putin stated that Russia was due to pay as much as $340 per thousand cubic metres ($340/tcm) for its purchases of gas from Central Asia, a price level which, when transit costs are taken into account, would compare favourably with $400–500/tcm that Russia was charging its hard cash customers in the EU and Turkey. However, there was never any actual confirmation that Russia was paying anything like $340 to its three Central Asian suppliers, Turkmenistan, Uzbekistan and Kazakhstan. Turkmen sources indicated during the Ukraine crisis of January 2009 that they were afraid the crisis would mean they would not receive as high a price from Russia as they had expected.

In the event, however, it was not prices but volumes that were to prove all-important in terms of Russia–Turkmenistan energy relations. At 01:30 on 9 April 2009, an explosion at Km 487 on the main Central Asia–Centre pipeline in Turkmenistan near the Uzbek border damaged one of the main pipes. The blast was the clearly foreseeable consequence of the action of Russian technicians on the line who, some 27 hours earlier, on the late evening of 7 April, gave the Turkmen authorities what they said was 24 hours' notice that they would be closing the valves in the pipe to reduce the rate of flow. Turkmenistan immediately asked for 72 hours' notice, saying this was necessary to give their technicians time to effect an orderly closure of gas input from the country's production wells. The Turkmen call was rejected. In the event, despite Turkmen efforts to close down their input into the system – the Turkmens eventually turned off 195 gas production fields, resulting in the suspension of 92 per cent of Turkmen gas exports to Russia – pressure built up at the approach to the Gazprom-controlled valve station and, at Km 487, ruptured the line. Within a few days, Turkmen technicians had repaired the line, but Russia declined to resume deliveries. The incident led to a furious war of words in which the Turkmen foreign ministry used unprecedentedly strong language to blame Russia for the explosion. Russia's main reason for forcing the line's closure was its own problem in selling gas to recession-hit markets in both the European Union and Ukraine. But the explosion also occurred after a summit meeting in Moscow between Berdymukhammedov and Russia's President Dmitry Medvedev which had been expected to lead to a firm contract for Russian companies to construct a major East–West pipeline project within Turkmenistan. Instead, probably as a result of disagreement on both the volume and price of Turkmen gas to Russia, Turkmenistan announced within days of the summit ending that it would host an open tender for the East–West pipeline, that it was specifically seeking participation by international companies, and that the line would also be capable of serving prospective Turkmen gas exports across the Caspian to Europe.

It was not until January 2010 that Turkmen gas exports to Russia were resumed. But while the Turkmen authorities believed that their negotiations with Moscow in autumn 2009 would lead to some 30 bcm in export volumes, Russia insisted that the 30 bcm/y figure was simply an upper limit and that it would only be purchasing around 11 bcm of gas from Turkmenistan in 2010. In contrast, before the 7 April 2009 explosion, it had anticipated that annual exports to Russia in both 2009 and 2010 would run at a rate of around 40–50 bcm/y.

Before the explosion, the corridor threatened Russia's monopsony in Central Asian gas purchases because it provided a potential way for both Azerbaijan and also the producers on the other side of the Caspian to access European markets in a manner that would enable them to maximize their gas export revenues; this would happen by means of a transit system which would be more transparent and reflect commercial terms for transit costs far better than any available from Russia. In the event, Moscow's own actions in curtailing Turkmen gas deliveries to Russia simply speeded up a process of export route diversification that was already under way, not least due to the start of construction work on the 30–40 bcm/y, 2,200-km Central Asian Gas Pipeline, which connects Turkmenistan to China by way of Uzbekistan and Kazakhstan.

One further point can be made: the more effective the corridor is in attracting Central Asian gas away from Russia, the more it forces Russia to address its own gas imbalance, an issue crucial for markets seeking to assess how much gas they might reasonably be able to import from Russia in years to come. Although the start-up of Turkmen gas exports to China at the end of 2009 increases the degree of competition posed to Russia, it is important to note that Turkmenistan has insisted it will meet its agreement to increase its export volumes to Russia at the same time as it develops its gas export trade with China. Indeed the problem is not Ashgabad's unwillingness to supply Russia, but Moscow's current unwillingness to receive large volumes of Turkmen gas it had previously contracted to purchase.

In turning to the Ukraine crisis of January 2009, it is worth taking as a starting point Russia's own gas balance, since this illustrates that Russia, as a supplier, faces constraints just as Ukraine, as a consumer, faces financial constraints in meeting its gas import bill.

The Russian gas balance is the product of a variety of factors, of which the main ones, as recorded or estimated for 2006, are:

Supply	709 bcm
• Production	646 bcm
• Imports from Central Asia	62 bcm
• Net withdrawals from underground storage	1 bcm
Demand	710 bcm
• Domestic consumption	444 bcm
• Exports to hard cash markets	152 bcm
• Exports to various FSU countries	112.5 bcm[10]
• Produced gas used for reinjection	2 bcm

(Sources: RusEnergy, Moscow and BP Statistical Summary 2007)

These figures are, to a certain extent, approximations – particularly where import and export levels are concerned. But they demonstrate how Russia has traditionally relied on Central Asian imports to help overcome the problem that its own production cannot satisfy its three main markets: Russian domestic consumers paying highly subsidized prices; the cluster of FSU countries paying a variety of prices that, overall, continue to entail considerable losses for Russia; and hard cash customers, essentially in the EU and Turkey. It should be noted that the Baltic republics of Latvia, Lithuania and (most probably) Estonia are included in the hard cash European customer section. The figures also demonstrated that Russia's supply of low- or mixed-cost exports still exceeded its import of relatively low-cost Central Asian gas. In sum, the more Central Asian gas gets sucked in other directions – initially to China (and with a little to Iran) and prospectively through the Southern Caucasus Energy Corridor to Europe – the more Russia has to face up to the fact that the bulk of its gas sales result in considerable foregone income by comparison with prices obtained through exports to hard cash customers in Europe.

Under such circumstances pressure increases on Russia to reform its own gas consumption patterns by raising domestic prices both faster than it would like and to higher levels than currently envisaged. Put simply, the more the Southern Caucasus Energy Corridor – and China – appear likely to attract Central Asian gas, the more Russia has to address the problem of its own profligacy in gas consumption, which averages three times the European per capita consumption. This is particularly true if Russia is to maintain its flow of hard cash gas exports and take full advantage of the dominant trend in the European gas market, which is characterized by rising consumption and falling production, thus leading to a stream of forecasts of steadily rising demand for imported gas. And although there are signs that Russia does want to address this problem, it is also clear that it wants to do so at its own pace.

The Ukraine crisis of January 2009

The Russia–Ukraine gas crisis of January 2009 carried a variety of implications for the Caucasus. The crisis itself provided fresh grounds for European customers to doubt both Russia's reliability as a supplier and Ukraine's reliability as a transit country. When Russia initially cut off gas supplies to Ukraine on 1 January, it had some justification since Ukraine was by then out of contract and had failed to agree terms on a price formula for 2009 deliveries. Moreover, Western analysts were concerned that Ukraine had failed to use the three years since the last major gas crisis with Russia to start reforming its own internal energy sector in order to curb energy use in general and reliance on imported gas in particular.

Yet when Russia cut off all supplies through the Ukrainian system on 7 January, including those bound for customers beyond Ukraine in the European Union, Turkey and the Balkans, it caused widespread consternation. While Ukraine had built up its gas storage so that it could counter a Russian cut-off, countries such as Bulgaria, Slovakia and the Czech Republic had made no such provision.

With energy security already at the top of the EU's agenda, the Russia–Ukraine crisis intensified the focus on developing the Southern Corridor as an alternative route for accessing Caspian gas, and one that would not be subject to either Russian or Ukrainian intervention. So while Russia's tough stance against Georgia in August contributed to concern as to just how strongly the European Union and Western security structures such as NATO might be committed to the concept of ensuring security in the southern Caucasus, its equally tough action in instituting a total cut-off via Ukraine has contributed both to Caspian producer concerns that they need a non-Russian outlet and to European and Turkish concerns that they have to address the Southern Corridor issue with greater purpose than ever before.

As of early 2011, all along the corridor there were increasing indications of progress, participation and cooperation in trying to expand the use of the Southern Caucasus as a conduit for Caspian gas. Turkmenistan informed the Nabucco meeting in Budapest on 26 January 2009 that it was ready and willing to export its gas to European markets by way of a Trans-Caspian gas pipeline; in November 2009, it specifically signalled its willingness for all gas produced in the Caspian by Malaysia's Petronas Carigali to be exported to Europe by means of Nabucco; Azerbaijan has voiced support for Nabucco, although in practice its concern is to ensure it can export its gas to the EU by whatever means seem most attractive, whether it be Nabucco, ITGI, TAP or even as liquefied natural gas (LNG) across the Black Sea to Romania; Turkey and the EU achieved a breakthrough on the vexed issue of transit terms when Baku and Ankara reached agreement on the terms under which gas should be carried through Nabucco in July 2009 and by other pipelines in a series of meetings in April, May and June 2010; and the proponents of the Nabucco project were promised serious money for further project development. On 13 January 2011, Azerbaijan President Ilam Aliev and European Commission President José Manuel Barroso signed a formal declaration in Baku in which Azerbaijan agreed to supply enough gas to the EU to open up the Southern Gas Corridor. Such statements and actions do not guarantee the further expansion of the Southern Corridor in general and the implementation of Nabucco in particular, but they do provide the strongest indication to date that Europe, including both the European Union and Turkey, are increasingly committed to opening up the Southern Energy Corridor to expanded gas transit with the kind of steely determination necessary to counter Russian objections.[11]

Conclusion

Over the next several years, three interrelated issues will dominate in terms of the region's energy relations: the role that Azerbaijan itself will play in the supply of both its own oil and gas to European and world markets; the role that Azerbaijan and Georgia will play in handling deliveries of oil and gas from the far side of the Caspian – essentially Kazakhstan and Turkmenistan – to hard cash world markets; and the role that internal and external forces will play, particularly in Georgia, in terms of seeking either to stabilize or to destabilize what is currently the most volatile section of the Southern Caucasus Energy Corridor.

By the end of 2010, further development of the Southern Corridor, ideally including a Trans-Caspian transportation system so that Turkmen gas can be piped to Europe via Azerbaijan and either the projected Nabucco or one of its Southern Corridor alternatives, has become a major strategic goal of the European Union. The question now is how much of a challenge the EU will have to pose to Russia in order to expand oil and gas flows through the Southern Caucasus Energy Corridor and whether the EU can provide a sufficiently compelling case to Caspian producers to enable them to develop further oil and gas resources to feed Europe by means of the corridor.

Appendix: Armenian energy issues

Armenia has its own issues with the twin pipelines that pass so close to its territory. When Baku–Ceyhan was first mooted in the early 1990s various options for the route were considered, including one from Armenia. This would have been the shortest and potentially the most direct route to Turkey but it suffered from the problem that to construct a line across Armenia would have required at least an interim settlement to the Nagorny Karabagh dispute. At times there were suggestions that construction of the pipeline would act as a factor to facilitate peace but it should be noted that pipelines tend to be built after peace has been established.

Today the Armenians argue that the developers of the BTC line and its gas twin took an uncommercial approach in that they bypassed Armenia and thus required extra kilometres of line to be built by passing them through Georgia. At present, the most that Armenia can now expect to secure from the BTC line is, in the event of peace, a short extension from the line to enable it to run its own refinery.

For the record, it should be noted that Armenia shares the prevailing assumption of its neighbour, Georgia, that it should diversify its energy sources. Armenia's Deputy Energy Minister, Areg Galstyan, has said that 'diversifying sources of gas supply for Armenia is a major priority'. However there are limits to the way in which Armenia can diversify. A new gas line from Iran has already reached the south-eastern corner of Armenia and will, eventually, reach as far as the capital of Yerevan, but the new pipeline is only capable of carrying around 350 to 400 mcm of gas per year, which is roughly only one-quarter of total Armenian annual consumption. The reason for this is thought to be the insistence of Gazprom, which is a shareholder in the project, that the line should remain small in effect to ensure that it does not compete with Russian supplies. This may be one instance in which Russia is using continued supply assurances as a carrot to encourage close relations.

In the long run, how Armenia fits into the overall picture of hydrocarbons development in the South Caucasus depends primarily on whether efforts to resolve the Mountainous Karabagh conflict prove successful. If they do not, then Armenia will continue to remain dependent on Russia and Iran, in energy as in security. If they do, then, for the record, it is worth stating that both Azerbaijani and Armenian officials have told the author that, in event of a Mountainous Karabagh peace settlement being reached, they would have no objection to Azerbaijan supplying both oil and gas to Armenia.

11 Brothers to neighbours

Russia–Belarus relations in transit

Folkert Garbe, Felix Hett and Rainer Lindner

Introduction

Energy conflicts between Russia and Belarus concerning the trade of natural gas and oil grabbed headlines in the winter of 2006–7.[1] These conflicts culminated in a complete cut-off of oil supply to Belarus for several days in January 2007. Previously, Russia had often been confronted with severe international criticism based on the assumption that it has intended to punish disagreeable governments (particularly in the cases of Ukraine and Georgia) and reward loyal regimes (Belarus for example) by pursuing a selective energy and foreign trade policy. Such behaviour was often referred to as Russia's use of an 'energy weapon'. implying that energy – and especially natural gas – is a foreign policy tool which the Russian leadership can freely deploy in order to maintain and increase its influence in the post-Soviet space. Bearing this assumption in mind, the Russian–Belarusian energy conflicts came as a surprise to many. Belarus, closely connected to its eastern neighbour via political, economic, military, historical and cultural linkages, was – and still is – widely considered to be Russia's closest ally on the international stage. If Minsk was principally loyal to Moscow, then why should it have been punished?

Instead of joining the widespread and often futile search for reasons why Russia decided to use its 'energy weapon' on Belarus, an alternative explanation for the conflicts might be found in the logic of the transformation of Russian foreign policy in the Putin era.[2] According to this line of argument, Putin's accession to power in 1999 marked a decisive shift away from his predecessor's foreign policy. Putin has realigned and rationalized Moscow's foreign policy based on increased appreciation of the country's political and economic weakness. Replacing the formerly unrealistic great power policy by pragmatism and economic modernization was considered a precondition for regaining a strong position on the international stage. These premises also became the driving factors of foreign policy, leading to the increased commercialization of relations with all countries, including those of the 'near abroad' within the Commonwealth of Independent States (CIS). According to this interpretation, the energy conflicts with Belarus do not constitute examples of the 'energy weapon' being used, but a further sign that Russia has gradually dropped its value-driven policy toward its western neighbour in exchange for a pragmatic approach based purely on national interests that are increasingly shaped by an economic rationale.

However, this argument, though certainly convincing, does not do justice to the full spectrum of Russia–Belarus relations, not least because Belarus remains an object of potential rivalry between Russia and the West. As such, Russian policy toward its 'Slavic brother' is constantly shaped by more than purely economic considerations. Minsk, for instance, remains the only partner Moscow has that openly and vigorously rejects further NATO enlargement. That this kind of support is not free became evident in the aftermath of the 2008 Georgian–Russian war, however. Belarus' reluctance to recognize the independence of Abkhazia and South Ossetia resulted in a diplomatic tug-of-war between Russia and the European Union during which Minsk refrained from making a decision, apparently trying to sell it at the highest possible price.

From such observations it follows that energy relations between Russia and Belarus need to be placed within the context of the two countries' broader economic and political relations. This chapter will attempt to do precisely that, comparing the energy conflicts of 2004 and 2006–7 as case studies able to shed light on the inner workings of this highly complicated but crucial area of mutual relations. The chapter will thus assess the interactions between a supplier of natural resources, Russia, and a hitherto indispensable transit country, Belarus, providing key insight into the idea of energy being used as a 'weapon'. The conclusion will therefore focus on the validity of referring to an 'energy weapon'.

The context: symbiotic economies, divergent polities

Virtual reintegration: the 'Union State' project

Between 1995 and 1999, Russia and Belarus signed numerous agreements and declarations of intent, stating their willingness to develop close political, economic and military cooperation. This process culminated in 1999 with the treaty on the creation of a Union State. However, the process of reintegration remained a merely declarative one in many aspects, as it was to a great extent based on the personal considerations of both heads of states. Lukashenko valued a close alliance with Russia not only as economically beneficial, but also as a foothold for his increasingly isolated regime. Lukashenko also viewed this treaty as a potential stepping stone for his own political career. His Russian counterpart, Boris Yeltsin, considered the popular idea of a post-Soviet reintegration an opportunity to shed his reputation as the 'destroyer of the USSR', thereby strengthening his weakened domestic position.[3]

Though Vladimir Putin did not suffer from such an association, the new Russian President initially continued Yeltsin's Belarus policy. The treaty of the Union State was ratified in January 2000. This document envisaged not only the adoption of a common constitution, but also common efforts to unify economic, financial and legal systems. An agreement on the introduction of a common currency was signed at the end of 2000,[4] further confirming the willingness of both sides to harmonize their economic laws. In particular, Belarus committed itself to substantial market reforms and to open its economy for Russian capital. In return, Moscow granted its neighbour a multitude of economic privileges.[5] First and foremost, the gas price paid by Belarus was gradually decreased to the level of Russian domestic prices. Thus, the Belarusian government managed to secure equal conditions and open access to the Russian markets for its manufactured goods.

However, differences about the function and structure of the Union State soon led to serious tensions. While Lukashenko insisted on the Union Treaty's wording granting both countries political equality, Putin bluntly rejected this idea in spring 2002, since it did not give enough consideration to Russia's economic, geographic and political preponderance. Nonetheless, the idea of post-Soviet reintegration still appeared attractive to the Russian leadership. Not only did the prospect of reunification with the 'brotherly Belarusian nation' promise political merit, but also the Belarusian state-directed economy – in particular assets like the highly developed petrochemical facilities – increasingly drew the Russian business community's attention (i.e. Russian business was interested in investing in Belarus' economy which had not yet undergone privatization). Thus, Putin clarified his point of view that integration was by all means possible, though on Russian conditions. What he was offering was, in reality, Belarus' accession to the Russian Federation. At this stage, the process of political integration reached a dead end, and has remained there ever since: the Belarusian concept of integration on equal grounds is as unacceptable to Russia as Putin's proposal is to Lukashenko. Moscow was not willing to accept any significant Belarusian leverage on its internal and external policies, while Lukashenko disliked the idea of being downgraded to the role of a Russian regional governor.[6]

When, from Lukashenko's point of view, the political integration was no longer beneficial, the Belarusian president changed his strategy. Instead of implementing existing agreements, he made additional demands concerning the Union State's foundation pillars – such as the common currency – and delayed them indefinitely. Furthermore, the Belarusian leadership increasingly sought to exclude Russian investors from Belarus.[7] Likewise, the proclaimed customs union remains incomplete to this day. Though a free trade area does exist on paper, Belarus has purposely and repeatedly undermined its full implementation.[8] The creation of a common economic space has stalled. The Union State's political embodiment finally turned out to be nothing but a virtual project.[9]

Officially, however, Lukashenko chose an entirely different course, repeatedly pledging his loyalty to Moscow while consistently spreading the idea of a confederation with Russia – emphasizing 'a union of two equal states'. This deliberately chosen rhetoric served economic as well as political purposes. On the one hand, it was aimed at upholding Russian diplomatic support for the Belarusian regime. On the other hand, the profit of economic privileges granted by Russia, in particular cheap gas and oil, crucially helped to further guarantee social-economic and political stability in Belarus – in a nutshell, to secure Lukashenko's position of power. This strategy proved to be remarkably successful, while Russia, lacking a consequent and decisive Belarus policy, failed to make significant progress concerning its reunification plans.

Economic and energy relations: asymmetric interdependence

Since 2003, the Belarusian economy has enjoyed annual growth rates of more than nine per cent on average, benefiting from Russia's dynamic development. Another factor was the increase of Belarusian exports to the EU, composed mainly of refined and petrochemical goods.[10] Hence Belarus, not possessing noteworthy oil or gas reserves itself, has paradoxically benefited from dramatically increased world market prices for energy resources.

This paradox evolved from the combination of two factors: advanced Belarusian petrochemical facilities and beneficial trading terms. Between 2001 and 2005, refiners in Belarus managed to nearly triple their profit margin.[11] The customs union concluded between Russia and Belarus in 1995 entitled Belarus to import Russian crude oil duty-free. When re-exporting the refined oil products, however, Belarus was required to yield 85 per cent of its export duties – that is, what the country charged its own exports to third countries – to Russia. However, Belarus stopped implementing this part of the agreement in 2001. Instead, Belarus imposed export duties that were far below the Russian level and withheld the entire revenue.[12] This made the processing of Russian crude in the Belarusian refineries especially attractive to Russian oil companies.[13] They increasingly set out to have their crude refined in Belarus, paying the Belarusian refineries for this service and exporting the finished oil products on their own account. The low Belarusian export duties allowed for significantly higher profit margins than would have been possible in the case of direct exports from Russia. At the same time, Belarusian refineries capitalized on the benefits of higher capacity utilization without having to worry about acquiring sufficient liquidity. The terms of trade in oil between Russia and Belarus were, therefore, not only beneficial for Belarus, but also for the Russian oil companies. In many ways, Belarus was turned into an 'offshore' tax haven for Russian businesses.[14] Thus, it can be concluded that one of the reasons that caused the Russian government to tolerate the non-observance of the customs union's provisions for five years was the intensive lobby from Russian oil companies and not merely 'additional support for the Belarusian economy as a whole, and, truth to tell, support during the election campaign [...] in Belarus,' as the Russian Energy Minister Khristenko put it.[15] Attempts to cut back this subsidy at the expense of Belarus and in favour of the Russian budget resulted in the oil transit conflict of 2007.

The Belarusian economy as a whole also benefits from special conditions concerning the import of natural gas.[16] Unlike petrochemical products, natural gas has never been one of Belarus' important exports. Nevertheless, natural gas, supplied exclusively by Russia, is the basis for Belarusian industry's functional capacity. Belarus' economy is highly energy intensive. Belarus expends twice as much energy to generate each per cent of GDP as neighbouring Lithuania. The share of natural gas in the Belarusian primary energy supply is, due to its importance for

Table 11.1 Belarus: oil import, refining and transit, 2005–7

	2005	*2006*	*2007*
Crude oil import (mln tons)	19.2	20.9	20.0
Oil products export (mln tons)	13.5	14.8	15.1
Crude oil transit via Druzhba (mln tons)	65.6	67.2	62.8
% of exports of Russian crude oil (carried out by Transneft)	29.0	29.4	27.5

Source: National Statistical Committee of the Republic of Belarus, Transneft, authors' own calculations.

Table 11.2 Belarus: natural gas import and transit, 2005–7

	2005	*2006*	*2007*
Import (bcm)	19.8	21.0	20.6
Price (USD/mcm)	46.7	46.7	100
Transit (bcm)	40.4	44.2	49.5
– via Beltransgaz Network (bcm)	18.7	n.a.	18.5
– transit fee (USD/100 km/mcm)	0.75	0.75	1.45
– via Yamal–Europe (bcm)	21.7	n.a.	31
– transit fee (USD/100 km/mcm)	0.46	0.46	n.a.
Gazprom's total exports to Europe (excl. CIS and Baltic States) (bcm)	156.1	161.5	168.5
Transit via Belarus as % of total exports to Europe	25.9	27.4	29.3

Source: Compiled from press releases of Gazprom and Beltransgaz, Gazprom's annual reports, authors' own calculations.

electricity generation, exceptionally high, reaching 64.6 per cent in 2005.[17] Thus, gas prices are a crucial factor in Belarusian industry's cost structure and, therefore, of its international competitiveness. Since Russia remains its most important export market, the Belarusian government has always sought to benefit from gas prices equal to those of Russia's internal market, particularly since the only selling point of tractors, trucks and television sets made in Belarus was their low price. Attaining this strategic goal, even temporarily, is one of the main explanations why the Belarusian leadership managed to deliver remarkable economic growth in recent years without undertaking major economic reforms.[18]

This fact raises a fundamental question: what motivation does Russia have in granting such privileges to its western neighbour? One answer, of course, is the Russian leadership's desire to sustain and to strengthen the diplomatic alliance and the public relations gains associated with a union state. Another important motive, however, lies in the fact that contrary to what is often perceived, dependency is mutual. Indeed, from Russia's perspective, Belarus, as a transit country for energy deliveries and other export products to Western Europe, is of particular importance. The shortest and, therefore, potentially cheapest transit routes from Russia to the European Union and Russia's Kaliningrad *oblast* pass through Belarus. This is especially significant in the area of energy exports: more than a quarter of Russia's gas deliveries to the EU and around 30 per cent of Russia's total oil exports (see Table 11.1) currently cross Belarusian territory each year.[19]

The oil transit that is carried out via the Druzhba ('friendship') pipeline is not without alternatives, as the transported volumes, though significant, could in principle also be brought to world markets by sea or rail. Gas transit, by contrast, is crucial because of Russia's very limited range of export routes; the bulk of gas transit travels via the Yamal–Europe pipeline, which began operating in 1999 but only reached full capacity in 2007. The rationale for this joint Russian–Polish–German project was to create an alternative to the notoriously unreliable Ukrainian gas transit network. The 575-km long Belarusian section is owned by Gazprom, but operated by the Belarusian state-owned Beltransgaz. It has an annual capacity of 33 bcm. Together with the older Northern Lights system owned by Beltransgaz, the theoretical transmission capacity of the combined Belarusian gas networks is 60 bcm per year.[20] These pipelines, however, are also used for internal distribution and are thus not fully available for exports. In 2007, approximately 46 bcm of natural gas transited to the European Union via Belarus. Beltransgaz was to earn around $360 million in transit fees in that year, apparently delivered in cash.[21]

Its geographical situation thus gives Belarus a rather strong bargaining position vis-à-vis Russia, as it is able to crucially intercept Russian energy deliveries to Western Europe. This holds especially true in the case of gas transit, since the EU remains Gazprom's most profitable market.[22] As delivery contracts are concluded on a long-term basis, delivery reliability is at once Gazprom's most important asset and, at the same time, potentially its weakest spot. Belarus and Ukraine as main transit states have the power to cast serious doubt on Gazprom's ability to fulfil its long-term obligations by interrupting transit. The gas giant tries to counter this threat with a twofold strategy. On the one hand, it aspires to gain control over the transmission infrastructure in transit countries by acquiring controlling stakes in the pipeline operators. On the other, it does its best to circumvent transit countries, most notably through new pipeline projects such as Yamal–Europe and Nord Stream through the Baltic Sea, scheduled for completion in 2011.[23] Thus, the Russian company is deliberately building over-capacity in order to be able to switch between different export routes.[24] In this context, the strategic value of Belarus and its gas pipeline operator Beltransgaz not only resides in the volumes transported. Until Nord Stream's completion, Belarus remains the one and only alternative in case of a Russian transit conflict with Ukraine. The preferential prices that Belarus enjoys are, therefore, not only a consequence of brotherly relations, but must be understood in terms of Russia's transit dependence on Belarus.[25] Economic relations between the two countries are thus best characterized as interdependent. This interdependence is nevertheless asymmetric, as Belarus in its current state is crucially dependent on Russia for its economic well-being, whereas Russia has not only theoretical alternatives to Belarus, but it is actually very much engaged in realizing them by further diversifying export routes.[26]

Within this relationship of economic interdependence, there are constant attempts to redistribute the costs and gains of bilateral relations. Thus, Gazprom is not only interested in gaining control over the Belarusian pipeline network, but also has a strong incentive to increase the gas price paid by Belarus. The most obvious of these is pure profit maximization. This seems even more convincing in the face of the challenges the Russian gas giant faces – gradually expiring gas fields, huge supply obligations and the subsequent need for large investments in order to keep production at the present level.[27] Consequently, the company is also pressuring for increased domestic gas prices within Russia. The Russian government in its turn has itself pledged to do the same in an agreement with the EU on WTO accession in 2004. This agreement was celebrated as a success by EU representatives, as it is expected that higher gas prices will lead to a more efficient use of resources, thus freeing more for export to Europe.[28] Attempts to reduce the cost of the Belarusian–Russian friendship are, in this regard, not only grounded in Gazprom's commercial interests, but are to a certain extent also in the officially declared interest of the European Union. However, this does not hold true when these attempts end in delivery interruptions, as it was the case during the 'gas war' in 2004.

In the following sections, we will analyse the Russian–Belarusian energy conflicts of 2004 and 2006–7 as manifestations of a broader attempt to redefine the cost and benefits of their energy relations.

The conflicts: rearranging relations

The 2004 gas crisis

In April 2002 Moscow and Minsk signed an agreement allowing Belarus to continue importing gas at Russian domestic prices. The document, however, contained an important condition for this concession on Russia's part, namely the creation of a joint venture between Russia's Gazprom and Belarus' Beltransgaz. Accordingly, the Russian company, which is state-owned, was to buy a 50 per cent stake of the Belarusian company by July of 2003. This agreement was of strategic interest not only for Gazprom but for the Russian leadership as well: while Gazprom would control gas flows to its central and Western European customers, the Kremlin would limit the Belarusian President's ability to use pipeline infrastructure as a bargaining chip.

This deal was never implemented, however. Having first determined Beltransgaz' total value to be $2.5 billion (while Gazprom had offered 700 million to $1 billion for a 50 per cent stake), the Belarusian leadership then doubled its claim and thus de facto denied the deal's implementation by deliberately fixing conditions that were unacceptable to Gazprom. The latter considered this move a breach of agreement and, as a consequence, announced, with the Kremlin's open approval, the doubling of gas prices to Belarus for the following year. At the same time, Moscow hinted at its willingness to negotiate, on the condition that Minsk proved to be more cooperative on the Beltransgaz issue.[29] However, Lukashenko did not give in, and reacted by alluding to the fact that Gazprom's intent to hike the gas price was contradictory to the Union State Treaty. Furthermore, he threatened to raise the transit tariff for Russian gas up to the considerably higher Ukrainian level and raised the issue of the so-far rent-free Russian military bases on Belarusian territory.[30]

As no compromise was found, Gazprom stopped its gas supply to Belarus on 1 January 2004. The country was then supplied by independent Russian gas companies until these contracts expired.[31] The conflict reached its climax on 18 February when Gazprom accused Belarus of stealing gas destined for Western Europe, and replied by cutting off gas transit through Belarus completely. This move led to an unprecedented political crisis between the two countries. A furious Lukashenko dubbed the gas cut-off 'an act of terrorism of the highest order'. He emphasized that relations with Moscow now would be 'poisoned by gas for a long time', threatening to abandon the Russian–Belarusian integration process and develop closer political ties with the EU.[32]

The Russian side decided to resume gas deliveries 10 hours later. Eventually, in addition to Lukashenko's aggressive rhetoric, another crucial factor came to the fore: the interruption of the gas transit had caused serious apprehension among Western European customer countries, putting into question Gazprom's reliability as a supplier. In fact, suspending deliveries came back to haunt Russia, as its reputation as a reliable trade partner suffered serious damage. Such circumstances encouraged Gazprom to compromise and settle the price issue with Beltransgaz. The new deal provided for Gazprom to continue its deliveries to Belarus at the slightly higher price of $46.68 per 1000 cubic metres (mcm) in 2005, while Belarus itself raised the transit tariff for Gazprom's use of the Belarusian pipeline infrastructure from $0.55 to $0.75 per mcm and per 100 kilometres ($/mcm/100km).

All things considered, Russia had to accept a bitter defeat. While the gas price was raised only slightly, the final compromise was rather about Gazprom saving face. With regard to Gazprom participating in Beltransgaz – the dispute's focal point – the Russian side did not achieve anything. Although agreement on Beltransgaz' partial sale remained officially valid, a clear and binding arrangement did not exist. Beltransgaz consequently remained entirely in the hands of the Belarusian state. What is more, Moscow's hard stance toward Belarus ensured that Russia lost considerable favour with the Belarusian people, thus jeopardizing the realization of the Union State.[33]

The 2004 Russian–Belarusian gas conflict thus revealed the extent of Lukashenko's political leverage over Moscow, this in spite of his country's obvious economic dependence on Russia. Understanding this, Gazprom thus pushed forward its plans to build a gas pipeline on the Baltic seabed as a means to bypass Belarus as a transit country.

Gas and oil – subsequent conflicts in 2006

Energy relations between Russia and Belarus eased temporarily in 2005. Despite the floundering integration process, in April Putin and Lukashenko concluded an agreement for Russian gas deliveries for the following year based on previous conditions. Russia's conciliatory position must be put in the context of the Belarusian presidential elections of March 2006 and the 'colour revolutions' in Georgia and Ukraine in 2003 and 2004. The rise to power of Western-orientated governments in the NIS caused great unease in the Kremlin. The perception that Russia was losing influence in the post-Soviet space encouraged Moscow to accept the subsidizing of Belarus, as long as the Lukashenko regime proved loyal. Gazprom was, meanwhile, also facing a severe gas crisis with Ukraine – eventually settled in January 2006.[34] Being confronted with two gas conflicts with critical transit countries at the same time would certainly have been perilous for Gazprom as well as for the Kremlin.

However, shortly after Belarus held its presidential election in March 2006, it became clear that the energy conflict had not been resolved: Gazprom announced the trebling of gas prices to be paid by Belarus in 2007. The company justified this price hike on economic imperatives making it necessary to adjust relations on a 'market basis'.[35] Similar to the conflict between Ukraine and Russia the preceding year, positions over the price issue were severely hardened. Gazprom demanded a price no lower than $105 per 1000 cubic metres of gas while Minsk was willing to pay a maximum of $75. The commonly held assumption that the conflict would escalate as had that between Russia and Ukraine was nevertheless wrong; after tedious negotiations, Gazprom president Aleksey Miller and Belarusian Prime Minister Sergei Sidorsky finally agreed to a new contract on 31 December, mere seconds before the old one expired. The new agreement stated that Gazprom was to supply gas to Belarus in 2007 at a price of $100 for 1000 cubic metres, conditions which were markedly more favourable than those offered to neighbouring countries.[36] However, it was agreed that until 2011 the price of gas should be gradually raised to Western European levels, while the fee that Belarus charged Gazprom for transiting gas through Beltransgaz's network was raised from 0.75 to $1.45/tcm/100km. Lastly, the conditions for creating the Gazprom–Beltrangaz joint venture were also decided: the Russian gas giant would purchase a 50 per cent stake in the Belarusian company for $2.5 billion, to be paid in four annual installments of $625 million.[37] Gazprom transferred the first two installments in June 2007 and February 2008 respectively, and has taken two out of eight seats on the company's board of directors.

Despite this new agreement, however, Russian–Belarusian energy relations did not ease up entirely, as the Kremlin soon put into question the existing conditions of the two countries' bilateral oil trade. Yet in May 2006 – when Gazprom hardened its stance on the price issue as well – President Putin reportedly ordered the fundamental alteration of Russia's trade, economic and financial policy toward Belarus, once again highlighting the close ties that exist between Russian politics and economics.[38] Any direct or indirect subsidy of the Belarusian economy was to be abandoned, particularly the duty-free delivery of crude oil so important to the Belarusian refining industry, with the objective of establishing the transition to Russian–Belarusian relations 'based on real benefit and genuine equality'. As a response to Minsk's repeated refusal to implement the arranged modalities of the bilateral trade union, the Kremlin decided in early December to end Belarus' exemption from export duty on crude oil. At the time, the corresponding duty amounted to $180.7 per ton.

In response, the Belarusian leadership adopted a hitherto non-existent 'transit fee' of $45 per ton of crude oil – of course at Russia's expense – shortly after New Year's Eve. As Russia refused to accept this fee, the Belarusian company Belneftekhim simply tapped the Druzhba pipeline, taking crude oil destined for Western Europe as compensation for the outstanding transit fee. However, the Russian oil transport network Transneft considered this illegal and brought oil transit to a complete halt. Crude oil deliveries via the Druzhba pipeline to Germany, Poland, Hungary, Slovakia and the Czech Republic were cut off for three days. Though the suspension of deliveries did not generate immediate shortages, it raised massive protests in Western Europe and was considered 'unacceptable' by the EU.[39] Significantly, international criticism was not only directed at Russia, but also to the Belarusian leadership and its actions.

In the face of such pressure, Belarus eventually withdrew the transit fee. Shortly after, Russian Prime Minister Fradkov and his Belarusian counterpart Sidorsky met in Moscow to discuss the oil transit issue. A compromise was eventually found after tough negotiations, and Belarus agreed to bring its export duties to Russian levels. Although it was obliged, on paper, to pass on 70 per cent of this revenue to Russia, Minsk could keep 100 per cent of the duty income. Russia was satisfied through the introduction of a duty on crude exports to Belarus that corresponded to 29.3 per cent of the standard Russian export duty.[40] Interestingly, a legal violation of the customs union thus became a practical measure aimed at enforcing its provisions.[41] Russia tried to limit the extent to which Belarus had so far been undermining the customs union to its own advantage.[42]

The consequences: a new modus vivendi, but further alienation

Compromise solutions and a cushioned price shock

In contrast to 2004, Gazprom enforced a crucial price hike for gas deliveries to Belarus after the 2006 dispute. Furthermore, the Russian company finally succeeded in realizing its long-sought objective of acquiring 50 per cent of Beltransgaz. To cut a long story short, the Russian company reached a more beneficial outcome than in 2004.

Nevertheless, it would be wrong to see this outcome as an all-out victory for the Russian side. In reality, Russia had to make considerable concessions to Belarus: not only did Gazprom agree to increase transit rates, but also to Belarus' high valuation of Beltransgaz's assets which it had previously deemed unacceptable. During the four subsequent years, Gazprom would supply $625 million to

Table 11.3 Estimation of additional costs of rising gas prices to the Belarusian economy in 2007 and 2008 as compared to 2006

Costs	*2007*	*2008*
Additional costs (USD/mcm)	53	79
Import volumes (bcm)	20	20
Additional costs, total (mln USD)	1060	1580
Offsets	*2007*	*2008*
Increased transit rates (mln USD)	70	0
Sale of a 12.5% stake in Beltransgaz (mln USD)	625	625
Balance (costs minus offsets (mln USD)	418	995

Sources: Gazprom, authors' own calculations.

the Belarusian budget annually via the purchase of four blocks of shares each representing 12.5 per cent of the company. Gazprom therefore agreed to push back its influential role in Beltransgaz to 2010 while beginning to pay for it in 2007. In addition, as Gazprom is scheduled to acquire a strict 50 per cent (as opposed to 50 per cent plus one share), decision-making in Beltransgaz will therefore rest on consensus between Gazprom and the Belarusian government.

The Beltransgaz deal, carried out fully on Belarusian terms, can also be considered as a partial transfer of Gazprom subsidies from one area to the other. While gas prices rose by \$53 per mcm in 2007 – representing a total increase of the Belarusian gas bill by more than \$1 billion – this was in part offset by the increased transit rates and the stock purchases for Beltransgaz (see Table 11.3 for details).[43] The 'price shock' faced by the Belarusian economy was thus substantially reduced.

Additionally, during 2008, severe doubts arouse as to whether the gas pricing contract struck in December 2006 had actually been implemented. In reality, this deal appeared to have suffered a fate similar to many other symbolic agreements in the context of the Union State.

According to the much quoted 'formula' that was never publicly spelled out in detail, Belarus was to pay 67 per cent of the 'European level' for a thousand cubic metres of imported natural gas – minus transit costs and Russian custom duties (netback price), said to reach 30 per cent. Thus, publicly emphasizing adherence to the formula, Gazprom fixed the Belarusian price at \$119.5 in the first quarter and later raised it to \$127.9, starting in the second quarter of 2008. At the same time, according to news agency reports, neighbouring Lithuania paid \$348 in January 2008,[44] while the average European price reached a high of \$500 in October – triggered by the record world oil prices.[45] It is obvious that the price paid by Belarus was far below 67 per cent of the so-called European level which is regularly referred to as the market price by Russian officials.[46]

Another interesting detail has occurred in the sphere of finance: the Russian government granted Belarus a state loan of $1.5 billion in December 2007 at preferential conditions, officially to cushion the effects of increased energy costs.[47] In November 2008, an agreement was reached granting Belarus another $2 billion, this time citing the looming world financial crisis as a reason, although the credit had been discussed long before the crisis became apparent.

A compromise was also struck over the oil dossier, with Russia forgoing over 70 per cent of its potential income from export duties. The reduction of the export duty toward Belarus not only implies higher import costs for Belarusian refiners, but also reduced profits for the Russian oil majors that had thus far profited from the Belarusian oil 'offshore'.[48] Their additional profit, realized thanks to duty-free imports of Russian crude oil into Belarus and the lower export duties for oil products from Belarus to the West, was substantially reduced. The real beneficiaries of this new arrangement were in fact both governments: while the Russian Ministry of Finance received approximately $1.2 billion of additional export duty income in 2007, the Belarusian Treasury also increased its revenues as the government had to bring export duties for oil products up to the Russian level.[49] However, beginning in March 2007, Belarus started to pass on parts of this new rent to Russian oil companies in order to partially offset the export duty they had to pay at the Russian–Belarusian border.[50] In this way Belarus apparently managed to sustain the competitive advantage its refineries had over Russian refineries for the custom of Russian oil companies: after a sharp slump in the first two months of 2007, Belarusian exports of oil products even surpassed their 2006 level by the end of the year.[51] Generally, fears – voiced by Lukashenko himself – that the new energy arrangements would lead to economic turmoil in Belarus did not materialize.[52] Belarus' 2007 bottom line further supports the argument that the price shocks were cushioned by other arrangements. Belarus' GDP grew 8.2 per cent while the budget ended with a surplus worth 0.6 per cent of GDP.[53] In other words, the conflict's outcome once again tended to demonstrate Belarus' rather strong position vis-à-vis Russia, able to limit Moscow's room for manoeuvre.

Nevertheless, though the overall environment regarding energy relations and the economic interdependence between Russia and Belarus did not change significantly from 2004 to 2006–7, a more decisive and coherent approach on the Russian side seems to have emerged in the latest energy conflicts. In order to test this hypothesis, it is necessary to grasp the larger picture, and come to grips with Russia's broader foreign and energy policy and its implications for Belarus.

'Allies' become 'just friends': normalizing relations

The Russia–Belarus energy conflicts included harsh rhetoric and diplomatic hostility on both sides. This seems quite remarkable, when one considers the proclaimed 'brotherly' aspect of Russian–Belarusian relations. In 2006 dialogue between Russia and Belarus even appeared close to an end. According to President Lukashenko, the increase in gas prices to the level of $200, as proposed by Russia's Gazprom at the beginning of the talks, would have caused 'the end of the relationship'. The Belarusian President accused Russia of having 'not only violated the norms of international law, but also destroyed the economic ties that had developed, and seriously harmed the prospects of the Belarusian oil refining industry'.[54] Similarly, the Russian side explicitly turned its back on Lukashenko. The delegation from Minsk travelling to Moscow on 9 January 2007 to negotiate the oil issue was at first not even granted a reception. Until recently, such demonstrative gestures of contempt would have been as unthinkable as Putin's distant references to the 'Belarusian colleagues'.

By significantly raising gas prices and imposing duty on oil exports to Belarus, Russia has taken decisive steps to alter the traditional, privileged relationship between the two countries. Russian policy-makers have obviously realized that a paternalistic, Soviet-style policy toward its neighbourhood is ineffectual. Instead, Russia has switched to a post-ideological policy strongly anchored in national and commercial interests.

This approach is underpinned by official discourse on Russian foreign policy. Thus, Russia's overall realignment of energy pricing policy becomes 'an element of mutual emancipation that allows to dispel all rudiments of the past and to establish pragmatic relations'.[55] Similarly, the significant change of Russia's policy vis-à-vis Belarus has found its *raison d'être* in the current foreign policy conception, adopted by the newly elected President Dmitry Medvedev in July 2008. Whereas the former, Putin-era conception still attributed 'front-ranking relevance' to 'the strengthening of the union between Russia and Belarus as in the current stage highest form of integration of two states', the new document soberly stated the goal of furthering 'agreed efforts to create favorable conditions for effective establishment of the Union State by gradually transforming relations between Russia and Belarus on the basis of market principles within the framework of developing a common economic space'.[56]

This new approach will have a considerable impact on the prospects of further political integration. Overall, the creation of a Union State has suffered a further crucial setback, though its final death knell has not yet tolled. The agreement on the oil dossier simply implemented provisions that had been ignored by Belarus since 2001. The aim of abandoning the special conditions on gas over the long term, however, shows that Russia has finally ceased to consider its eastern neighbour as its 'little brother' but instead as an actual sovereign country. Thus understood, Russia's emergence as an energy power defending its own national interests can be considered the final step of the USSR's dismantling. While Gazprom is – as demonstrated above – acting more and more like a 'normal' monopolist, Russia is itself acting – from a neo-realist perspective – more and more as a 'normal' nation state.

Yet Russian officials' emphasis on the commercialization of bilateral relations is not altogether coherent, taking into account the outcome of the latest energy conflicts. The partial implementation of commercial relations and Russia's reluctance to cause Belarus significant economic damage point to another relevant factor in contemporary Russian foreign policy: international competition and the 'display' of power. The Kremlin's perception of political and economic competition between the USA, the EU and Russia in eastern Europe stresses Belarus' political value. As Moscow takes into account Belarus' military and transit infrastructure in a potential 'arms race' with the West, the military factor may grow in prominence in Belarus–Russia relations. One must not forget Kaliningrad's geostrategic role and the possibility of cooperation between Belarus and Russia in response to placing elements of the US missile defence system in central Europe. Accordingly, it is no coincidence that Russian officials have noted Belarus' hesitant stance in the recent Caucasus conflict with suspicion. However, the mere fact that Belarus approved Russia's actions in Georgia only in a lukewarm manner and did not openly support it highlights the fact that Russian–Belarusian relations are in transition; while Belarus remains a valuable 'friend', it is no longer a close ally of Russia.

Conclusion

The recent Belarusian–Russian energy conflicts are the result of Moscow's attempt to reconstruct relations with the states of the post-Soviet region based on pragmatism and devoid of ideology. This implies a deliberate turning away from the traditional carrot-and-stick pattern which, in the eyes of Russian policy-makers, has failed to fulfil Russia's national interests. The Kremlin's priority is the same as that of Russia's energy industry: gain access to the distribution networks in all countries which are transiting Russian gas and maximize profits.

This gradual commercialization of relations is accompanied by a somewhat weakened – and weakening – Belarusian position. Although geopolitical realities, namely Russia's dependency on Belarus for transit, have so far remained the same, change is already visible on the horizon. With the completion of new gas and oil pipeline projects bypassing Belarus from 2011, Minsk's bargaining power will shrink. However, from a Belarusian perspective, selling Beltransgaz when they did might have been the most viable option for Belarus, since the country prevailed in pushing through its high valuation of the company. This move has also helped preserve Belarus' continued relevance as a transit country. Gazprom now has increased incentives to remain in Belarus, and to be able to modernize its pipeline infrastructure there in a safer investment climate. If Gazprom's strategy of building excess capacity for exporting gas to Europe succeeds, Ukraine and Belarus might soon compete against each other, as transit countries, for gas volumes pumped through their territory. If such competition comes to pass, Belarus will have a clear advantage, not only because its pipeline infrastructure is more recent, but also because of Gazprom's long-term commitment to this infrastructure, given the investment it has already made.

It has also become obvious that Moscow is not interested in having its neighbour undergo abrupt economic and political destabilization, not only because Belarus is a crucial transit country, but also out of concern for losing a close partner on the international stage. By granting its neighbour extensive state loans to soften the energy price hike's effects, Russia has proved to be conciliatory. Lukashenko has been clearly told that Russian subsidies cannot be counted on forever, however. In the long term, the Belarusian economy will only be able to cope with new market conditions if the country is modernized structurally and opened for foreign capital. It will therefore be difficult for Lukashenko to further resist the penetration of Russian capital into Belarus. Increased Russian investments or takeovers of Belarusian state-directed companies like Beltransgaz are thus highly probable in the future. It is also very likely that Russian business circles, which for a long time expressed interest in the privatization of vital assets in Belarus, have pushed the Kremlin to harden its stance toward Belarus. Whereas the prospects for further political rapprochement are rather bleak, economic integration is nevertheless likely to proceed.

The issue of Russia's bilateral energy relations with Belarus is therefore of vital importance for both countries. With this in mind, it is not appropriate to accuse Russia of using of its 'energy weapon' vis-à-vis Belarus. Such a concept only serves to obscure the facts even more, rather than shed light on an economic and political relationship that is already notoriously untransparent. Closer analysis of the energy crises between Moscow and Minsk rather proves the impotency of any supposed 'energy weapon' Russia might wield; Russia, be it embodied in Gazprom or in the Kremlin, simply has not been in a position to unilaterally force its conditions on Belarus, but has had to seek compromise in order to ensure the transit of its natural gas to Europe, thus preserving its reputation as a reliable supplier of energy. The specificity, and drawback, of the 'energy weapon' lies in the fact it immediately backfires on he who wields it.

Conclusion

Adrian Dellecker and Thomas Gomart

This book has tried to analyse Russia's foreign and energy policies together rather than separately, in order to better grasp their correlation and gain deeper understanding of broader geopolitical issues in Eurasia. As editors, we encouraged the contributors not only to explore special factors through global, regional and bilateral case studies, but also through the time factor – specifically, the 'Putin era'. Interestingly, the level of research interest on energy issues seems almost directly related to the level of energy prices: the higher these are, the more researchers and experts deal with energy issues. When prices decline, research interest wanes in equal proportion. Here lies a sizable paradox which deserves to be highlighted: while volatility and reliability are inherent to energy policy, foreign policy requires perseverance and flexibility. As a result, it is a considerable gamble for a great power like Russia to pursue a foreign policy that is narrowly dependent on energy and its associated revenues.

That being said, it would have been a significantly missed opportunity had Russia not exploited the rise in international energy prices between 2000 and 2008. On balance, and focusing on the last decade, it is fair to say that Russian authorities were far more successful than unsuccessful in leveraging the opportunity opened by the global demand for energy in that window. Even if there are considerable uncertainties as to the sustainability of its development, there is no doubt that Russia is today far more influential on the international stage than it was in 2000. And this is not just true for the Russian state: there is also no doubt that Russian energy companies have reinforced their positions on the global markets over the same period, at both global and regional levels.

Identify the political factor

This book has sought to give added fuel to the often-partisan debate on the geopolitics of energy by discussing both Russia's attempt to 'lock in' CIS net exporters into its energy network and the efforts of various external players to 'lock Russia out'.

In the book's first part, the chapters explored the specific link between energy and foreign policy in Russian strategic thinking, helping to set the stage for a better understanding of the mechanisms and stakes involved in the complex

Eurasian energy equation. This exercise included the study of institution and policies as well as reserve/production data and the realities of infrastructure financing. The notion that there is a perceived competition between Russia's foreign ambitions and the actions of other players has been seriously nuanced, notably by exposing the significant constraints inherent to energy investment projects, constraints which in reality focus minds together. Indeed, this first part has mainly concluded on the need not to overestimate the role of geopolitics in what are often pure business decisions. This part helped to understand and highlight how different actors, be they companies or states, simply tend to the maximizing of profits, undermining the more traditional view of Eurasian energy relations, which can often over-emphasize the 'political factor' in the domain of energy relations. Given the required significant upfront costs of energy investments, pipeline projects cannot afford to be solely motivated by politics, let alone by grand strategic considerations. Political factors do come into play and are inherent to each project, but these are rather symptoms of a weak institutional framework, pervasive corruption or opaque policy-making within the ruling elites, rather than grand design.

The chapters of the second part explored the extent to which Russian energy companies are dependent on Central Asian oil and gas in order to honour their contractual obligations to consuming markets. This is particularly true for Gazprom, given the nature of the natural gas market. This is because natural gas is far more difficult and costly to extract, store and distribute than other hydrocarbons: unlike oil and coal, natural gas cannot easily be diverted from one consuming market to another. As a result, the vast majority of global trade in natural gas is still conducted by dedicated pipelines in segmented markets, limiting choices. Even if this may be somewhat altered in the coming decades as more and more liquefied natural gas (LNG) projects come on stream, the natural gas trade can be expected to continue to be centralized around big players until the considerable investments in LNG are amortized.

The chapters of the second part thus demonstrated why this basic model of Russian dependence on Central Asian gas will only be reinforced in the coming decades, with obvious consequences for the 'East Eurasian' market. This second part also highlighted the fundamental differences between each of the respective relationships: between Russia and Turkmenistan, Russia and Kazakhstan, Russia and Uzbekistan, as well as Russia and Azerbaijan. Taken as whole, the chapters seem to confirm that there has in fact been a 'Russian comeback' in terms of influence, relative to the 1990s, though this comeback differs significantly from country to country in terms of both intensity and efficiency.

Kazakhstan, for example, will continue to be Moscow's top priority in the coming decade. One the other hand, Kazakhstan has successfully expanded its options in terms of foreign policy; the succession of Nazarbaev is clearly a critical unknown factor of the coming equation, however. Uzbekistan, which has never been an easy partner for Moscow, is an ever greater unknown: while the energy environment certainly suffers from the nature of the regime, there is also a question surrounding Karimov's succession, adding to the uncertainty. Azerbaijan, benefiting from a highly valuable geopolitical location, has been courted by external players since the 1990s. It has attracted much attention as a key country by the US and the EU for establishing the so-called South Corridor, but despite this Russia and Azerbaijan seem to have reached a sort of balance in their relations. Turkmenistan is a particular unknown of the Eurasian energy equation, given the lack of reliable data on its reserves and production as well as the lack of institutional framework to develop its energy efficiently. The recent inauguration of the Turkmenistan–China pipeline illustrates a key new factor of the coming decade, however: China's increasing involvement in the CACR.

Finally, the chapters of the last part shed light on Russia's relations with transit countries, which are also net importers and dependent, to varying degrees, on Russia for their own domestic consumption. Indeed, compared to 1990s, when the BTC and the SCP were negotiated, the period 2000–8 saw Russia reclaim its central role as a transit hub and reinforce its advantage in terms of transit infrastructure. One should not underestimate the value and crucial importance of the CAC, inherited from the Soviet period, as the main transportation system in Eurasia. Its fate will also be one of the main factors in the future Eurasian energy equation: if the necessary investments for its maintenance and modernization are made, it will be hard to rival. But rival projects exist, often backed by powerful interests, and could yet be built from scratch.

This issue has been on the table for some years now. The so-called South Corridor, particularly, has long been presented by Europe and the US as a secure route that bypasses both Russia and Iran, and therefore as the best way to embed Europe in the South Caucasus for the long term. The projects and counter-projects which gravitate around this idea of parallel routes include Nabucco, the Trans-Caspian and South Stream, which were examined in detail in this part's chapters. In the end, the credibility of each project seems to be in the political eye of the beholder. This highlights an important finding: while business rationale is constantly invoked to justify one project rather than another, preferences for one project over another nevertheless seem to reflect political interpretations. Some contributors seem more inclined to think in terms of regional integration; others tend more for the fragmentation of the FSU space. While pipeline projects cannot be solely motivated by geopolitics, as already discussed, mental maps and political preferences do impact the business mindset as the expert point of view. Even in business, after all, there are rival dogmas: there is no single path to corporate prosperity.

There has also been a clear shift in perception since the war in Georgia in 2008. The conceived stability of the South Corridor is now more than ever highly questionable, and there can now be no doubt that Russia intends to stay the dominating military power in this region. There is also no doubt that Baku has drawn the inevitable political and military conclusion of the Georgian defeat, while Tbilisi can no longer promote itself as a key actor of stability – which is precious in hydrocarbon trade and necessary in that of natural gas. For understandable reasons, expert production in transit issues has been mainly focused on Georgia, Azerbaijan and Ukraine. This book was also keen to draw attention to Belarus, which helpfully completes the picture: because of its political structure and key geographic location, Belarus is sure to remain an important part of Russia's future energy and foreign policies.

Two main conclusions

Taken together, these chapters lead us to two main conclusions. The first is related to the notion of energy power broadly speaking; the second concerns the overall transformation of the FSU.

While there is no doubt that Russia strives to fuel its power project by energy, it would be a mistake to reduce Russian power to this alone. Russia's power project must be understood in the long term as trying to maintain Russian strategic autonomy in an increasingly interdependent international system. This power project is national in essence and global in ambition, and consequently unclear at the regional level. Furthermore, unlike other countries in Europe, or in Asia for that matter, Russia does not see its future in terms of the integration or grouping of states. Unlike the many countries that make up the EU, for example, Russia intends to go it alone. In such a context, it would be misleading to explain Russia's power project by its energy policy alone. Its energy policy is in fact included into a much broader strategic culture, which remains mainly defensive and territorial. In the Russian mindset, energy policy has been a tool of control over space to be protected against external predations.

In fact, energy policy has become a tool for the projection of power only quite recently for Russia. In the 1990s, due to domestic turmoil, the Russian authorities were in an inward-looking and defensive posture. In the 2000s, after the renationalization of the energy landscape, led by Vladimir Putin, Russia has been much more eager to use its energy policy as a tool of global influence. The level of Russia's international ambitions has risen proportionately to energy prices, which helped fuel the engine of power. These years also produced a positive feedback loop: thanks to energy, Russia reinforced its power, which it used in turn to reaffirm its dominance of the energy landscape, both domestically and internationally. Russian authorities practised a concentrated form of resource nationalism to strongly limit foreign interference and to actively project Russian power abroad by merging the approaches of energy companies and the state apparel. Yet a basic but significant problem remains in this strategy: company and state interests do not always converge.

As explained in the Preface, one of the roundtables at the beginning of this editorial project was entitled: 'The Restoration of Russian Power: Toward an Energy-Based Deterrence?'. Yet we quickly learned that energy conflicts cannot be understood solely by looking at the level of the nation-state and statist relations. In fact, a multiplicity of competing interest groups – public and private, domestic and foreign – interact at different levels, all trying to make a profit out of the energy crunch. For this reason the notion of an energy-based deterrent is, at the end, inappropriate to properly account for the complexity of the overall energy question in Eurasia.

On the one hand, the Kremlin's involvement in all major energy agreements was a fact during Putin's tenure. Yet it is still too early to say whether Putin has durably transformed Russia's traditional geopolitical approach in foreign policy into a more innovative geoeconomic approach. At the regional level, however, the connection between geopolitics and geoeconomics has been particularly visible. Energy resources and infrastructure were embedded in a policy dedicated to the promotion of Russia's economic and security interests, mainly in the FSU but also at the global level. Added to this, there has been what may be called 'the Putin factor'. Under him, the Kremlin deliberately politicized energy issues to the extent that it became seen from abroad as the sole gate of entry. The conclusion was that a contract could be signed only if President Putin was in the loop; if he was not, the business risk for foreign investors was considerably augmented. In other words, neither market forces nor legal frameworks secured investments, but only presidential endorsement and personal connections. Despite this, it would be misleading to assume that presidential endorsement necessarily fits into an overarching power project and geopolitical vision. The increased importance of purely commercial considerations, internal power bargains, not to mention personal favours involving money in Russian decision-making processes has been continuously underestimated by Western expertise.

The fact is that to be consistent in geopolitics and geoeconomics requires strong leadership and vision. Yet commercial considerations and trade agreements require flexibility and cooperation. For Russia, this paradox is of consequence, as energy supplies will continue to be Russia's chief source of external revenue for some time.

Russian authorities understand well that exercising leverage through energy supplies can become very self-negating by encouraging customers to look to alternative suppliers. For this reason they are used to saying that they prize Russia's reputation as a reliable and sustainable supplier. In Putin's years as president from 2000 to 2008, Russia held a dual discourse: while insisting on its ability to deliver energy as pledged, it was not afraid to push its weight around in order to be recognized as an energy power. This dual approach can be seen to have succeeded at least in part, since Russia has clearly become a central player in the geopolitics of energy. Russia's January 2009 gas crisis with Ukraine, however, significantly undermined Russia's reputation on the European market (the same can be said about Ukraine), since it revealed that aligning Kremlin and Gazprom interests, as the brand it has been touted, is increasingly complicated and sensitive to manage. Russia remains uncomfortable in terms of global marketing. As a promoter of the BRIC (Brazil, Russia, India and China) grouping, Russia is likely to be increasingly compared to Brazil in terms of energy and foreign policies. The natural consequence would be for the link between the Kremlin and Gazprom to be compared to that which exists between Brasilia and Petrobras, which could be very challenging for Russia's reputation in the coming decade.

Russia faces a significant problem when trying to reconcile its bilateral, regional and global priorities. Its international policy is directly influenced by the vast expanse of Russian territory. Within the FSU, it is well known that Russia favours bilateral approaches, of which it is always the dominant partner and where it holds the balance of power when dealing with each capital, even if no two relations are alike in terms of real influence. As a result, the FSU space remains considerably fragmented; there is a lack not only of political integration, but also of a desire to integrate. The customs union between Moscow, Astana and Minsk appears to be the maximum possibility of integration for now, given Russia's lack of interest in the matter. From the outside, this only confirms that the FSU is ever more fragmented.

Yet the FSU space can also be seen as having been partly reintegrated between 2000 and 2008. The war in Georgia in 2008 also marked a turning point in terms of perceptions of Russia within the FSU. By using force in the FSU, Moscow has conspicuously added a military component to its foreign and energy policies. This added dimension, coming on top of energy dependency and infrastructure such as the CAC, may be the bonding agent that maintains the notion of the FSU for years to come. In the end, Russian strategy regarding the CACR remains ambivalent. On the one hand, Russia wants to appear as the main security player, able to deliver security guarantees and to use force to defend its interests, while striving to be a credible energy partner.

This ambivalence is particularly visible in Central Asia with regard to the increasing Chinese involvement. In it for the long run, Russia certainly feels much more threatened by Chinese influence than by Western military presence in Afghanistan and Kyrgyzstan. Russia will try to maximize the potential of the CAC for diplomatic use in the relevant countries and try to attract investment. It is important to understand that the CACR countries have no intention to break with Moscow, as was often thought in Western strategic circles in the past two decades. What is more, dealing with Moscow in a non-transparent way in fact remains a significant asset when negotiating with Europe. For them, showing an interest in the European market in order to increase revenues does not mean they intend to cease to interact with Moscow. For these countries, this dual approach is not contradictory – a point which remains difficult to understand by Western expertise. Taken together with the threat of China, these elements in fact contribute to an integration of sorts within the FSU, even if it may not be understood as such from the outside, or may not have formal legal or institutional frameworks.

In the end, true strategy, as contrasted to tactic, involves going beyond geography alone and considering the factor of time. It has already been shown that Russia's energy policy has been used to recast the power structure domestically and to project power abroad. However, as a key component of Russian foreign policy, Russia's energy policy has also been to project itself into the future. In this regard, the period from 2000 to 2008 must be read as a very particular sequence in Russian recent history. On the one hand, these eight years were used to restore the Russian energy industry after the turmoil of the 1990s. On the other hand, they were also characterized by short-, medium- and long-term considerations. Many official documents or statements, for example, refer openly to 2020 as a deadline for this transformation to occur. It transpires that the Russian leadership has tried to shape a power project fuelled by energy for a single generation. It is this ability to project itself in both space and time that drives Russian power and foreign policy; energy policy is also a timely tool.

Notes

Introduction

1 Energeticheskaya strategiya Rossii na period do 2020 goda, Pravitel'stvo Rossijskoi Federatsii, Ukaz 1234, 28 August 2003.
2 Jeffrey Mankoff, 'Eurasian Energy Security', Council on Foreign Relations, *Council Special Report*, no. 43, February 2009, p. 4–5.
3 Jonathan Stern, *The Future of Russian Gas and Gazprom*, Oxford, Oxford Institute of Energy Studies, 2005, pp. 33–5.
4 Thomas Gomart, 'Vladimir Poutine ou les avatars de la politique étrangère russe', *Politique étrangère*, no. 3–4, 2003, pp. 789–802; 'Politique étrangère russe: l'étrange inconstance', *Politique étrangère*, no. 1, 2006, pp. 25–36; *Russian Civil-Military Relations: Putin's Legacy*, Washington DC, Carnegie Endowment for International Peace, 2008.
5 Vladimir Putin, 'Mineral'no-syr'evye resursy v strategii razvitiâ rossijskojèkonomiki', in N. Marfenin, *Rossiâ v okruzausem mire: 2000*, Moscow MNEPY, 2000.
6 Bobo Lo, *Vladimir Putin and the Evolution of Russian Foreign Policy*, London, RIIA/ Blackwell Publishing, 2003, p. 54.
7 Jim O'Neill, 'Building Better Global Economic BRICs', Goldman Sachs, Global Economic Paper, no. 60, November 2001; Dominic Wilson and Roopa Purushothaman, 'Dreaming with BRICs: The Path to 2050', Goldman Sachs, Global Economics Paper, no. 99, October 2003.
8 Thomas Gomart, *Russian Civil–Military Relations: Putin's Legacy*, op. cit., pp. 2–3.
9 Pavel Baev, *Russian Energy Policy and Military Power*, London, Routledge, 2008, p. 2.
10 Fiona Hill, *Energy Empire: Oil, Gas and Russia's Revival*, London, FPC, 2004; Mickael Klare, *Rising Powers, Shrinking Planet*, New York, Metropolitan Books, 2008; Ian Bremmer, *The End of the Free Market*, New York, Portfolio, 2010.
11 Margarita Balmaceda, *Energy Dependency, Politics and Corruption in the Former Soviet Union*, London, Routledge, 2008.

1 Energy security in Eurasia: clashing interests

1 J. A. Stanislaw, Power Play. Resource Nationalism, the Global Scramble for Energy, and the Need for Mutual Interdependence. Deloitte, Item #8127, 2008, p. 16.
2 See European Commission, *Green Paper, Towards a European Strategy for the Security of Energy Supply* (adopted 29 November 2000), Luxembourg 2001.
3 See European Commission, *An EU Energy Security and Solidarity Action Plan. Second Strategic Energy Review 2008*. Commission Working Staff Document: Europe's Current and Future Energy Position. Demand–Resources–Investments. Communication from the Commission to the European Parliament, the Council, the

European Economic and Social Committee of the Regions, Brussels, November, COM(2008) 744.

4 See also J. H. Keppeler, *International Relations and Security of Energy Supply: Risks to Continuity and Geopolitical Risks.* External Study for the Directorate-General for External Policies of the Union. Policy Department, Brussels: European Parliament, February 2007; D. Helm, *The Russian Dimension and Europe's External Energy Policy*, Oxford: University of Oxford, 3 September 2007; J. Stern, *The New Security Environment for European Gas: Worsening Geopolitics and Increasing Global Competition for LNG*, Oxford: Oxford Institute for Energy Studies, NG 15, October 2006 and F. Umbach, *Globale Energiesicherheit. Herausforderungen für die europäische und deutsche Außenpolitik*, Munich: Oldenbourg Verlag, 2003, pp. 166 ff.

5 See European Commission, *An EU Energy Security and Solidarity Action Plan. Second Strategic Energy Review 2008*, p. 7.

6 See European Council (2007), *Presidency Conclusions*. 7224/1/07 REV 1, CONCL 1, Brussels, 8–9 March 2007, 26 pp.; to the EAP, pp. 11–26.

7 See F. Umbach, *Zielkonflikte der europäischen Energiesicherheit. Dilemmata zwischen Russland und Zentralasien*, DGAPanalyse, No. 3, Berlin, November 2007 and ibid., Europe's Next Cold War, *Internationale Politik (Global Edition)*, Summer 2006, pp. 64–71.

8 See European Council, *The EU and Central Asia: Strategy for a New Partnership*, Brussels, 22 June 2007. Online. Available at: http://www.auswaertiges-amt.de/diplo/en/Europa/Aussenpolitik/Regionalabkommen/EU–CentralAsia-Strategy.pdf (accessed 5 July 2007).

9 See also F. Umbach, *Memorandum: The European Union and Russia – Perspectives of the Common 'Strategic Energy Partnership'*. Personal Analysis for the Sub-Committee on Foreign Affairs, Defence and Development Policy, House of Lords/Great Britain, in: House of Lords and its European Union Committee (Eds.), The European Union and Russia, Report with Evidence, 14th Report of Session 2007–8. HL Paper 98, Norwich–London: The Stationery Office, pp. 185–8.

10 See IEA, *World Energy Outlook 2008*, Paris 2008, pp. 37 ff. (Executive Summary).

11 See Joseph A. Stanislaw op. cit., p. 6 f.

12 See also F. Umbach, The Legs of the Triangle – The EU–China Relations, in W. Jung (Ed.) *The New Strategic Triangle: China, Europe and the United States in a Changing International System*, Konrad-Adenauer-Stiftung, KAS-Schriftenreihe No. 76, Beijing: KAS, pp. 36–45.

13 See Sergei Guriev/Anton Kolotilin/Konstantin Sonin, *High Oil Prices and the Return of 'Resource Nationalism'*. VoxEU.org, 12 April 2008. Online. Available at: http://www.voxeu.org/index.php?q=nod/1050 (accessed 18 December 2008). To the debate on resource nationalism see D. G. Victor, What Resource Wars?, *The National Interest*, 11 December 2007; S. W. Goodman and P. J. Kern, Bad Tidings, ibid., 2 January 2008; M. T. Klare, Clearing the Air, ibid., 1 February 2008.

14 See Joseph A. Stanislaw, op. cit.

15 See ibid., p. 3 f.

16 Javier Solana, *Toward an EU External Energy Policy*, Address at the EU Energy Conference, S324/06Brussels, 20 November 2006.

17 See BP, *Statistical Review of World Energy 2008*, June 2008, p. 22.

18 Islamic extremism is nevertheless not as severe a problem as it is in many Arab states such as Morocco, Algeria, Iraq and others. This threat also varies partly because Islam in North Africa or in the Persian Gulf states dates further back and is thus more deeply rooted. In some parts of Central Asia, interest in Islamic extremism in fact seems to be declining with improving economic conditions (such as in Kazakhstan).

19 See European Council, *'Baku Initiative'. An Energy Dialogue between the EU and the countries of the Black Sea, Caspian Basin and their Neighbours*, Brussels 2004.

Online. Available at: http://ec.europa.eu/external_relations/energy/baku_initiative/index.htm (accessed on 22 January 2008).

20 See International Crisis Group, *Central Asia: What Role for the European Union?*, Asia Report, no. 113, 10 April 2006.

21 See also Anna Mateeva, *EU Stakes in Central Asia,* Chaillot Paper, No. 91, July 2006; Vladimir Socor, Central Asia–Europe Energy Projects: Itemizing What Went Wrong, *Eurasia Daily Montor*, Vol. 4, No. 106, 31 May 2007 and Svante E. Cornell, Commentary: So Far, Europe's Approach to Central Asia has been Moralistic and Counter-Productive, *Europe's World*, Spring 2007.

22 See Brenda Shaffer, From Pipedream to Pipeline: A Caspian Success Story, *Current History*, October 2005, pp. 343–6.

23 See F. Umbach, *Zielkonflikte der europäischen Energiesicherheit*, p. 21.

24 Gazprom Annual Report, 2007.

25 See also 'Kazakhstan, EU to Sign Energy Deals in December', *RFE/RL Feature Article*, 30 November 2006; Kazakh Leader's EU Visit Yields Energy Steps, ibid., 4 December 2006; EU, Kazakhstan: The Geopolitics of Energy Cooperation, *Strafor. Com*, 4 December 2006; Energie-Dialog der EU mit Kasachstan, *Neue Züricher Zeitung*, 5 December 2006.

26 See European Council, *The EU and Central Asia: Strategy for a New Partnership.*

27 See F. Umbach, *Zielkonflikte der europäischen Energiesicherheit* , pp. 10 ff.

28 See Frank-Walter Steinmeier, Reviving the Silk Road, *Frankfurter Allgemeine Zeitung*, 30 June 2007. Online. Available at: http://www.auswaertiges-amt.de/diplo/en/Infoservice/Presse/Interview/2007/070630-Seidenstrasse.html (accessed 12 August 2007) and Johannes F. Linn, *Central Asia: A New Hub of Global Integration*, Brookings Institution, 12 December 2007.

29 To the latter point see also Nargis Kassenova, *The New EU Strategy Towards Central Asia: A View of the Region*, CEPS–Policy Brief, No. 148, Brussels, January 2008.

30 So China's foreign minister, Tang Jiaxuan – see *NAPSNet–Daily Report*, 6 March 2003.

31 Wu Lei/Shen Qinyu, Will China Go to War over Oil?, *Far Eastern Economic Review*, April 2006, p. 39 f.

32 See also D. Zweig and B. Jianhai, China's Global Hunt for Energy, *Foreign Affairs*, September–October 2005, pp. 25–38; F. Umbach, China's Energy Insecurity in Context of Growing Geopolitical Competition: Implications for the Future EU–China Relations, in: Klaus Lange (Ed.), *European–Chinese Security Cooperation: Possibilities and Limits*, Akademie für Politik und Zeitgeschehen der Hanns-Seidel-Stiftung, Studies and Comments, No. 6, Munich 2008, pp. 43–59; ibid., Globale Energiesicherheit, pp. 103 ff.; and ibid., Future Impacts of Chinese and Asian Dependency on Energy Imports from the Middle East and Central Asia, in Erich Reiter and Peter Hazdra (Eds.), *The Impact of Asian Powers on Global Developments*, Heidelberg–New York 2004, 143–163.

33 See F. Umbach, The Wounded Bear and the Rising Dragon – The Sino–Russian Relationship at the Beginning of the 21st Century: A View from Europe, *Asia–Europe Journal* 2/2004, pp. 43–62.

34 See Sebastien Peyrouse, Sino–Kazakh Relations: A Nascent Strategic Partnership, *China Brief*, Vol. 8, Issue 21, 7 November 2008.

35 See ibid., p. 1 ff.

36 See 'China, Kazakhstan Ink Deals', *China-Daily*, 10 April 2008.

37 See Stephen Blank, China's Recent Energy Gains in Central Asia: What Do They Portend, *CACI-Analyst*, 31 October 2007.

38 See F.Umbach, *Zielkonflikte der europäischen Energiesicherheit*, p. 18 ff. To the background see also Svante E. Cornell/Niklas Nilsson (Eds.), *Europe's Energy Security. Gazprom's Dominance and Caspian Supply Alternatives,* Washington–Stockholm–Nacka: Central Asia–Caucasus Institute/Silk Road Studies Program, 2008.

39 To the Nabucco project see also Nicklas Norling, *Gazprom's Monopoly and Nabucco's Potentials: Strategic Decisions for Europe. Silk Road Paper,* Washington–Stockholm–Nacka: Central Asia–Caucasus Institute/Silk Road Studies Program, November 2007.
40 The international pipeline consortium of six companies is equally owned with 16.67 per cent by Austria's OMV, Hungary's MOL, Turkey's state-owned BOTAS Petroleum Pipeline Corporation, Bulgaria's state-owned Bulgargaz, Romania's state-owned Transgaz and Germany's privately owned RWE.
41 Two European companies involved in the Nabucco project are also participating at the South Stream project: Bulgaria's state-owned Bulgargaz and Austria's state-owned OMV, which is the most engaged. Romania explicitly declined any Russian offers to participate, whereas Turkish and German companies have not been approached by Russia.
42 See South Stream Pipeline Costs Doubling, *Vedomosti*, cited in *RIA Novosti*, 30 July 2008. See also Zeyno Baran, *Security Aspects of the South Stream Project.* Hudson Institute/Center for Eurasian Policy/European Parliament, 2008, pp. 19–22.
43 See Vladimir Socor, OMV Joins with Gazprom to Undercut Nabucco, *EDM*, Vol. 5, 29 January 2008 and Roman Kupchinsky, The Shadowy Side of Gazprom's Expanding Central European Gas Hub, ibid., Issue 217, 12 November 2008.
44 See European Council (2007), *Presidency Conclusions*, p. 19.
45 See also Zeyno Baran op. cit., p. 21 f.
46 See Ahto Lobjakas, EU: Nabucco Looks to Middle East for Gas, *RFE/RL*, 6 May 2008.
47 See Aladimir Socor, Romania Firmly Aboard Nabucco Projects, Hints at Leverage vis-à-vis South Stream, *EDM*, Vol. 5, Issue 210, 3 November 2008 and Zeyno Baran, op. cit., p. 18.
48 See Vladimir Socor, Gazprom Seducing Romanian Government with South Stream, *EDM*, Vol. 5, Issue 204, 24 October 2008.
49 See Vladimir Milov, *The EU–Russia Energy Dialogue: Competition versus Monopolies*, *Russie.Nei.Visions*, No. 13, IFRI , Paris, September 2006, p. 7; ibid., *Russia and the West. The Energy Factor,* Paris–Washington DC: IFRI and CSIS, July 2008; V. Milov, L. L. Coburn and I. Danchenko, Russian Energy Policy 1992–2005, *Eurasian Geography and Economics*, Vol. 47, 206, No. 3, pp. 285–313; A. Riley and F. Umbach, Out of Gas. Looming Russian Gas Deficits Demand Readjustment of European Energy Policy, *Internationale Politik (Global Edition)*, Spring Issue, No. 1, pp. 83–90 and ibid., *Zielkonflikte der europäischen Energiesicherheit*, pp. 16 ff.
50 According to the well-known and independent Russian military expert Pavel Felgenhauer and some Western experts and journalists, the war had been deliberately been pre-planned in Moscow, albeit the time-point may have taken by surprise the Kremlin leadership – see Pavel Felgenhauer, The Russian–Georgian War was Preplanned in Moscow, *EDM*, 14 August 2008 and Brian Whitmore, Did Russia Plan Its War in Georgia?, *RFE/RL*, 15 August 2008.
51 See, for instance, Bruce Pannier, Georgia–Russia Conflict Changes the Energy Equation, *RFE/RL*, 2 September 2008.
52 See Andrew E. Kramer, Russia Claims Its Sphere of Influence in the World, *New York Times*, 1 September 2008.
53 European Council, *Conclusions of the Council. Emergency Meeting*, 12594/08 DE, CONCL 3, Brussels, 31 September 2008.
54 See Bruce Pannier, Nabucco Gets a Boost in Baku, *RFE/RL*, 10 September 2008.
55 See Fariz Ismailzade, The Georgian–Russian Conflict Through the Eyes of Baku, ibid., No. 154, 12 August 2008.
56 To China see Jing Huang, Beijing's Perspectives on the Russo–Georgian Conflict: Dilemma and Choices, *China Brief*, Vol. 8, Issue 17, 3 September 2008.
57 Eric Marat, CSTO Summit Disappoints Moscow, Unites Central Asia, *EDM*, 12 September 2008.

58 See Pavel K. Baev, Russia Assembles Its Allies and Ponders the Content of its 'Privileged Interests', *EDM*, Vol. 5, 9 September 2008.
59 See Farkhad Sharip, Nazarbayev Plays Double Game on South Ossetia, ibid., No. 167, 2 September 2008 and Are Kazakhstan's Economic Bonds with Georgia the Price of Stronger Ties with Russia?, ibid., No. 190, 3 October 2008.
60 See Vladimir Socor, Oil Development in Kazakhstan Underscores Significance of Azerbaijani–Georgia Transit Corridor, ibid., Issue 212, 5 November 2008.
61 See Bruce Pannier, Gazprom Works to Advance Russia's Interests in Central Asia, *RFE/RL*, 19 October 2008 and Farangis Najibullah, Moscow Seeking Alliances in Energy-Rich Central Asia, ibid., 4 September 2008.
62 See 'Caspian: Turkmen, Azerbaijani Deal Could Raise Energy Hopes', *Eurasianet.org*, 6 March 2008.
63 See Vladimir Socor, Azerbaijan Quietly Lending Impetus to Nabucco Project, *EDM*, Vol. 5, Issue 228, 2 December 2008 and F. Umbach, Diversifizierung statt Protektorat. Energiepartnerschaft zwischen Russland und der EU, *Die politische Meinung*, September 2008, pp. 25–30.
64 Originally the Socialist Hungarian government supported the South Stream pipeline with an agreement with Russia of January 2008. At that time, it called the Nabucco project a 'distant dream', which had provoked harsh criticism by the political opposition in the parliament as well as many energy and foreign policy experts in Hungary. The MOL company, by contrast, did not support the South Stream project. Meanwhile, the Hungarian government changed its opinion, inter alia by appointing one of its most senior diplomats, Mihaly Bayer, as a full-time special envoy for the Nabucco project – see Vladimir Socor, Hungary Doubling Efforts for Nabucco Ahead of Budapest Summit, *EDM*, 12 December 2008.
65 See John C. K. Daly, Iran Gains from Georgian Confrontation, *EDM*, Vol. 5, No. 199, 17 October 2008; Emrullah Uslu, Turkey and Iran Sign Accord on Natural Gas Cooperation, ibid., Issue 224, 21 November 2008, and Bruce Pannier, Kazakhstan Set to Increase Energy Cooperation with Iran, *RFE/RL*,16 October 2008.
66 The 'Peak oil' theory is based on the assumption and the indisputable fact that the oil resource base is finite. Once the world has depleted already half of all the oil that can ever be recovered technically and economically, production will enter a period of long-term decline and lead to very high oil prices. But most of the peak oil discussions focus exclusively on conventional oil resources but dismiss unconventional oil resources and natural gas liquids (NGL), which may keep total oil production much longer, but is also much more expensive. Thus the era of 'cheap oil' has already ended.
67 See European Commission, *An EU Energy Security and Solidarity Action Plan. Second Strategic Energy Review.* Communication from the Commission to the European Parliament, the Council, the European Economic and Social Committee of the Regions, Brussels, November 2008.
68 See also Roman Kupchinsky, The Price of Gas and Russian Democracy, *EDM*, Vol. 5, Issue 240, 17 December 2008.
69 According to Vladimir Milov and Western experts, the total Yamal fields development budget runs as high as US$ 200 billion (estimated in 2002 just at US$ 45) and Shtokman at around US$ 70 billion – V. Milov, *Russia and the West*, p. 10 f.
70 See Sergei Blagov, Russia's Gazprom Eyes Governmental Support, *EDM*, Vol. 5, Issue 233, 8 December 2008.

2 Hydrocarbon production and exports in Central Asia: the impact of institutions and policies

1 IEA, *World Energy Outlook*, Paris: International Energy Agency, 2006.
2 *Financial Times, 27 November 2006.*

3 Including both crude oil and natural gas liquids.
4 The exception to this trend was Uzbekistan, which managed to sustain production growth through the 1990s, with output nearly tripling to 191 kbd in 1998–9 before falling just over 20 per cent over the subsequent five years. However, in addition to being a minor producer, Uzbekistan was a significant oil consumer. Oil exports were limited, as rising output was needed to cover domestic consumption.
5 From 1995 through to 2002, each of the IEA's successive global projections raised forecasts for non-OPEC supply to 2010, yet each was low of the mark, with the NIS accounting for most of the difference. The IEA was not alone: most forecasts, including those of NIS governments, pointed to slower growth than was actually achieved. See IEA, *World Energy Outlook*, Paris: International Energy Agency, 2004, p. 525.
6 'Those quantities that geological and engineering information indicates with reasonable certainty can be recovered in the future from known reservoirs under existing economic and operating conditions.' BP (2008) is the BP Statistical Review of World Energy (2008). Available at: http://www.bp.com/genericarticle.do?categoryId=4003391&contentId=7045863.
7 While all NIS reserves data must be treated with caution, estimates for Uzbekistan and Turkmenistan are particularly problematic, since foreign involvement is so limited and the regimes in question are so secretive. In the case of Turkmenistan, alternative estimates suggest that reserves may be close to triple the levels reported by BP. The IEA cites estimates running to around 11 billion tons for the Turkmen section of the Caspian alone. See IEA, *World Energy Outlook*, Paris: International Energy Agency, 2006.
8 USGS, *World Petroleum Assessment 2000, Washington, DC: US Geological Survey Digital Data Series 60, 2000.*
9 It should also be noted that the USGS estimates cited here are for undiscovered oil; they do not include potential reserves growth via increases in the share of oil known to be in the ground that is assessed as economically recoverable. There is considerable scope for growth in the NIS here, too, although a great deal of the easiest growth in this respect has already taken place.
10 Given that larger volumes of oil have been located in all the Caspian basin states except Turkmenistan, this is hardly surprising: it seems unlikely that Caspian geology would conform so exactly to post-Soviet borders.
11 US Energy Information Administration, 'Country Analysis Briefs: Azerbaijan', US Department of Energy, February 2007. Online. Available at: http://www.eia.doe.gov/emeu/cabs/Azerbaijan/pdf.
12 On the reasons for these divergent choices, see Pauline Jones Luong and Erika Weinthal, 'Prelude to the Resource Curse: Explaining Energy Development Strategies in the Soviet Successor States and Beyond', *Comparative Political Studies 34:4, May 2001; and Pauline Jones Luong, 'Rethinking the Resource Curse: Ownership Structure and Institutional Capacity', Paper prepared for the Conference on Globalization and Self-Determination, Yale University, 14–15 May 2004.*
13 In recent years, former Soviet republics have consistently ranked below most other major oil exporters in Transparency International's Corruption Perceptions Index. The major exceptions are Angola, Indonesia (now, in any case, a net oil importer) and Nigeria (which consistently occupies second-to-last place in the index).
14 In 2000, Azerbaijan decided to abolish joint ventures and convert them to the more investor-friendly PSAs in an effort to spur increased development.
15 In addition to SOCAR, the AIOC consortium includes BP, Unocal, Inpex, Statoil, ExxonMobil, TPAO, Devon Energy, Itochu and Delta/Hess.
16 This refers primarily to oil. Azerbaijan has lately enjoyed more success in finding new gas deposits.
17 World Bank, 'Azerbaijan: Issues and Options Associated with Energy Sector Reform', Report No. 32371-AZ, Washington: The World Bank, 31 March 2005. Online. Available

at: http://wdsbeta.worldbank.org/external/default/WDSContentServer/IW3P/IB/2005/06/20/000160016_20050620104718/Rendered/PDF/323710AZ.pdf (accessed 1 January 2009).

18 Jones Luong and Weinthal, op. cit.

19 During 1996–7, the government sold off the Chymkent Oil Refinery, Yuzhneftegaz, Mangistaumunaigaz, Aktyubinskneft and the Uzen oil fields.

20 Unlike Azerbaijan, moreover, Kazakhstan has been relatively successful in attracting FDI to other sectors – the total stock of FDI is estimated to have reached $29.75 billion by the end of the third quarter of 2004.

21 Calls to revise long-standing contracts were sometimes justified with reference to allegations of corruption on the part of Western advisors involved in negotiating the original deals; see Martha Brill Olcott, Kazakhstan: Unfulfilled Promise, Washington, DC: Carnegie Endowment for International Peace, 2002; and Richard Pomfret, 'Kazakhstan's Economy since Independence: Does the Oil Boom Offer a Second Chance for Sustainable Development?', Europe–Asia 57:6, September 2006.

22 The fine for the delay was reportedly around $150 million.

23 President Nursultan Nazarbayev himself spoke on several occasions of the possible renegotiation of terms for the North Caspian PSA (i.e. Kashagan), arguing that Western companies had taken advantage of Kazakhstan in the early years after independence.

24 It is worth noting that BG originally agreed to sell the stake to China's CNOOC and China Petrochemical Group, but neither the consortium partners nor the authorities in Astana wished to bring the Chinese into the project.

25 The state's pre-emptive rights even extend to the purchase of a stake in a legal entity which is in a position to influence, directly or indirectly, the decisions of a subsoil use company.

26 See, e.g 'Roundtable with MEBP Kelimbetov', *International Tax & Investment Centre Bulletin*, July–August 2005, p. 3.

27 The legal status of this requirement is at least open to question, as the relevant decree states that a tender winner must form a consortium or joint venture with Kazmunaigaz in order to conclude a petroleum operations contract. However, other legislation, including the Petroleum and Subsoil Laws, require that such a contract be concluded directly with the winner of a tender, and not with any other person or entity. (See Decree of the Government of the Republic of Kazakhstan No. 708 [29 June 2002], 'On Approval of Regulations for Representing the State Interest by the National Company in Service Contracts for Petroleum Operations through a Mandatory Share.')

28 In mid-2005, Minister for the Economy and Budget Planning Kairat Kelimbetov publicly acknowledged that the 2004 tax changes were indeed linked in part to a desire on the part of his ministry and the Ministry of Finance to delay tenders for the development of the remaining blocks of the North Caspian. He indicated that this desire reflected macroeconomic concerns rather than any belief that an optimal depletion strategy would involve leaving more oil in the ground for future generations. See 'Government's "Thinking" on Future Oil Tenders', International Tax & Investment Centre Bulletin, May–June 2005; and 'Roundtable with MEBP Kelimbetov', op. cit.

29 Gerard Anderson, 'The Kazakhstan Upstream Tax Regime: A Comparative Analysis', *InvestKazakhstan*, March 2004, p. 70.

30 The project entailed the development of fields in Uzbekistan's central Ustyurt and Southwest Gissar regions, with the partners forecasting natural gas production of roughly 2bcm and oil production of 2.6 kbd by 2006.

31 These appear to have been partially offset by smuggling from Turkmenistan, where prices are lower still. The situation in the gas sector in Uzbekistan is similar, since Uzbekneftegaz has to supply gas at tariffs that are well below cost-recovery levels, payment discipline is sometimes patchy, and there is a great deal of cross-subsidy from industry to households.

32 This is a clear example of the 'obsolescing bargain model' developed in Ray Vernon, Sovereignty at Bay: The Multinational Spread of US Enterprise, New York: Basic

Books, 1971. See also Jones Luong op. cit., p. 6; as well as T. H. Moran, Multinational Corporations and the Politics of Dependence: Copper in Chile, Princeton, NJ: Princeton University Press, 1974; and Franklin Tugwell, The Politics of Oil in Venezuela, Stanford, CA: Stanford University Press, 1975.

33 Bargains between states and multinationals in manufacturing sectors are far less likely to obsolesce, not least because their investments tend to be smaller, more mobile and more closely tied to knowledge-based, firm-specific advantages. See Lorraine Eden, Stefanie Lenway and Douglas A. Schuler, 'From the Obsolescing Bargain to the Political Bargaining Model', Bush School Working Paper No. 403, Texas A&M University, January 2004. Online. Available at: bush.tamu.edu/research/working_papers/leden/Eden-Lenway-Schuler-FINAL-GBS.pdf (accessed 1 January 2009).

34 Ironically, the obsolescing bargain model anticipates that the host government's perception of the cost-benefit ratio offered by the relationship with the foreign investor will be more likely to deteriorate if the investment turns out to be much more profitable than anticipated – oil booms are thus highly likely to cause oil-sector investment bargains to come under strain.

35 This may be one reason why there has been no parallel toughening of the state's position in Azerbaijan: investors have been exiting – albeit because they have failed to find sufficient recoverable reserves rather than because of mistreatment at the hands of the authorities. Azerbaijan's bargaining position now looks somewhat weaker, owing to growing scepticism about how much oil remains to be found there.

36 These conclusions are unlikely to sound in any way surprising or controversial to observers of Russia's gas monopoly or other large, state-owned oil and gas companies in the NIS.

37 See, e.g. D. Michael Shafer, Winners and Losers: How Sectors Shape the Developmental Prospects of States, Ithaca: Cornell University Press, 1994; Michael L. Ross, 'The Political Economy of the Resource Curse', World Politics 51, 1999, pp. 297–322; Richard Auty, 'Patterns of Rent-Extraction and Deployment in Developing Countries: Implications for Governance, Economic Policy and Performance', Paper prepared for the Poverty Reduction and Economic Management Unit Seminar, World Bank, April 2004; and Jones Luong, op. cit. This proposition has not undergone much empirical analysis, for the simple reason that most of the literature focuses on minerals sectors in the period from the 1960s through the 1990s – a time during which the vast majority of mineral-rich countries opted for state ownership and control of mineral reserves. There were simply not very many cases with which to contrast the performance of state-owned enterprises.

38 It has also been suggested that the NOCs will tend to use less sophisticated technologies. However, this problem should, in principle, be relatively easy to overcome, given the potential for relying on Western oil service companies, which employ the most advanced technologies available.

3 How to get a pipeline built: myth and reality

1 The most spectacular example, and a dominant factor in Europe's increasing need for imports, is the UK turning from self-sufficiency in gas (and even exporting for a few years) into a major importer as its reserves are depleted and its domestic production declines. Thus, the new Langeled pipeline from Norway to the UK was put onstream in 2007 as the UK turned from a gas exporter into a net gas importer. See a description at http://www.statoil.com/statoilcom/svg00990.nsf?opendatabase&lang=en&artid=F5255D55E1E78319C1256FEF0044704B (Statoil website) and a map at http://www.npd.no/English/Produkter+og+tjenester/Publikasjoner/Faktaheftet/Faktaheftet+2007/Kapittel+14.htm (from the Norwegian Petroleum Directorate).

2 Nabucco is explicitly described as being able to bring Iranian gas (among others) to Europe. See for instance, the presentation 'Security of Gas Supplies in Europe',

William C. Ramsay, Deputy Executive Director, International Energy Agency, Vienna, 26 June 2006 (http://iea.org/textbase/speech/2006/ramsay/nabucco.pdf). America is more hostile to Europe's desire to diversify supplies in this way. See 'Iran gas to Europe a problem', Derek Sands, UPI, http://www.upi.com/International_Security/Energy/Analysis/2007/08/13/analysis_iran_gas_to_europe_a_problem/3110/ (13 August 2007) or a worried view from Israel in this article: 'An Iranian pipeline through Europe', Zvi Ba'rel, Haaretz, http://www.haaretz.com/hasen/spages/888446.html (2 August 2007).

3 Azeri reserves include the Shah Deniz gas field in the Caspian Sea, developed by BP and partners. In this case, the pipeline concept has already been brought to reality, with first gas exports to Turkey via the South Caspian Pipeline (SCP) in July 2007, a 700-km, 10 bcm/y pipeline going from Baku to the Turkish border: see BP's press release with more details on both the gas field and the pipeline, 'Shah Deniz gas flows to Turkey': http://www.bp.com/genericarticle.do?categoryId=9006615&contentId=7034735).

4 Widely touted to hold up to 200 bn barrels of oil and gas, the region is now expected to hold no more than 30 bn barrels. See for instance for some more background on regional resources: 'Caspian Oil: a primer', Jérôme Guillet, The Oil Drum http://europe.theoildrum.com/story/2006/11/18/102426/08 (22 November 2006).

5 Turkmenistan is one of the countries with the largest gas reserves in the world (60–100 trillion cubic feet, according to various estimates), but these reserves are only connected to Russia via an old Soviet pipeline and thus have no direct outlet to world markets. Building another export route has thus long been seen as a priority for the country as well as for consumers keen to diversify away from Russian gas. See the US Department of Energy summary at this link: http://www.eia.doe.gov/emeu/cabs/Centasia/NaturalGas.html.

6 As described on the project's website, the much-publicized 30 bcm/y Nabucco would connect the Caspian region to Central Europe via Turkey, for a total length of up to 3,300 km. See http://www.nabucco-pipeline.com/project/project-description-pipeline-route/index.html.

7 The Blue Stream, which includes a Russian portion, an underwater portion below the Back Sea, and a Turkish section, is a 1,200 km pipeline which was completed in late 2002, for a maximum capacity of 16 bcm/y. A map and a construction history can be found on Gazprom's website: http://www.gazprom.com/eng/articles/article8895.shtml

8 The Iran–India gas pipeline, also known as the Peace pipeline (or IPI), is a proposed 2,775 km gas pipeline project to deliver natural gas from Iran to Pakistan and India. It was first mooted in 1989. See the Wikipedia article: http://en.wikipedia.org/wiki/Iran-Pakistan-India_gas_pipeline for some additional background.

9 The Turkmenistan–Afghanistan–Pakistan pipeline (or Trans-Afghanistan Pipeline/TAP) is a 1,600 km gas pipeline that would transport natural gas from Turkmenistan through Afghanistan into Pakistan. It is supported by the Asian Development Bank (see this review by the ADB: 'Gas Pipeline Race', October 2005), at http://www.adb.org/Documents/Periodicals/ADB_Review/2005/vol37-5/gas-pipeline.asp as well as the following Wikipedia article: http://en.wikipedia.org/wiki/Trans-Afghanistan_Pipeline which provides more historical background.

10 See, for instance, 'Ukraine vs Russia: Tales of pipelines and dependence', Jérôme Guillet, European Tribune, http://www.eurotrib.com/story/2005/12/30/173336/17 (30 December 2005).

11 See for instance this violent editorial in the Washington Post about Gerhard Schröder taking the role of Chairman of the Russian–German Nordstream pipeline project company: 'Gerhard Schroeder's Sellout', 13 December 2005 (http://www.washingtonpost.com/wp-dyn/content/article/2005/12/12/AR2005121201060.html/12dec2005), or Tom Lantos calling Schröder a 'prostitute' – see 'US Lawmaker's "Prostitute" Remarks Provokes Germany' Deutsche Welle, 13 June 2007 (http://www.dw-world.de/dw/article/0,2144,2607360,00.html).

12 And indeed, the most recent announcement suggests that the project has been postponed: this article in the *International Herald Tribune* suggests that the project is still beyond the horizon: http://www.iht.com/articles/ap/2008/11/26/business/EU-Serbia-Russia-South-Stream.php.

13 See, for instance, with respect to the TransCaspian pipeline, various announcements: 'Shell Exploration B.V. Signs a Strategic Alliance Agreement with the Government of Turkmenistan http://www.shell.com/home/content/media-en/news_and_library/press_releases/1999/shellexplorationbv_10101756.html (6 August 1999), or the signing of the Istanbul Protocol as flagged by the White House Press Office: 'Caspian Energy Pipelines' http://www.clintonpresidentialcenter.org/legacy/111799-fact-sheet-on-pipeline-a.htm (17 November 1999).

14 Production or transit flows are also often expressed in million cubic feet per day (mcfd). 10 bcm/y is almost equal to 1 mcfd. For comparison purposes, Blue Stream has 16 bcm/y capacity, Langeled 20 bcm/y, Nabucco 30 bcm/y (proposed), the existing Transmed pipeline from Algeria to Italy 24 bcm/y, the under-construction Medgaz pipeline from Algeria to Spain 8 bcm/y to be increased to 16 bcm/y.

15 Wikipedia, 'List of natural gas fields', http://en.wikipedia.org/wiki/List_of_natural_gas_fields (accessed 10 December 2008).

16 In May 2007, Vladimir Putin triumphantly announced an agreement with Turkmenistan and Kazakhstan to build a new pipeline connecting them together, and to be used to send Central Asian gas to Russia. See 'Moscow gets Central Asian agreement on pipeline to Russia', *International Herald Tribune*, http://www.iht.com/articles/2007/05/13/business/pipeline.php (13 May 2007). Given the weak likelihood of this agreement actually leading to actual investment, given that it is not needed, its massive impact in the Western media can only be seen as a highly successful diplomatic move from the perspective of the perceived competition for Turkmen gas.

17 A typical price for the transport of gas will be $2/100 km/tcm, i.e. $50/tcm (1 tcm = 1000 cubic metres) to be compared to a typical price today of $250/tcm for gas delivered to Europe – after several years of increases.

18 The USA nominated a special roving ambassador to the region, whose main role was to ensure that the energy resources of the Caspian were developed in a way that guaranteed to the maximum extent their availability on the open market. Richard Morningstar was the first such Special Envoy in 1998, followed by John Wolf in 1999 and Steve Mann in 2002. See for instance the briefings provided by the White House upon the signing of framework agreements on 17 November 1999: 'Background on U.S. Caspian Energy Policy' (http://www.clintonfoundation.org/legacy/111799-fact-sheet-on-pipeline-b.htm); 'Caspian Energy Pipelines' op. cit. See also 'L'évolution de la politique américaine dans la région de la mer Caspienne', Annie Jafalian, *Annuaire français de relations internationales*, 2003, no. 4, pp. 532–54, (http://www.frstrategie.org/barreCompetences/energiesSecu/afri.pdf).

19 As late as the middle of 1999, BP was openly expressing its hostility to the Baku–Ceyhan route. See for instance 'Caspian Energy Accords: a Job Half Done', Simon Henderson, http://www.washingtoninstitute.org/templateC05.php?CID=1302 (December 1999). There are better, more up-to-date references here.

20 See the links above referring to the framework agreements signed on 17 November 1999 – they covered the TransCapsian project and the BTC equally.

21 'Turkish Power Market: A crossroad of power', *Power Engineering International* (October 2000) http://pepei.pennnet.com/articles/article_display.cfm?article_id=83970. Botas, the Turkish national gas company, was even forecasting demand above 50 bcm/y in 2010 and reaching 80 bcm/y in 2020. See 'How will Turkey's gas deal with Iran affect other initiatives?', Charles Coe, Alexander Oil and Gas, http://www.gasandoil.com/goc/news/ntc03653.htm (8 August 2000).

22 See Charles Coe, op. cit. and 'Turkey Country Brief – Natural Gas', US Department of Energy – Energy Information Agency, http://www.eia.doe.gov/emeu/cabs/Turkey/NaturalGas.html (October 2006 update).
23 'Turkey: Economy Spells Trouble for Russia, Iran, Azerbaijan', Michael Lelyveld; Eurasianet.com (4 April 2002) http://www.eurasianet.org/departments/business/articles/pp040702.shtml.
24 'Russia: Gazprom Secures Pipeline Financing', Michael Lelyveld, RFE/RL (8 April 2000) http://www.b-info.com/tools/miva/newsview.mv?url=news/2000-04/text/apr18a.rfe.
25 Tcm = thousand cubic metres. To use imperial units, it costs roughly 10 cents to transport a MBTU over 100 miles.
26 For instance, the cost of energy is one of the major business inputs in industries like aluminium smelting or petrochemicals, and the cost of procurement of gas may be the main factor in deciding where to locate a factory.
27 That issue is the single biggest difference between oil and gas pipelines – once oil is on the market (i.e. on a boat on open ocean, or in a big, open network), there is no need to care about who will buy it. Oil cargoes will always be sold, at or close to the prevailing market prices, and paid for in hard currency. So finding a committed buyer (unless the end destination of the pipeline is a narrow market with only one or a few buyers) is not usually as burning an issue for oil pipelines. But for gas, it is a fundamental issue. Gas is fully dependent on highly specific infrastructure, and it is not possible to count on selling gas on the open market and be done with it – there is a requirement to know that it will be handled properly, on a consistent basis, for all the relevant volumes pumped day in and day out.
28 'Take or Pay contracts as investment drivers for gas development projects in developing countries: are there co pilots?' http://www.dundee.ac.uk/cepmlp/car/html/CAR10_ARTICLE4.pdf (Kunle Adegun, Centre for Energy, Petroleum and Mineral Law and Policy, 20 February 2007).
29 'Iran–Pakistan–India Gas Pipeline Imperiled', Bill Samii, http://www.rferl.org/featuresarticle/2005/03/d107f257-10bc-4b2a-83e7-bb87c3533ea5.html (18 March 2005).
30 If gas proves to be far more expensive than coal, the government might decide to revert to solid fuel. 'Russian Oil and Gas Report', http://www.uofaweb.ualberta.ca/chinainstitute/nav03.cfm?nav03=61796&nav02=58139&nav01=57272 (University of Alberta, 20 June 2007).

4 The role of Central Asian gas: is it possible to bypass Russia?

1 It should be noted that the crisis did not develop all of a sudden on 1 January 2006; rather, during the second half of 2005, negotiations between Russia and Ukraine deteriorated as Ukraine resisted Russia's drive to receive international market prices for its gas. Disputes over debts and prices occur annually.
2 We define Central Asia as Kazakhstan, Turkmenistan and Uzbekistan.
3 Stephen Bierman, 'Supply may be the least of Nabucco's challenges,' *Nefte Compass*, 21 June 2007. The writer reports that according to Socar, the revised estimates of the reserves of Shah Deniz may be as high as 1 trillion cubic metres, which (even if not that high) could support, in the view of then US Deputy Assistant Secretary of State Matthew Bryza, production of 50 bcm/y from the field by 2015.
4 Anonymous, 'US and Azerbaijan to sign agreement on TransCaspian pipeline – report,' *European Spot Gas Markets*, 15 August 2007.
5 Anonymous, 'Central Asia: Russian, Turkmen, Kazakh leaders agree on Caspian pipeline,' *RFERL*, 12 May 2007. Online. Available at: http://www.rferl.org/featuresarticle/2007/05/09bf853d-a94b-48be-a3dd-937138d0b299.html (accessed 15 May 2007). See also Anonymous, 'Turkmen president gives cautious backing to TransCaspian

pipeline: reports,' *International Herald Tribune*, 26 June 2007. Online. Available at: http://www.iht.com/articles/ap/2007/06/26/asia/AS-GEN-Turkmenistan-US.php (accessed 30 June 2007), and Anonymous, 'The TransCaspian gas pipeline will go ahead,' *The Russian Oil and Gas Report*, 16 May 2007.

6 Vladimir Socor, 'Russia surging farther ahead in race for Central Asian gas,' *Eurasia Daily Monitor*, 16 May 2007. Online. Available at: http://www.jamestown.org/edm/article.php?article_id=2372168 (accessed 16 May 2007).

7 Anonymous, 'Azerbaijan: looming gas power?' *World Gas Intelligence*, 10 October 2007.

8 Anonymous, 'Independent audit shows Turkmen gas field "world class",' *RFERL*, 21 October 2008. Online. Available at: http://www.rferl.org/content/Independent_Audit_Shows_Turkmen_Gas_Field_WorldClass_/1329822.html (accessed 27 October 2008).

9 Anonymous, 'US official upbeat on Nabucco project,' *Assa-Irada – News*, 16 October 2008.

10 The ownership/unbundling provisions of the EU third liberalization package make it a requirement that producer companies have no equity interest.

11 Conversation with industry expert, 12 September 2007.

12 The US seems to have changed its views on Russian participation in Nabucco, driven by Washington's desire to restrict possibilities for Iranian gas exports. See Michael Kuser, 'US flip-flops on Caspian gas,' *Business Week Online*, 26 September 2007, p. 17. Online. Available at: http://www.businessweek.com/globalbiz/content/sep2007/gb20070925_280972.htm?chan=top+news_top+news+index_global+business (accessed 30 September 2007).

13 Anonymous, 'Attacking Iran's route to EU,' *World Gas Intelligence*, 14 March 2007.

14 Jill Junnola, 'Turkey–Greece–Italy gas pipe deal signed,' *International Oil Daily*, 27 July 2007.

15 Online. www.upi.com/Science_News/Resource-Wars/2010/03/11/ENI-calls-for-South-Stream-Nabucco-links/UPI-96591268317232/

16 Online. www.bloomberg.com/apps/news?pid=newsarchive&sid=ae4.eb4lPqjk

17 Tehran has proposed to supply gas to Nabucco and this was backed by Turkey; however, due to the political context this is rejected by both the EU and the US.

18 Iraqi gas would be imported via the Arab gas pipeline from the Ekas field, in Iraqi Kurdistan.

19 Anonymous, 'Russia joins in Central Asian pipeline game,' *Nefte Compass*, 30 August 2007. Online. Available at: http://www.energyintel.com/DocumentDetail.asp?document_id=210949 (accessed 15 September 2007).

20 Anonymous, 'Gazprom's eastern strategy,' *World Gas Intelligence*, 12 September 2007, and Anonymous, 'Turkmenistan grants China gas license,' *Kommersant*, 31 August 2007. Online. Available at: http://www.kommersant.com/p800095/hydrocarbon_sales/ (accessed 15 September 2007).

21 Jing Yang, 'Price terms of Sino-Turkmenistan gas import project finalized – NDRC official,' *Interfax China*, 5 September 2007. Online. Available at: http://www.interfax.cn/displayarticle.asp?aid=27613&slug=CHINA-ENERGY-GAS (accessed 10September 2007). With these prices at the Turkmen border, delivered prices to China's far-distant eastern consuming regions are likely to be far higher, and may make gas uncompetitive with coal.

22 Anonymous, 'Gazprom's pipe politics,' *World Gas Intelligence*, 27 June 2007. These remarks suggest the illogical conclusion that the Europeans plan to achieve less dependence on Russian gas by becoming more dependent on Russian pipelines.

23 Gerhard Schroeder, as German Chancellor, signed the agreement. The German government guaranteed a credit for the project of 1 billion Euros shortly before Schroeder left office. When he was subsequently hired as head of the Nord Stream shareholders' committee, he aroused controversy in Germany, with the media accusing him of pushing the pipeline project not for reasons of greater energy security for Germany, but rather for personal gain.

24 Anonymous, 'Gazprom increases projected cost of Nord Stream project,' *DataMonitor Newswire*, 20 August 2007. See also Vladimir Socor, 'More problems for Russo-German gas pipeline project,' *Eurasia Daily Monitor*, 16 August 2007. Online. Available at: http://www.jamestown.org/edm/article.php?article_id=2372380 (accessed 20 August 2007). Some observers suggest that costs were deliberately underestimated to sell the project and win German financial support.

25 Judy Dempsey, 'Gazprom plans to re-route controversial European pipeline,' *International Herald Tribune*, 23 August 2007. Online. Available at: http://www.iht.com/articles/2007/08/23/news/pipeline.php?page=1 (accessed 29 August 2007), and Tom Kaeckenhoff and Tanya Mosolova, 'Update 1 – Nord Stream to hike cost estimates in early 2008,' Reuters UK, 13 December 2007. Online. Available at: http://uk.reuters.com/article/companyNews/idUKL1368466120071213?symbol=EONG.DE (accessed 19 December 2007).

26 I. Tomberg, 'The North European gas pipeline project in the geopolitical context,' *International Affairs*, Vol. 52, Iss. 2, March 2006, pp. 95–101.

27 Jonathan P. Stern, *The Future of Russian Gas and Gazprom*, Oxford, UK: Oxford University Press, 2005, pp. 74, 103.

28 Some observers believe that the credit crunch together with Gazprom's high level of indebtedness may be responsible for the date slipping from 2010–13 to 2012–15, and that the company has postponed both the start date and the detailed planning and engineering. See Vladimir Socor, 'France ready to leap on the South Stream bandwagon,' *Eurasia Daily Monitor*, 31 October 2008. Online. Available at: http://www.jamestown.org/edm/article.php?article_id=2373495 (accessed 31 October 2008).

29 Anonymous, 'Gazprom's pipe politics,' op. cit., and Judy Dempsey, 'Eni of Italy signs a pipeline deal With Gazprom,' *New York Times*, 23 November 2007, p. C5.

30 Zeyno Baran, 'Security aspects of the South Stream project,' Hudson Institute, October 2008.

31 Originally, South Stream's route called for landfall in Bulgaria, which involved Turkish waters. Recent press reports suggest landfall in Romania may now be under consideration.

32 The word 'encircle' is used frequently in discussions of actual or proposed Russian gas exports to Europe. See 'Gazprom's proposed Nord Stream pipeline to Germany and the Blue Stream pipeline expansion to Turkey will essentially encircle Europe from the north and south and will help diversify Russia's export routes, but will do nothing to help Europe lessen its dependency on Russia', as excerpted form Michael Cohen, 'Russia and the European Union: an outlook for collaboration and competition in European natural gas markets,' *Demokratisatsiya,* Vol. 15, No. 4, Autumn 2007, p. 7; 'Expressing concern at Gazprom's apparent bid to "encircle" Europe by securing upstream assets in Algeria and expanding the Gazprom- and Eni-owned Blue Stream pipeline…, Bryza gave his full support to the Nabucco project,' as excerpted from Anonymous, 'US pitching for Nabucco – minus Iran', *World Gas Intelligence*, 4 April 2007; and Sohbet Karbuz, 'Putin's battle over Caspian energy resources and transport routes', *Alexander's Gas and Oil,* 17 July 2007, where the author states that 'Putin's strategy for Caspian oil and gas involves the following goals:
'– encircle Europe by pipelines from the northwest, the southeast and the middle (for instance, the Nord Stream bypasses Belarus and Poland, the South Stream bypasses Ukraine and Turkey)…'

33 Russia has gas contracts in place with various Western European countries that do not expire until between 2020 and 2040.

34 Anonymous, 'EU leaders put necks further into Russian energy noose', *The New American*, Vol. 23, Iss. 15, 23 July 2007, p. 7.

35 Vladimir Socor, 'South Stream: Gazprom's new mega project,' *Eurasia Daily Monitor*, 25 June 2007. Online. Available at: http://www.jamestown.org/edm/article.php?article_id=2372249 (accessed 26 June 2007).

36 Renata Goldirova, 'EU weakens "Gazprom clause" on foreign energy investors'. *EUobserver.com*, 13 October 2008. Online. Available at: http://euobserver.com/9/26914 (accessed 5 November 2008).
37 Zeyno Baran, op. cit.
38 Ben Hirschler, 'Exporters to discuss gas OPEC in June: Qatar'. Reuters, 25 January 2008. Online. Available at: http://www.reuters.com/article/idUSWLB630420080125 (accessed 31 January 2008).

5 Ups and downs of the Russia–Turkmenistan relationship

1 Hereafter 'prices of gas supply from Turkmenistan to Russia' are referred to as FOB prices on Turkmenistan–Uzbekistan border.
2 Russian daily newspaper *Vremya Novostej*, 21 November 2005. Available at: www.vremya.ru/2005/216/8/139473.html.
3 News Central Asia. Available at: www.newscentralasia.com/modules.php?name=News&file=article&sid=1628.
4 This estimate is based on assessment of the extent of gas supply cuts to the Russian power stations during January–February 2006, as well as some additional information about Gazprom's inability to supply gas to Europe above minimum contract volumes. Real evaluation of the actual deficit of gas is complicated due to non-transparency of data on the actual gas demand applications from consumers – this data has never been publicly released by Gazprom.
5 For more on the role of Itera on the post-Soviet gas market, see Stern, Jonathan, Oxford Institute for Energy Studies, *The Future of Russian Gas and Gazprom*, Oxford: Oxford University Press, 2005, pp.72–7.
6 *Gazprombank Financial Report 2005*, p.31. Available at: www.gazprombank.ru/media/papers/financial_report/2005/gpb_ar2005eng_ias.pdf.

6 Kazakhstan's oil and gas development: views from Russia and Kazakhstan

1 George W. Bush and Nursultan Nazarbayev, 'Joint Statement between the United States of America and the Republic of Kazakhstan,' The White House: President George W. Bush, 29 September 2006. Online. Available at: http://georgewbush-whitehouse.archives.gov/news/releases/2006/09/20060929-1.html (accessed 1 February 2009).
2 Energy Information Agency, 'Kazakhstan: Natural Gas,' *Country Analysis Brief*, 2008. Online. Available at: http://www.eia.doe.gov/emeu/cabs/Kazakhstan/NaturalGas.html (accessed 5 February 2009).
3 Energy Information Agency, 'Kazakhstan: Oil,' *Country Analysis Brief*, 2008. Online. Available at: http://www.eia.doe.gov/emeu/cabs/Kazakhstan/Oil.html (accessed 5 February 2009).
4 Ibid.
5 Chevron was initially delayed in the development of the Tengiz by Russian interests in the project. In fact, it took them almost two years to finally come to an agreement on the field. Chevron currently owns 50 per cent of the venture formed to mine the field TengizChevroil, followed by ExxonMobil with 25 per cent and the state-run KazMunaiGas with 20 per cent. See M. Olcott, *Kazakhstan: Unfulfilled Promise*, Washington DC: Carnegie Endowment for International Peace, 2002, pp.145–6.
6 Olcott, op. cit., pp. 149–51.
7 CPC ownership: Russia, 24 per cent; JSC National Company–KazMunaiGaz, 19 per cent; Chevron Caspian Pipeline Consortium Co.,15 per cent; Lukarco BV, 12.5 per cent; Mobil Caspian Pipeline Co.., 7.5 per cent; Rosneft–Shell Caspian Ventures Ltd, 7.5 per cent; CPC Company, 7 per cent; BG Overseas Holding Limited, 2 per cent; Eni International (NA) NVS.ar.l., 2 per cent; Kazakhstan Pipeline Ventures LLC, 1.75 per cent; Oryx Caspian Pipeline LLC, 1.75 per cent. See 'Caspian Pipeline

Consortium,' Online. Available at: http://www.cpc.ru/portal/alias!press/lang!en-US/tabID!3360/DesktopDefault.aspx (accessed 5 February 2009).

8 Adrian Dellecker, 'Caspian Pipeline Consortium, Bellwether of Russia's Investment Climate?', *Russie.Nei.Visions*, No. 31, June 2008.

9 Gulmira Kenjegalieva, 'Russia gets Kazakh Gas,' *Gazeta.kz*, 9 October 2006. Online. Available at: http://eng.gazeta.kz/art.asp?aid=81754 (accessed 5 June 2009).

10 'Gazprom, Rosneft Sign Caspian Oil Deal,' *Eurasianet.org*, 20 June 2006. Online. Available at: http://www.eurasianet.org/resource/kazakhstan/hypermail/200206/0035.shtml (accessed 5 February 2009).

11 'Kazakhstan Joins BTC Pipeline,' *Bank Information Center*, June 27, 2006. Online. Available at: http://www.bicusa.org/en/Article.2862.aspx (accessed 8 February 2009).

12 Even at the summit meeting there was evidence that Nazarbayev and Putin had come away with differing impressions on what they had discussed regarding CPC expansion. See, Andrey Kolesnikov, 'Consortium without Consensus,' *Kommersant*, 11 May 2007. Online. Available at: http://212.248.33.60/page.asp?id=764456 (accessed 6 February 2009).

13 Nargis Kassenova 'Kazakhstan and the South Caucasus Corridor in the Wake of the Georgia–Russia War', *EU–Central Asia Monitoring*, no. 3, January 2009, p.3.

14 Reuters, 'Putin Pledges to Clarify Mystery of Pipeline Sale,' Forbes.com, 30 October 2008. Online. Available at: http://www.forbes.com/feeds/afx/2008/10/30/afx5627698.html (accessed 9 February 2009).

15 'Caspian Pipeline Consortium 1st Phase expansion due 2011: Chevron', Platts, 2 February 2009. Available at: http://platts.com/Oil/News/836117.xml.

16 Caspian Pipeline Consortium press release, 14 January 2009. Available at: http://www.cpc.ru/desktopdefault.aspx?alias=press&lang=En-US&tabid=3552.

17 I. Gorst and G. Dyer, 'Pipeline Brings Asian Gas to China', *Financial Times*, 14 December 2009.

18 Dinara Sarsenova, 'Program to Develop Kazakh Sector of the Caspian Shelf Presented,' *The Times of Central Asia*, 16 July 2003.

19 Information available online at 'KazMunayGas Group of Companies,' *KazMunaiGaz,* 2009. Online. Available at: http://www.kmg.kz/page.php?page_id=1020&lang=2&parent_id=334 (accessed 5 February 2009).

20 Lukoil assets in Kazakhstan: Karachaganak, 15 per cent; Kumkol, 50 per cent; Tengiz, 2.7 per cent; Karakuduk, 50 per cent; North Buzachi, 25 per cent; Arman, 25 per cent; Kazakhoil Aktobe, 25 per cent; Tyub Karagan, 50 per cent; Atashsky, 50 per cent; South Zhambai and South Zaburnuye, 12.5 per cent. *Lukoil Factbook 2008.* Online. Available at: http://www.lukoil.com/materials/doc/DataBook/DBP/2008/FactBook/facts2008eng.pdf (accessed 10 February 2009).

21 BBC Monitoring, 'Russia, Iran, Kazakhstan, Azerbaijan, Turkmenistan still Negotiating Caspian', 14 April 2004; Gazeta.kg, 'Legal Status of the Caspian Sea to be Discussed in Baku', 15 March 2004; Pravdaonline, 'Five Countries to Discuss Legal Status of Caspian Sea in Tehran', 4 December 2003.

22 Lukoil is currently participating in three Kazakh oil and gas projects. Lukoil holds a 15 per cent share in the Karachaganak Integrated Organization, which is developing the Karachaganak field, and a 50 per cent share in the joint venture Turgai Petroleum, which is developing the Kumkol field in the Kyzyl-Orda region of central Kazakhstan. Lukoil also has a share in the Russia–US joint venture Lukarco, which in turn owns a 5 per cent share in the Tengiz field. Lukoil also owns a 12.5 per cent share in the Caspian Pipeline Consortium, which exports oil from Tengiz to Novorossiisk. Interfax Central Asia, 10 February 2003.

23 Kazakhstan and Russia signed a protocol in May 2002, which declares Kazakh jurisdiction over the Kurmangazy oil field and Russian jurisdiction over the Tsentralnoye and Khvalynskoye fields in the north of the Caspian. Interfax Central Asia, 2 October 2002.

24 Reuters, 'Kazakhs Turn up Heat on Rosneft', 14 April 2004.

25 Interfax, *Central Asia and Caucasus Business Report*, 22–28 March 2004.
26 Bloomberg, 'Eni's Kashagan Project Delayed', 4 October 2006.
27 Sergei Blagov, 'New Deal will Process Kazakh Gas at Nearby Russian Facility', *Eurasia Daily Monitor* 3, no. 185, 6 October 2006. Online. Available at: http://www.jamestown.org/edm/article.php?article_id=2371524 (accessed 20 February 2009).
28 This, of course, is a highly complicated thing to negotiate, as it involves the Karachaganak joint venture partners, the Kazakh government, the Russian gas industry as well as the Russian government, which stands behind them. Lukoil Overseas started talks with Gazprom in 2003 over increasing the price of gas produced in Karachaganak field. Gazprom has been paying only about $14.5 per tcm, one-third of what it pays for Turkmen and Uzbek gas. Interfax Central Asia, 28 August 2003.
29 Vladimir Socor, 'Trans-Caspian Oil Pipeline Planned in Kazakhstan,' *Eurasia Daily Monitor* 2, no. 95, 16 May 2005.

7 Uzbekistan: central Asian key

1 Online. Available at: www.gov.uz/ru/content.scm?contentId=30549 (accessed 10 December 2008); 'Russian PM to Visit Uzbekistan for Energy Talks', *RIA Novosti,* 1 November 2007. Online. Available at: http://en.rian.ru/business/20071101/86193441.html (accessed 10 December 2008).
2 R. Cutler, 'The Caspian Energy Conundrum', *Journal of International Affairs*, 56, 2. Spring 2003.
3 Interview with European institution official, London, October 2007.
4 There are also significant Uzbek minorities in the neighbouring states in the region, particularly in Kyrgyzstan and Tajikistan, 13.8 per cent and 15.3 per cent of the respective populations.
5 R. Pomfret, 'Resource Abundance, Governance and Economic Performance in Turkmenistan and Uzbekistan', in R. Auty and I de Soysa (eds) *Energy, Wealth and Governance in the Caucasus and Central Asia. Lessons Not Learned*, London: Routledge, 2006, pp. 80–2, 89.
6 'In-Depth Report on the Investment Climate and Market Structure in the Energy Sector of Uzbekistan' October 2005. Online. Available at: www.encharter.org (accessed 10 December 2008), p. 6. This brief overview of the domestic situation draws on discussions with experts on Central Asia in London in October 2007 and particularly on B. Pannier, 'Uzbekistan', in *Nations in Transit*, Freedom House, 2007, pp. 755–60.
7 Equally, the government has withheld gas supplies from domestic customers to allow it to be exported – in October 2006 the government announced that 60 per cent of homes in Andijan as well as most of the city's schools and hospitals would receive no gas that winter. Gas shortages were a reason contributing to the protests in May 2005 and they have begun to appear again, according to the International Crisis Group, in *Central Asia's Energy Risks*, Asia Report, 133, 24 May 2007, pp. 31–2. Online. Available at: www.icg.org (accessed 10 December 2008).
8 *Central Asia's Energy Risks*, pp v, 5, 20, 24.
9 For an assessment of the range of governmental responses to this threat, see V. Naumkin, 'Uzbekistan's State Building Fatigue', *The Washington Quarterly*, 29, 3, Summer 2006.
10 EurAsEC is a collective economic organization comprising Belarus, Kazakhstan, Kyrgyzstan, Russia and Tajikistan.
11 Brill Olcott, M. ''Druzhba narodov' v mire energetiki', *Pro & Contra*, 2–3 (32), Summer 2006.
12 'Uzbekistan: Tashkent Strives to Diversify its Trade Partners'. Online Available at: www.eurasianet.org/departments/insight/articles/eav031908.shtml (accessed 10 December 2008).
13 IEA document, pp. 275, 281; Pomfret, p.78.

14 IEA document, p.287.
15 *BP Statistical Review of World Energy*, June 2008. Online. Available at: www.bp.com (accessed 9 December 2008). Despite the lack of clarity regarding Uzbekistan's figures, BP's *Statistical Review* is considered to provide a reliable 'ball-park figure'.
16 Another estimate is slightly higher, putting production at approximately 59–62 bcm annually between 2004 and 2006. A. Ilkhamov, 'Russia Lures Uzbekistan as its Strategic Satellite in Central Asia', *Russian Analytical Digest*, 29/07, p. 9. This brings it closer to Turkmenistan's production level.
17 *Central Asia's Energy Risks*, p.16.
18 'In-Depth Report on the Investment Climate', October 2005. Online. Available at: www.energycharter.org (accessed 10 December 2008).
19 *Central Asia's Energy Risks*, p. 16; V. Paramonov and A. Strokov, *Structural Independence of Russia and Central Asia in the Oil and Gas Sectors*, Conflict Studies Research Centre Paper, 7. 16 E. Swindon: Defence Academy of the UK, June 2007, p. 4; *BP Statistical Review*.
20 M. Kenisarin, 'Addressing the Problem of Uzbekistan's New Energy Policy Formulation', *Central Asia and the Caucasus*, 3, 39, 2006. 141 and *passim*.
21 Barry, pp. 48–9.
22 Ibid, pp. 52–3.
23 'In-Depth Report on the Investment Climate', pp. 3–4.
24 IEA document, pp. 270, 287.
25 Barry, pp. 49–50; 'In-Depth Report on the Investment Climate', p. 3.
26 The extent of this expansion remains somewhat unclear: one analyst suggests that the Central Asia–Centre III line will be expanded from 45 bcm to 65 bcm. Interview with Dr N Mehdiyeva, gas consultant and Russia/CIS specialist at Poyry Energy Consulting, London, autumn 2007. Others suggest that Gazprom seeks to increase the capacity to 80 bcm, which in combination with the 19-bcm capacity of the Bukhara–Khiva line would mean an overall capacity of nearly 100 bcm. Paramonov and Strokov, p.4. Where all this gas will come from is another question, but as Mehdiyeva argues, it seems likely that this pipeline network will compete for the region's gas, drawing it away from the Nabucco line so favoured by the EU.
27 The ICG notes that the funding for this remains unidentified. *Central Asia's Energy Risks*, p.18. It might also be asked where the additional gas will come from.
28 Paramonov and Strokov, p. 4.
29 'KazTransGas starts negotiations with Uzbekistan to buy additional volume of gas', *Uzreport*, 10 September 2007. Online. Available at: http://business.uzreport.com/mir.cgi?lan=e&id=37037 (accessed 10 December 2008).
30 Paramonov & Strokov, p. 6; Cutler, p. 93.
31 Russian policy towards the area has passed through three phases. First, in the early 1990s, a complete refusal to maintain links with the region, followed by a reactive policy of random responses to local developments in the late 1990s, to a policy of more coherence and a more pragmatic 'voluntarism' under then-president Putin which resulted in a gradual improvement in Russian policy. For Russian views of this evolution, see S. Lounev, 'Central Asia and the Southern Caucasus: Geopolitical Value for Russia', *Central Asia and the Caucasus*, 3, 39, 2006 (especially pp 22–5); I. Iniutin, 'Russian Strategy in Central Asia', *Central Asia and the Caucasus*, 2, 38, 2006.
32 For instance, the Moscow Telephone Network has become involved in mobile communications in Uzbekistan, acquiring a 74 per cent share in Uzdurobat. Other major Russian companies investing in Uzbekistan are Vimm-Bill-Dann which is investing $7.3 million in modernisation of the Tashkent Dairy Combine, Transaero and the Cherkiziv Agro-Industrial complex. I. Iniutin, 'Russian Strategy in Central Asia', *Central Asia and the Caucasus*, 2. 38, 2006, p. 32; M. Laumulin, 'Gazprom as a Transnational Corporation and Central Asia. Part I', *Central Asia and the Caucasus*, 5, 41, 2006, p. 44.

33 Iniutin, p. 32.
34 Online. Available at: www.gov.uz/ru/content.scm?contentId=30549 (accessed 10 December 2008). This reflects a significant increase: in 2005, there were 360. Iniutin, p. 33.
35 S. Luzyanin, *Vostochnaya politika Vladimira Putina. Vozvrasheniye rossii na 'Bolshoi Vostok', 2004–2008*. Moscow: Vostok-Zapad, 2007, pp 212-3; Laumulin, p. 44.
36 In August 2007, LUKoil completed the construction of a section of trunk gas pipeline. Other estimates put the reserves much higher, at some 329 bcm, of which industrial production is due to begin in the last quarter of 2007. The maximum annual production is estimated to be 10 bcm, with a total accumulated production being 207 bcm. 'LUKoil completes construction of gas pipe section in Uzbekistan', 25 August 2007. Online. at http://bbjonline.hu/index.php?id=30570 (accessed 10 December 2008).
37 I. Tomberg, 'Central Asia and the Caspian: New Stage in the Great Energy Game', *Central Asia and the Caucasus*, 5, 41, 2006, p. 31.
38 *Central Asia's Energy Risks,* p. 16.
39 Interfax, 12 October 2007.
40 This was taken over by LUKoil in 2008, giving it access to seven fields with some estimated 100 bcm of reserves.
41 President Karimov signed the summit document in May 2007. *Central Asia's Energy Risks*, p. 17.
42 Iniutin, p. 36.
43 Pomfret, p. 84.
44 Paramonov & Strokov, p. 3.
45 Central Asia's Energy Risks, p. 16.
46 'Russia's Central Asia Energy Strategy Experiences a Few Set Backs', *Eurasianet*, 22 July 2007. Online. Available at: http://www.eurasianet.org/departments/insight/articles/eav051107_.shtml (accessed 10 December 2008).
47 *Central Asia's Energy Risks*, p. 16.
48 'Russia's Central Asia Energy Strategy', op. cit.
49 'Uzbekistan: Stalemate Forces Alignment with Russia', *Oxford Analytica Daily Brief*, 22 May 2006.
50 'Uzbekistan continues to drift away from Russia, but not necessarily toward the West', *Eurasianet*, 15 July 2008. Online. Available at: www.eurasianet.org/departments/insight/articles/eav071508f.shtml (accessed 10 December 2008).
51 'Russia's Central Asia Energy Strategy'.
52 C. Zeigler, 'China, Russia and Energy in the Caucasus and Central Asia Region and East Asia', in Auty and de Soysa (eds), pp. 240, 243.
53 'Uzbekistan: Tashkent is up to its old geopolitical tricks', *Eurasianet*, 20 May 2008. Online. Available at: www.eurasianet.org/departments/insight/articles/eav052008.shtml
54 Ilkhamov, p. 10.
55 This price increase to Gazprom may well be passed on to Gazprom's customers in Europe.
56 Interview with Western expert, London, October 2007.
57 Brill Olcott, p. 35.
58 Interview with European expert on Central Asia, London, June 2007.

8 The uncertain trajectory of Russia–Azerbaijan relations in the multiple-pipeline era

1 Sound analysis of this phenomenon can be found in John Roberts, 'Pipeline politics', pp 77–89 in Shirin Akiner (ed.), *The Caspian: Politics, Energy and Security*, London and New York, RoutledgeCurzon, 2004. Willy Olsen's chapter in that volume 'The role of oil in the development of Azerbaijan', pp. 126–46 provides an insightful assessment of the oil–politics interplay in Azerbaijan. Another noteworthy analysis

is Jan H. Kaliciki & Jonathan Elkind, 'Eurasian transportation futures', pp. 149–174 in Jan H. Kaliciki and David L. Goldwyn (eds), *Energy and Security: Towards a New Foreign Policy Strategy*, Washington, Woodrow Wilson Center Press, 2005.

2 A concise evaluation of that decision is in Douglas W. Blum, 'Why did Lukoil really pull out of the Azeri–Chirac–Guneshli oilfield?', *PONARS Memo* 286, Washington: CSIS, December 2003.

3 One example of these assessment is Yuri Aleksandrov, 'Baku–Ceyhan: Where is the oil?', *Politcom.ru*, 12 September 2002 (http://www.politcom.ru/2002/p_neft11.php); my analysis is in Pavel Baev, 'Russia's happiness in multiple pipelines', *CACI Analyst*, 16 June 2004 (http://www.cacianalyst.org/view_article.php?articleid=2452).

4 In February 2006, Viktor Khristenko, Russian Industry and Energy Minister, noted that in 2005 the amount of oil exported by Azerbaijan through this pipeline increased to 4.1 million tons compared with the average of 2.7 million tons in the previous years, and suggested a new deal for expanding this route; see 'Moscow proposes oil transportation deal to Azerbaijan', RIA–*Novosti*, 22 February 2006 (http://en.rian.ru/world/20060222/43702295.html).

5 An extensive if slightly exaggerated evaluation of the impact of this event can be found in S. Frederick Starr and Svante E. Cornell, *The Baku–Tbilisi–Ceyhan Pipeline: Oil Window to the West*, Washington, Johns Hopkins University, May 2005 (http://www.silkroadstudies.org/BTC.htm).

6 President Ilham Aliyev opening the 14th International Oil and Gas Exhibition in Baku in June 2007, announced that the estimates of oil reserves from the mid–1990s of about 500 million tons were increased to 1–1.35 billion tons, and for natural gas the estimates were increased from 600 bcm to 1200–1330 bcm; see Vasily Kashin and Aleksandr Tutushkin, 'Russian oil will see competition', *Vedomosti*, 8 June 2007.

7 Huge enthusiasm among Russian experts stood in sharp contrast with many critical comments in the US; see, Ali Aliyev, 'From Bulgaria to Greece', *Expert*, 15 March 2007; Vlad Socor, 'Agreement signed on Trans-Balkan pipeline, rival to Trans-Caspian project', *Eurasia Daily Monitor*, 16 March 2007; Stephen Blank, 'The Burgas–Alexandroupolis pipeline and its implications', *CACI Analyst*, 5 April 2007 (http://www.cacianalyst.org/?q=node/4584).

8 One solid upbeat analysis is Ivan Rubanov, 'Not by the right of force', *Expert*, 21 May 2007; a more cautious view is Vladimir Milov, 'The Great Uncertainty: Russian–Central Asian Energy Relations', presentation at the Carnegie Endowment, Washington, 13 June 2007 (http://www.carnegieendowment.org/events/index.cfm?fa=eventDetail&id=1013&&prog=zru).

9 See Ali Aliev 'The inertia of expensive oil', *Expert*, 18 December 2008. See also A. Dellecker, 'Caspian Pipeline Consortium, bellwether of Russia's investment climate', *Russie.Nei.Visions*, No. 31, June 2008.

10 On the Tengiz–Baku tanker traffic, see John C.K. Daly, 'With no pipelines in sight, tankers roam the Caspian', *CACI Analyst*, 16 May 2007 (http://www.cacianalyst.org/?q=node/4615); Sohbet Mamedov, 'Kazakh oil is ready to come to the world market', *Nezavisimaya gazeta*, 17 December 2008.

11 As of mid-2010, Bulgaria is still wavering about this project; see Ali Aliev, 'The pipe business', *Expert*, 11 June 2010.

12 For a balanced view on the fallout of that mini-crisis, see Jerome Guillet, 'Gazprom as a predictable partner: Another reading of the Russian–Ukrainian and Russian–Belarussian energy crises', *Russie.Nei.Visions*, No. 18, March 2007.

13 In 2006, Azerbaijan imported some 4.5 bcm of gas from Russia, but in 2007, it increased production to 17 bcm (of which 11 bcm was oil-associated gas and 3.25 bcm came from Shah Deniz) and exported 5 bcm while fully covering its own needs estimated at 11–12 bcm; in 2008, production is expected to increase to 27.5 bcm (of which 8 bcm at Shah Deniz). See Vladimir Mishin, 'Azerbaijan's hydrocarbon resources are still uncertain', *Nezavisimaya gazeta*, 10 June 2008.

14 Visiting Washington in March 2007, Azerbaijan's Foreign Minister Elmar Mammadyarov signed a Memorandum of Understanding on Energy Security Cooperation; see Yuri Simonyan, 'Not a friendly advice', *Nezavisimaya gazeta*, 22 March 2007.

15 The prospects for Nabucco have not been improved at all by Berdymuhammedov's commitment for exporting gas to China; see on that Kirill Martynov, 'Asian prisoner', *Kommersant*, 24 June 2006, and Pavel Burmistrov and Igor Grebtsov, 'Gas dances around the Caspian', *Expert*, 11 September 2008; and on manoeuvring for rescuing Nabucco, Vladimir Socor, 'Nabucco gas project retains political and business momentum', *Eurasia Daily Monitor*, 5 February 2010.

16 See Sohbet Mamedov, 'Baku cultivates the European direction', *Nezavisimaya gazeta*, 22 June 2010.

17 On Putin's courting the Aliyev family, see Sergei Pravosudov, 'Aliyev turns to the northern neighbour', *Nezavisimaya gazeta*, 13 January 2001. The proceedings of the Key West talks still remain confidential; insightful reflections can be found in Thomas De Waal, *The Black Garden: Armenia and Azerbaijan Through Peace and War*, New York and London, New York University Press, 2003, pp. 279–80. Heydar Aliyev died in late 2003, and his son carefully maintains his cult of a 'great statesman'; see Thomas De Waal, 'Heidar Aliyev: A political colossus', *IWPR Caucasus Reporting Service*, 18 December 2003 (http://www.iwpr.net/?p=crs&s=f&o=158717&apc_state=henicrs2003).

18 The conclusion that the elections 'failed to meet OSCE commitments and other international standards for democratic elections' supported by a long list of violations was one of the strongest in the records of OSCE election monitoring; see OSCE/ODIHR Election Observation Mission Final Report, 12 November 2003 (available at http://www.osce.org/documents/odihr/2003/11/1151_en.pdf).

19 Interview with Vasily Istratov, Russian Ambassador to Azerbaijan, September 2006; see also Sergei Blagov, 'Russia moves to repair ties with Azerbaijan', *Eurasia Insight*, 2 April 2007 (http://eurasianet.org/departments/insight/articles/eav040207a.shtml).

20 See Vladimir Belousov, 'The right for the truth', *Rossiiskaya gazeta*, 9 November 2005; Andrei Gromov, 'There will be no revolution', *Expert*, 24 October 2005.

21 On the mediation efforts, see Svetlana Gamova, Sohbet Mamedov, 'Moscow has come to the Karabakh dead-end', *Nezavisimaya gazeta*, 22 June 2009; on a possible peace operation, see Sergei Markedonov, 'Whether Russian peacekeepers in Nagorno Karabakh', *Polit.ru*, 2 June 2006 (http://www.polit.ru/analytics/2006/06/02/nkr.html).

22 On the reorganization of Azerbaijan's armed forces by NATO standards, see Sohbet Mamedov, 'The Caucasian North-Atlantic district', *Nezavisimaya gazeta*, 15 December 2006; and also 'Azerbaijan: Defence Sector Management and Reform', *Europe Briefing* 50, Brussels: International Crisis Group, 29 October 2008.

23 On the scant results of State Secretary Clinton's visit to the area in summer 2010, see Aleksandr Karavaev, 'Reset for the South Caucasus', *Nezavisimaya gazeta*, 7 July 2010.

24 See Maria Rozhkova, 'Rusal will come to Azerbaijan', *Vedomosti*, 26 February 2006; Andrei Gromov, 'Oil modernization', *Expert*, 25 December 2006; Mina Muradova and Khazri Bakinsky, 'Building bridges for President Aliyev's re-election?', *Eurasia Insight*, 30 May 2007 (http://eurasianet.org/departments/insight/articles/eav053007.shtml).

25 The IMF warned in mid-2010 that the oil production in Azerbaijan was set to peak so growth could be generated only by other sectors; the report 'Regional Economic Outlook: The Caucasus and Central Asia' is available at (http://www.imf.org/external/pubs/ft/reo/2010/mcd/eng/mreo0510cca.pdf). See also Fariz Ismailzade, 'Azerbaijan seeks new partners to diversify its economy', *Eurasia Daily Monitor*, 29 March 2010.

26 The trigger for these legislative measures was the anti-Chechen pogrom in Kondopoga, Karelia in August 2006; see 'Russian Babylon', *Expert*, 29 January 2007; on the impact of the recession, see Christian Lowe, 'Financial crisis hits migrant workers in Russia', *New York Times*, 14 January 2009.

27 Solid analysis of the maritime borders question can be found in Ali Granmayeh, 'Legal history of the Caspian Sea', pp 17–47 in Shirin Akiner (ed.), *The Caspian: Politics, Energy and Security* (note 1). On the Ashgabat summit, see Lyudmila Romanova, 'Five president, six opinions and one scandal', *Nezavisimaya gazeta*, 25 April 2002.

28 One good description of those exercises is Sergei Sokut, 'The military return to the Caspian', *Nezavisimoe voennoe obozrenie*, 16 August 2002. I first made the argument on Moscow's interest in preserving the maritime dispute in Pavel Baev, 'Gunboats in the great anti-terrorist game', *CACI Analyst*, 28 August 2002 (http://cacianalyst.org/view_article.php?articleid=17&SMSESSION=NO).

29 As one British military expert argued: 'Whether we in the West like to admit it or not, Russia's military presence in the Caspian Sea not only ensures that the Russian interests in the region are safeguarded, but also guarantees that the other littoral states derive security benefits from having a creditable military force in the region, large enough to act as a deterrent, which they do not have to pay for and which ensures a degree of stability in the region as a whole.' See Steven J. Main, 'The Bear, the Peacock, the Sturgeon, the Eagle and the Black Black Oil: Contemporary Regional Power Politics in the Caspian Sea', CSRC Defence Academy of the UK, December 2005 (http://www.defac.ac.uk/colleges/csrc/document-listings/caucasus/). On the recent developments in Russian military capabilities see Vladimir Muhin, 'Military–Caspian metamorphosis', *Nezavisimoe voennoe obozrenie*, 4 June 2009.

30 See Andrei Kolesnikov, 'Disagreements are thrown into the sea', *Kommersant*, 17 October 2007.

31 My reflections on this pledge are in Pavel Baev, 'Putin protects Islam and praises democracy in Grozny', *Eurasia Daily Monitor*, 14 December 2005.

32 See Viktor Myasnikov, 'Terrorist war on the basis of cost-efficiency', *Nezavisimoe voennoe obozrenie*, 30 July 2010.

33 An elaborate assessment of the Islamic factor can be found in Svante Cornell, 'The Politicization of Islam in Azerbaijan', *Silk Road Paper*, Johns Hopkins University–SAIS, October 2006 (available at http://www.silkroadstudies.org/new/docs/Silkroadpapers/0610Azer.pdf). For an insightful view on the situation in Lenkoran, see Idrak Abbasov, 'Talysh of Azerbaijan look South and North', IWPR *Caucasus Reporting Service*, 12 July 2007 (http://iwpr.net/?p=crs&s=f&o=337022&apc_state=henpcrs).

34 See on that Fariz Ismailzade, 'Azerbaijan moves to shape image of Islamic world', *Eurasia Daily Monitor*, 30 April 2007.

35 One informed analysis is Anvar Valiyev, 'Al Qaeda in Azerbaijan: Myths and Realities', *Terrorism Monitor*, 18 May 2006 (http://jamestown.org/terrorism/news/article.php?articleid=2370001).

36 This message is reinforced by the significant increase of migration from Iran, particularly of relatively wealthy Azeri families who seek to establish 'safe heavens' in Baku; see Kemal Ali, 'Iranian Azerbaijanis move North', IWPR *Caucasian Reporting Service*, 13 April 2006 (http://www.iwpr.net/?p=crs&s=f&o=261051&apc_state=henicrs2006).

37 On the issue of bases during the visit of US Defense Minister Robert Gates to Azerbaijan in mid-2010, see Sohbet Mamedov, 'Gates has reassured Aliyev', *Nezavisimaya gazeta*, 8 June 2010.

38 This deal was much discussed in mid-2010; see Vladimir Muhin, 'New geopolitical "Favorit"', *Nezavisimaya gazeta*, 2 August 2010. Moscow officially cancelled the contract on delivering the S–300 system to Iran; see Henry Meyer, 'Iran demands Russian S–300 to make it "invincible" to Israel', *Business Week*, 13 June 2010.

39 See Taleh Ziyadov, 'Putin's Gabala offer gets mixed reaction in Azerbaijan', *Eurasia Daily Monitor*, 22 June 2007; Richard Weitz, 'The Gabala gambit and Azerbaijan's geopolitics', *CACI Analyst*, 13 June 2007 (http://www.cacianalyst.org/?q=node/4633).

40 See Andrei Terehov, 'US open door to the anti-missile system for Russia', *Nezavisimaya gazeta*, 31 May 2010.

41 Informed analysis of this step is in Alexander Pikayev, 'Why Russia supported sanctions against Iran', *CNS Report*, June 2010 (http://cns.miis.edu/stories/100623_russia_iran_pikayev.htm).
42 See Fariz Ismailzade, Obama's missile defense decision renews discussion on Gabala in Azerbaijan', *Eurasia Daily Monitor*, 29 September 2009.
43 One good analysis of this evolution is Richard Sakwa, 'Russia and Turkey: Rethinking Europe to Contest Outsider Status', *Russie.Nei.Visions* No. 51, IFRI, May 2010.
44 As one keen analyst argued, despite the divides over oil, 'agreements on gas served to bind Turkey and Russia together'. See Fiona Hill, 'Seismic shifts in Eurasia: The changing relationship between Turkey and Russia and its implication for the South Caucasus', *Journal of Southeast European and Black Sea Studies*, vol. 3, no. 3, September 2003, pp 55–75, especially p. 63.
45 On the Putin–Erdogan dialogue, see Igor Torbakov, 'Russia and Turkey forge new ties on security, trade', *Eurasia Insight*, 8 August 2005 (http://eurasianet.org/departments/insight/articles/eav080805.shtml).
46 See Mamuka Tsereteli, 'The Blue Stream pipeline and the geopolitics of natural gas in Eurasia', *CACI Analyst*, 30 November 2005 (http://www.cacianalyst.org/view_article.php?articleid=3848&SMSESSION=NO).
47 The memorandum on energy partnership between the EU and Azerbaijan signed on that occasion was certainly not a binding contract but nevertheless an important commitment to open an alternative route for Caspian gas to Europe; see Vlad Socor, 'Azerbaijan at the forefront of Caspian basin energy initiatives', *Eurasia Daily Monitor*, 22 March 2007.
48 Back in early 2002, General Tuncer Kilinc, secretary of the powerful National Security Council, triggered a political earthquake in Turkey with the suggestion that Ankara should start looking eastward for new allies, singling out Russia as the most promising partner; see Igor Torbakov, 'Eurasian idea could bring together erstwhile enemies Turkey and Russia', *Eurasia Insight*, 18 March 2002 (http://www.eurasianet.org/departments/insight/articles/eav031802.shtml).
49 Italian ENI is the main partner in the latter project, as it was in Blue Stream; see Evgeny Utkin, 'Settling the South Stream', *Expert*, 12 April 2010.
50 Erdogan also made a short visit to Moscow in mid-August 2008 on the background of the Russian–Georgian war and received reassurances that the energy infrastructure would not be targeted; my comment is Pavel Baev, 'Russia values oil more than war', *Moscow Times*, 22 August 2008.
51 My more elaborate evaluation of this re-configuration is in Pavel Baev, 'Energy intrigues on the EU's southern flank: Applying game theory', *Problems of Post-Communism*, vol. 57, no. 3, May/June 2010, pp 11–22.
52 See Joshua Kucera, 'Washington is getting more involved in reconciliation process', *Eurasianet*, 18 February 2010 (http://www.eurasianet.org/departments/insight/articles/eav021910a.shtml).
53 See Sohbet Mamedov and Svetlana Gamova, 'Baku yet again talks about war', *Nezavisimaya gazeta*, 24 June 2010.
54 Moscow found the Platform interesting because it did not include the US and Iran, but Medvedev nevertheless clarified that support for that initiative would not limit Russia's freedom of political manoeuvre in the region; see Vladimir Kazimirov, 'To use either force or reason', *Nezavisimaya gazeta*, 25 March 2010.
55 One noteworthy collection of analyses is Svante E. Cornell and S. Frederick Starr (eds), *The Guns of August 2008: Russia's War in Georgia*. New York and London: M.E. Sharpe, 2009.
56 See Dmitri Butrin, 'Ilham Aliyev changes the routes', *Kommersant*, 9 November 2006; Aidyn Gadzhiev, 'What Aliyev brings to Georgia', RIA–*Novosti*, 6 February 2007 (http://www.rian.ru/analytics/20070206/60302991.html).

57 See Fariz Ismailzade, 'The Georgian–Russian conflict through the eyes of Baku', *Eurasia Daily Monitor*, 2008; Sabrina Tavernise, 'A Northern neighbour growls, and Azerbaijan reassesses its options', *New York Times*, 22 October 2008.
58 See Judy Dempsey, 'Georgia crisis could thwart EU project to bypass Russia for natural gas', *International Herald Tribune*, 28 August 2008.
59 Sharp argument in favour can be found in Katinka Barysch, 'Should the Nabucco pipeline project be shelved?', *CER Policy Brief*, May 2010 (http://www.cer.org.uk/pdf/pb_Nabucco_5may10.pdf).
60 The ruling of the International Court of Justice on the Kosovo case in July 2010 was taken in Armenia for an important precedent; see Thomas de Waal, 'A forever smoldering conflict in the Caucasus', *Moscow Times*, 5 July 2010.
61 See Igor Torbakov, 'The Georgia crisis and Russia–Turkey relations', *Jamestown Foundation Report*, Washington DC, 26 November 2008 (http://www.jamestown.org/programs/recentreports).

9 Russia, Gazprom and the CAC: interests and relations

1 E. Pirani (ed.), Russian and CIS Gas Markets and their Impact on Europe. Oxford University Press, 2009.
2 The more time passes after the dissolution of the USSR, the less clear for outsiders the original design of the pipeline system, its convenience and capability to ensure delivery on contracts.
3 http://www.rambler.ru/news/economy/10344/12617370.html (accessed 22 April 2008).
4 http://www.kaztransgas.kz/article.php?article_id=669 (accessed September 2007).
5 See 'Oil and Gas of Turkmenistan', November 2007, p.9.
6 See Article 7 Transit (Final Act of the European Energy Charter Conference), Declarations, no. 3. with respect to Article 7, p. 31.
(1) Each Contracting Party shall take the necessary measures to facilitate the Transit of Energy Materials and Products consistent with the principle of freedom of transit and without distinction as to the origin, destination or ownership of such Energy Materials and Products or discrimination as to pricing on the basis of such distinctions, and without imposing any unreasonable delays, restrictions or charges.
7 A high-level meeting on Nabucco in Budapest on 27 January 2009 was devoted to seeking solutions for the pipeline. Political support for Nabucco was very strong, but financing of $8 billion for the project was not identified. See: http://www.regnum.ru/news/1117000.html (accessed 28 January 2009).
8 L. Grigoriev and M. Salikhov, *GUUAM 15 Years Later*, Moscow, Regnum, 2007.
9 Presentation of Didier Houssin (IEA), 'The Challenges of the Changing International Energy Markets', 8 November 2007, Vienna.
10 The GECF is more akin to the IEA than OPEC.
11 This set of Gazprom's objectives is the author's responsibility, but the author would argue that it covers most actual events and decisions with and of Gazprom.
12 Political instability in Ukraine is obvious and for Russian gas interests there is no simple way to avoid conflicts.
13 President Vladimir Putin at the Kremlin reception for the diplomatic corpus mentioned: '…for us the CIS space – not the chessboard for geopolitical games', cited in *Kommersant*, 29 November 2007.
14 Глаголев А. И. (ООО «ИРЦ Газпром») Орлов Ю.Н. (ИПМ им. М. В. Келдыша РАН) «Газотранспортные проекты ОАО «Газпром» в Средней Азии», стр.40, Москва, 2004.
15 *Kazpravda*, 29 October 2007.

10 After the war: the Southern Corridor

1 The corridor goes by various names. The one I prefer is the Southern Caucasus Energy Corridor, since this covers both oil and gas pipelines and specifies the geographical component linking Caspian producers with European markets. The EU, in its Second Strategic Energy Review, refers to a 'southern gas corridor' to be developed 'for the supply of gas from Caspian and Middle Eastern sources'. The Southern Caucasus Energy Corridor constitutes one element of such a system: other elements might include connections from Iran, Iraq and Egypt, as well as the transit systems through Turkey and the Balkans. Alternative terms for the corridor include: The Southern Caucasus Corridor, The East–West Energy Corridor and The Fourth European Gas Corridor. The latter refers to three other 'corridors' serving Europe: the pipelines from Russia; the pipelines from North Africa; and the pipelines from Norway. In practice, Europe also enjoys the fruits of two other major gas sources: the North Sea and liquefied natural gas (LNG) imports.

2 An EU Energy Security and Solidarity Action Plan; Communication from the Commission to the European Parliament, the Council, the European Economic and Social Committee and the Committee of the Regions. Second Strategic Energy Review (SEC-2008-2794; SEC-2008-2795).

3 Every so often the passage of crude oil in a pipeline has to be slowed down and this is done by making the crude go around three sides of a rectangle before returning to the original route. In this instance, western defence analysts said the bombs landed along the one side of the rectangle that did not contain part of the BTC pipeline. The analysts and diplomats briefed the author in Baku in September 2008.

4 The explosion that shut down the BTC 48 hours before the Georgian war erupted remains a subject of concern. The Kurdish PKK claimed responsibility for a blast at Valve 30 on the BTC which prompted an immediate line shutdown. Western military analysts see no reason to dispute this, even though officials from Turkey's state energy company, Botas, were continuing to argue in late 2008 that the blast was an accident. But both Western defence analysts and diplomats remain perturbed at the coincidence of timing, prompting them to consider whether the PKK might itself have been prompted to make such an attack at this time. The PKK has said that it regards the BTC as a legitimate target and, indeed, the writer was warned by Turkish security forces in May 2006, ironically after stopping at a BTC valve station in southern Turkey, that PKK forces were moving into the area and that safety could not be guaranteed after nightfall. But the Valve 30 blast was the first open attack on the line in two years of operation and its closure before the Georgian war broke out was one of the reasons why the markets subsequently reacted with relative equanimity to the actual outbreak of fighting in Georgia.

5 Roberts, 'Azerbaijan looks set to extend plateau production', Platts, London, 11 June 2007.

6 Bryza, Keynote address to the 14th Annual Caspian Oil and Gas Conference, Baku, 6 June 2007. Bryza was in mid-2010 nominated as the next US Ambassador to Azerbaijan.

7 Projections from charts presented by BP Azerbaijan President Bill Schrader at the 14th Annual Caspian Oil and Gas Conference, Baku, 6–7 June 2007.

8 'It's now or never for Nabucco', *New Europe*, Issue 819, 1–7 February 2009. Nabucco is not the only beneficiary of EU largesse; the TGI Interconnector pipeline is to receive €150 million.

9 A. Dellecker, 'The Caspian Pipeline Consortium: Bellwether of Russia's Investment Climate ?' *Russie.Nei.Visions*, No. 31, June 2008.

10 The 62 bcm import figure derives from RusEnergy's figure of 56 bcm for transit and 6 bcm for imports; the 112.5 bcm FSU export figure is derived from RusEnergy's figure of 208 bcm for total exports plus 56 bcm for transit less the BP Statistical Summary figure of 151.5 bcm for exports to identifiable hard cash markets. BP cites lower

figures for both production and consumption: 612.1 bcm for production and 432.1 bcm for consumption. The RusEnergy figures are preferred in this context because they make specific allowance for Russia's import and transit trade. These figures should be used as illustrations of the overall situation, rather than as precise data.

11 The fact that Azerbaijan is considering LNG is significant since it would require liquefaction facilities to be located on the Georgian Black Sea coast, an investment that would cost several billion dollars. The fact that Azerbaijan is even contemplating such a project is testimony to its belief that there will be no repetition of the Georgia–Russia war of August 2008.

11 Brothers to neighbours: Russia–Belarus relations in transit

1 This chapter was concluded in late autumn 2008. More recent developments in Belarus–Russia relations, such as various conflicts on trade, oil and gas repeat in principle the patterns outlined below, but are not the object of our research. Yet it is worth mentioning that the discussed pursuit of narrowly defined economic self-interest in Minsk and Moscow also hinders the coming into full force of the renewed customs union, this time between Belarus, Russia and Kazahkstan, originally scheduled for 1 July 2010.

2 For an early account, see: B. Lo, *Vladimir Putin and the Evolution of Russian Foreign Policy*, Malden: Blackwell 2003. See also: D. Lynch (ed.), *What Russia Sees*, Paris: Chaillot Papers No. 74, 2005; G. Mangott (ed.), *Russlands Rückkehr. Außenpolitik unter Vladimir Putin*, Baden-Baden: Nomos, 2005.

3 Y. Drakokhrust, D. Furman, 'Belarus and Russia: The Game of Virtual Integration,' in: M. Balmaceda *et al.* (eds), *Independent Belarus. Domestic Determinants, Regional Dynamics, and Implications for the West*, Cambridge, MA: Harvard University Press, 2002, pp.232–55; K. Koktysh, 'Belorussiya v evropeiskom kontekste,' in: ibid., A. Moshes, 'Mezhdu Vostokom i Zapadom: Ukraina i Belorussiya na evropeiskom prostranstve,' Moscow: Centr Karnegi, 2003, pp.9–50; ibid., 'The Belarusian Policy of Russia: the Era of Pragmatism', *International Issues and Slovak Foreign Policy Affairs* 2, 2006, pp.18–29.

4 *Russian Ministry of Foreign Affairs' Bulletin*, 4 December 2000. Online. Available at: http://www.mid.ru/Ns-rsng.nsf/arh/432569D80022146643256 9AB004B57BF?Open Document (accessed 16 October 2008).

5 Next to low energy prices, these included regular debt reductions, beneficial conditions for Belarusian industry products on the Russian market, and Russian domestic tariffs concerning rail transport.

6 It would neither be feasible to grant Belarus any special status within the fragile construction of Russian federalism, as this would most likely lead to similar demands for more autonomy by other federal republics such as Bashkortostan, Chechnya or Tatarstan.

7 In spite of the close economic links, Russia only comes in fourth place concerning foreign direct investments in Belarus. See Ministerstvo inostrannych del RF, *Torgovo-ekonomicheskoe sotrudnichestvo Rossii i Belorussii*, 13 May 2008. Online. Available at: http://www.ln.mid.ru/ns-rsng.nsf/6bc38aceada6e44b432569e700419ef5/b8551e14 99a0895243256a540030d695?OpenDocument (accessed 16 October 2008).

8 Thus Minsk in total disagreement to a customs union's purpose strives to squeeze out Russian goods from the Belarusian market by implementing strict quotas, see L. Zaiko, 'Russia and Belarus: Between Wishing and Reality', *Russia in Global Affairs* 1, 2006, pp. 108–117.

9 For details see R. Deyermond, 'The State of the Union: Military Success, Economic and Political Failure in the Russia–Belarus Union,' *Europe–Asia Studies* 8, 2004, pp. 1191–1205.

10 Zaiko 2006, op. cit.

11 Reportedly, the refining margin increased from $5.8 to $15.4 per barrel in 2001 and 2005 respectively. See International Monetary Fund, *Republic of Belarus: Selected Issues*, June 2005, IMF Country Report No. 05/217, p. 9 ff. Online. Available at: http://www.imf.org/external/pubs/ft/scr/2005/cr05217.pdf (accessed 16 October 2008). See also 'Nikakikh chudes,' *Ekspert*, 12 December 2005.

12 According to a statement of the Russian Ministry of Economics, the Russian national budget annually suffered losses to the amount of $3–4 billion from the non-transfer of export duties. See: 'Russia, Belarus End Bruising Dispute,' *The Moscow Times*, 15 January 2007.

13 Cf. W. Kononczuk, *Belarusian–Russian Energy Conflict: The Game is Not Over*, Warsaw: Batory Foundation, 2007. Online. Available at: http://www.batory.org.pl/doc/belarusian–russian-energy-conflict.pdf (accessed 16 October 2008).

14 'Lost in Exportation. Russia closes the Belarusian oil offshore', *Kommersant*, 13 December 2006. Online. Availabe at http://www.kommersant.com/p729812/Russian_Belarus_Oil_Duties/ (accessed 16 October 2008).

15 This remark was made in a televised cabinet meeting with President Putin. Generally, Khristenko's explanations as to why the negotiations with Belarus on the non-observance of the customs treaty lasted for five years 'without tangible success' are rather cryptic. Kononczuk (op. cit.) cites him as having stated that 'all Russian oil companies pushed through to be able to send oil through Belarus in order to avoid higher taxes' (p. 6). Presidential elections took place in Belarus in March 2006. Excerpts from the meeting's transcript are published on the presidential website, http://www.kremlin.ru/eng/text/speeches/2007/01/09/1012_type82913type82917_116543.shtml (accessed 16 October 2008).

16 For Belarusian–Russian gas relations, see C. Bruce, *Fraternal Friction or Fraternal Fiction? The Gas Factor in Russian–Belarusian Relations*, Oxford: Oxford Institute for Energy Studies, 2005. Online. Available at: http://www.oxfordenergy.org/pdfs/NG8.pdf (accessed 16 October 2008). See also K. Yafimava, *Post–Soviet Russian–Belarusian Relationships. The Role of Gas Transit Pipelines*, Stuttgart: Ibibid, 2007.

17 Whereas the energy intensity of the Belarusian economy is slightly lower than in Ukraine or in Russia, the share of natural gas in the primary energy supply is higher than in those countries as Belarus so far does not possess any nuclear power plants. See International Energy Agency (IEA) Statistics, http://www.iea.org/Textbase/stats/country/maps/EUROPE/tpes_gdp_ppp.htmindex.asp (accessed 16 December 2008).

18 For this argument, see D.M. Nuti, 'The Belarus Economy: Suspended Animation between State and Markets', in: S. White *et al.* (eds), *Postcommunist Belarus*, Oxford: Rowman and Littlefield, 2005, pp. 97–122.

19 R. Götz, *Energietransit von Russland durch die Ukraine und Belarus. Ein Risiko für die europäische Energiesicherheit?* Berlin: Stiftung Wissenschaft und Politik, 2006. Online. Available at: http://www.swp-berlin.org/de/common/get_document.php?asset_id=3554 (accessed 16 October 2008).

20 Cf. Yafimava (*op. cit.*), pp.70 ff.

21 For this figure, see Gazprom's press release, http://www.gazprom.com/eng/news/2007/08/24810.shtml (accessed 17 October 2008).

22 According to Gazprom's Annual Report 2007, average net revenues from the sale of 1000 cubic metres of natural gas to the so-called 'far abroad' (all European countries except the Baltic States and CIS members) were 5,183 rubles in 2007 as opposed to 2,672 rubles earned in the 'near abroad' (former Soviet Union) and 982 rubles in the Russian Federation.

23 The Nord Stream pipeline, which is supposed to have a capacity of 55 bcm per year in its final stage, will go from Vyborg in Russia to Greifswald in Germany. It will consist of two parallel lines, of which the first one is to be completed at the end of 2011. Although doubts are frequently raised with regard to this date due to extensive

environmental audits requested by Baltic Sea states, Nord Stream officials maintain that the schedule will be kept. The pipeline is realized by a joint venture of Gazprom, BASF/Wintershall, E.ON Ruhrgas and N.V. Nederlandse Gasunie. For a game theory approach demonstrating how Gazprom's bargaining power will increase after completion of the project, see: F. Hubert, 'Strategic investment in international gas transport systems,' *EIB Papers* 2, 2007, pp. 62–81.

24 R. Götz, *Gasproms Diversifizierungsstrategie der Exportpipelines und Exportrichtungen*, Berlin: Stiftung Wissenschaft und Politik, 2007. Online. Available at: http://www.swp-berlin.org/de/common/get_document.php?asset_id=3982 (accessed 16 October 2008).

25 Cf. J. Guillet, 'Gazprom as a Predictable Partner. Another Reading of the Russian–Ukrainian and Russian–Belarusian Energy Crises,' Paris: Institut français des relations internationales, 2007. Online. Available at: http://www.ifri.org/files/Russie/ifri_Gazprom_guillet_anglais_mars2007.pdf (accessed 16 October 2008).

26 On 26 November 2008, Russian Prime Minister Putin signed a decree to construct the Baltic Pipeline System 2, a connector between the Druzhba pipeline and the Russian oil export terminal in Ust-Luga near St Petersburg which will bypass Belarus. The first section with a capacity of 30 million tons of oil per year is due for completion in autumn 2012. Until December 2013 the pipeline is to be upgraded to 50 million tons annually. Initial plans for the project were released immediately during the escalation of the Russian–Belarusian oil conflict in January 2007. Construction is to be carried out by Transneft, the state-owned pipeline operator. See Transneft's press release, 'Putin podpisal rasporjazhenie o stroitel'stve BTS-2', http://www.transneft.ru/news/newsitem/?id=7303 (accessed 3 December 2008).

27 Cf. R. Götz, *Russlands Erdgas und Europas Energiesicherheit*, Berlin: Stiftung Wissenschaft und Politik, 2007.

28 See EC Press Release, 'Russia–WTO. European Union–Russia deal brings Russia a step closer to WTO membership', 21 May 2004. Online. Available at: http://trade.ec.europa.eu/doclib/docs/2004/may/tradoc_117299.pdf (accessed 16 October 2008). Cf. A. Spanjer, 'Russian Gas Price Reform and the EU–Russia Gas Relationship: Incentives, Consequences and European Security of Supply,' *Energy Policy* 5, 2007, pp. 2889–2898; H.-J. Spanger, 'EU–Russland: Was bleibt von der strategischen Partnerschaft?', *Internationale Politik und Gesellschaft* 2, 2007, pp. 91–113.

29 For instance, then Prime Minister Mikheil Kasyanov stated that if Beltransgaz was sold to Gazprom under 'acceptable conditions', Belarus would continue to further purchase Russian gas at Russian domestic prices, see 'Putin ne pomog Lukashenko', *Nezavisimaya gazeta*, 6 December 2003.

30 Russia operates several important military facilities on Belarusian territory. Most prominent are the missile early warning station near Baranovichi and the naval communication system in Vileika.

31 At the time, monopoly export rights were not yet granted to Gazprom. For a detailed account of the conflict see Yafimava 2007 (*op. cit.*), pp. 90 ff.

32 'Lukashenko otozval iz Rossii posla i zayavil, chto otnosheniya teper nadolgo otravleny gazom', *NEWS-ru.com*, 19 February 2004. Online. Available at: http://www.newsru.com/finance/19feb2004/gaz23.html; 'Rossiya pobedila v 'gazovoi voine': Lukashenko gotov platit, skolko Putin poprosit', *NEWSru.com*, 19 February 2004. Online. Available at: http://www.newsru.com/world/19Feb2004/otrava.html (both accessed 16 October 2008).

33 'Minsk menyaet vector', *Nezavisimaya gazeta*, 3 November 2003.

34 J. Stern, *The Russian–Ukrainian gas crisis of January 2006*, Oxford: Oxford Institute for Energy Studies, 2006. Online. Available at: http://www.oxfordenergy.org/pdfs/comment_0106.pdf (accessed 16 October 2008).

35 'Gazprom to triple Price for Belarus', *The Moscow Times*, 5 April 2006.

36 The Ukraine in 2007 paid $130, Moldavia $170 and the Baltic states $240 for 1000 cubic metres of Russian gas.

37 For an early account of the gas conflict, see K. Yafimava, J. Stern, *The 2007 Russia–Belarus Gas Agreement*, Oxford: Oxford Institute for Energy Studies, 2007. Online. Available at: http://www.oxfordenergy.org/pdfs/comment_0107-3.pdf (accessed 16 October 2008).
38 'Aleksandra Lukashenko gotovyat k nedruzhestvennomu pogolosheniyu – Moskva prekrashchaet subsidirovanie belorusskoi ekonomiki', *Kommersant*, 12 May 2006.
39 'Putin Stands Tough on Oil Dispute', *The Moscow Times*, 10 January 2007.
40 This percentage rate was supposed to be increased to 33.5 and 35.6 per cent in 2008 and 2009.
41 IPM Research Center: *Small and Medium Business in Belarus. Quarterly Review 1/2007*. Online. Available at: http://www.research.by/pdf/SME2007e01.pdf (accessed on 16 October 2008).
42 On the conflict see also F. Garbe, 'Energische Integration? Russlands Energiekonflikt mit Belarus', *Osteuropa* 4, 2007, pp. 65–76.
43 The argument that the Beltransgaz takeover has more the character of a subsidy than of a commercially viable acquisition is supported by the apparently low profits of the Belarusian pipeline operator. A Belarusian newspaper cited the figure of roughly 6,830 years that Gazprom would need to regain the investment of $1.25 billion already made, assuming dividend payments would remain at the 2008 level of approximately $183,000 (*Belgazeta*, 27 October 2008, p. 2).
44 'Russian Jan gas price to Lithuania up 78 pct yr/yr', *Reuters*, 25. January 2008.
45 'V gazovom voprose Rossiya primenyaet 'grubye' formulirovki', *Belorusskie novosti*, 2 October 2008. Online. Available at: http://naviny.by/rubrics/economic/2008/10/02/ic_articles_113_159272/ (accessed 17 October 2008).
46 Rumour has it that the Belarusian government refused to pay the increased price as of the second quarter of 2008, continuing to pay $119.5. This was confirmed in late June by the Belarusian First Deputy Energy Minister, whereas Gazprom stated that Belarus had not accumulated any debts. See 'Gazprom confirms that Belarus owes nothing to it', *Belorusskie novosti*, 26 July 2008. Online. Available at: http://naviny.by/rubrics/inter/2008/07/26/ic_news_259_294496/ (accessed 18 October 2008).
47 IPM Research Center: *Belarusian Monthly Economic Review* 1, January 2008. Online. Available at: http://www.research.by/pdf/BMER2008e01.pdf (accessed 17 October 2008).
48 F. Hett, *Belarus unter Druck? Die belarussisch–russischen Energiekonflikte und ihre Folgen*, Berlin: Stiftung Wissenschaft und Politik, 2007. Online. Available at: http://www.swp-berlin.org/commom/get_document.php?asset_id=4515 (accessed 16 October 2008).
49 Revenues from export duties on oil products are said to account for 10 per cent of overall Belarusian budget revenues in 2008. See 'Krizis i Belarus' obyazatel'no peresekutsya?' *Belorusskie Novosti*, 14 October 2008. Online. Available at: http://naviny.by/rubrics/economic/2008/10/14/ic_articles_113_159462/ (accessed 14 October 2008).
50 IPM Research Center: *Belarusian Monthly Economic Review*, 3, March 2007. Online. Available at: http://research.by/pdf/BMER2007e03.pdf (accessed 18 October 2008).
51 National Statistical Committee of the Republic of Belarus, http://belstat.gov.by/homep/en/indicators/ftrade1.php (accessed 18 October 2008).
52 Lukashenko stated shortly after the conflicts had been resolved that the new arrangements would cost Belarus 3.5 billion dollars alone in 2007. Later on, he raised this estimate to 5 billion dollars. The Belarusian leader apparently over-exaggerated the consequences of the energy conflicts, most likely in order to appeal to Russian solidarity and to create fear in Moscow of causing a severe economic crisis which could lead to instability in its immediate neighbourhood. See 'Lukashenko Returns the Favor,' *Kommersant*, 24 January 2007. Online. Available at: http://www.kommersant.com/p736297/oill_natural_gas_duties_Belarus. 'Belarus to Annually Lose $5 Billion from Higher Prices for Gas,

Oil,' *Kommersant*, 7 February 2007. Online. Available at: http://www.kommersant.com/p-10052/transit_energy_prices/ (both accessed 16 October 2008).

53 Data according to the IPM Research Center, Minsk.

54 See R. Lindner, *'Friendship' Blockaded. The Russia/Belarus Conflict is a Post-Soviet Turning Point*, Berlin: Stiftung Wissenschaft und Politik, 2007. Online. Available at: http://www.swp-berlin.org/en/common/get_document.php?asset_id=3704 (accessed 16 October 2008).

55 Ministerstvo inostrannych del RF: *Obzor vneshnei politiki Rossiyskoj Federatsii*, 27 March 2007. Online. Available at: http://www.ln.mid.ru/ns-osndoc.nsf/0e9272befa34209743256c630042d1aa/d925d1dd235d3ec7c32573060039aea4?OpenDocument (accessed 18 October 2008).

56 *The Foreign Policy Concept of the Russian Federation*, 12 July 2008. Online. Available at: http://www.kremlin.ru/eng/text/docs/2008/07/204750.shtml (accessed 10 October 2008).

Selected bibliography

Books

Akiner, S. (ed.), *The Caspian: Politics, Energy and Security,* London: Routledge, 2004.

Baev, P.K. *Russian Energy Policy and Military Power: Putin's Quest for Greatness*, London: Routledge, 2008.

Balmaceda, M. *Energy Dependency, Politics and Corruption in the Former Soviet Union*, London: Routledge, 2008.

Balmaceda, M. *et al.* (eds), *Independent Belarus. Domestic Determinants, Regional Dynamics, and Implications for the West*, Cambridge, MA: Harvard University Press, 2002.

Barysch, K. (ed.), *Pipelines, Politics and Power*, London: Centre for European Reform, 2008.

Bremmer, I. *The End of the Free Market*, New York: Portfolio, 2010.

Cornell, S.E. and Nilsson, N. (eds), *Europe's Energy Security. Gazprom's Dominance and Caspian Supply Alternatives,* Washington–Stockholm–Nacka: Central Asia–Caucasus Institute/Silk Road Studies Program, 2008.

De Waal, T. *The Black Garden: Armenia and Azerbaijan Through Peace and War*, New York: New York University Press, 2003.

Drakokhrust. Y. and Furman, D. 'Belarus and Russia: The Game of Virtual Integration', in M. Balmaceda *et al.* (eds), *Independent Belarus. Domestic Determinants, Regional Dynamics, and Implications for the West*, Cambridge, MA: Harvard University Press, 2002.

Ebel. R. and Menon, R. (eds) *Energy and Conflict in Central Asia and the Caucasus*, New York: Roman & Littlefield Publishers Inc., 2000.

Fortescue, S. *Russia's Oil Barons and Metal Magnets: Oligarchs and the State in Transition*, London: Palgrave Macmillian, 2006.

Goldman, M.I. *Petrostate: Putin, Power, and the New Russia*, New York: Oxford University Press, 2008.

Gomart, T. *Russian Civil–Military Relations: Putin's Legacy*, Washington, DC: Carnegie Endowment for International Peace, 2008.

Grigoriev, L. and Salikhov, M. *GUUAM 15 Years Later*, Moscow: Regnum, 2007.

Helm, D. *The Russian Dimension and Europe's External Energy Policy*, Oxford: Oxford University Press, 3 September 2007.

Hill, F. *Energy Empire: Oil, Gas and Russia's Revival*, London: FPC, 2004.

Kaliciki, J.H. and Goldwyn, D.L. (eds), *Energy & Security: Towards a New Foreign Policy Strategy*, Washington, DC: Woodrow Wilson Center Press, 2005.

Klare, M.T. *Rising Powers, Shrinking Planet: The New Geopolitics of Energy*, New York: Metropolitan Books, 2008.

Kononczuk, W. *Belarusian–Russian Energy Conflict: The Game is Not Over*, Warsaw: Batory Foundation, 2007.

Larsson, R. *Russia's Energy Policy: Security Dimensions and Russia's Reliability as an Energy Supplier*, Stockholm: Swedish Defense Research Agency, 2006.

Lo, B. *Vladimir Putin and the Evolution of Russian Foreign Policy*, London: Blackwell/ Chatham House, 2003.

Luzyanin, S. *Vostochnaya politika Vladimira Putina. Vozvrasheniye Rossii na 'Bolshoi Vostok' 2004–2008,* [*Vladimir Putin's Eastern Policy. Russia's Come-Back to the 'Great East' 2004–2008*], Moscow: Vostok–Zapad, 2007.

Mangott, G. (ed.), *Russlands Rückkehr. Außenpolitik unter Vladimir Putin*, Baden-Baden: Nomos, 2005.

Moran, T.H. *Multinational Corporations and the Politics of Dependence: Copper in Chile*, Princeton, NJ: Princeton University Press, 1974.

Nuti, D.M. 'The Belarus Economy: Suspended Animation between State and Markets', in S. White *et al.* (eds), *Postcommunist Belarus*, Oxford: Rowman and Littlefield, 2005.

Olcott, M.B. *Kazakhstan: Unfulfilled Promise*, Washington, DC: Carnegie Endowment for International Peace, 2002.

Pomfret, R. 'Resource Abundance, Governance and Economic Performance in Turkmenistan and Uzbekistan', in R. Auty, and I. de Soysa (eds) *Energy, Wealth and Governance in the Caucasus and Central Asia. Lessons Not Learned*, London: Routledge, 2006.

Shafer, D.M. *Winners and Losers: How Sectors Shape the Developmental Prospects of States*, Ithaca, NY: Cornell University Press, 1994.

Starr, S.F. and Cornell, S.E. *The Baku–Tbilisi–Ceyhan Pipeline: Oil Window to the West,* Washington, DC: Johns Hopkins University, May 2005.

Stern, J. *The Future of Russian Gas and Gazprom*, Oxford: Oxford University Press, 2005.

Stulberg, A.N. *Well-Oiled Diplomacy: Strategic Manipulation and Russia's Energy Statecraft in Eurasia*, New York: State University of New York Press, 2007.

Tugwell, F. *The Politics of Oil in Venezuela*, Stanford, CA: Stanford University Press, 1975.

Umbach, F. 'China's Energy Insecurity in Context of Growing Geopolitical Competition: Implications for the Future EU–China Relations', in K. Lange (ed.), *European–Chinese Security Cooperation: Possibilities and Limits*, Munich: Akademie für Akademie für Politik und Zeitgeschehen der Hanns-Seidel-Stiftung, Studies and Comments, No. 6, 2008.

Umbach, F. *Globale Energiesicherheit. Herausforderungen für die europäische und deutsche Außenpolitik*, Munich: Oldenbourg Verlag, 2003.

Vernon, R. *Sovereignty at Bay: The Multinational Spread of US Enterprise*, New York: Basic Books, 1971.

Wenger, A., Perovic, J. and Orttung, R.W. *Russian Business Power: The Role of Russian Business in Foreign and Security Relations*, London: Routledge, 2006.

Yafimava, K. *Post-Soviet Russian–Belarusian Relationships. The Role of Gas Transit Pipelines*, Stuttgart: Ibidem-Verlag, 2007.

Ziegler, C.E. 'China, Russia and Energy in the Caucasus and Central Asia Region and East Asia', in R. Auty and I. de Soysa (eds), *Energy, Wealth and Governance in the Caucasus and Central Asia*, London: Routledge, 2005.

Periodicals

Baran, Z. 'Security Aspects of the South Stream Project', Center for Eurasian Policy, Hudson Institute, *Report*, October 2008.

Barry, M. 'The Development and Use of Production Sharing Agreement Law in Uzbekistan Oil and Gas', *Central Asia and the Caucasus*, Vol. 5, No. 41, 2006.

Blum, D.W. 'Why did *Lukoil* really pull out of the Azeri–Chirac–Guneshli oilfield?', CSIS, *PONARS Policy Memo, No.* 286, December 2003.

Bruce, C. 'Fraternal Friction or Fraternal Fiction? The Gas Factor in Russian–Belarusian Relations', Oxford: Oxford Institute for Energy Studies, *NG8*, March 2005.

Chow, E. and Elkind, E. 'Where East Meets West: European Gas and Ukrainian Reality', *The Washington Quarterly*, Vol. 32, No. 1, January 2009.

Cohen, M. 'Russia and the European Union: An Outlook for Collaboration and Competition in European Natural Gas Markets', *Demokratisatsiya,* Vol. 15, No. 4, Fall 2007.

Cornell, S.E. 'The Politicization of Islam in Azerbaijan', Johns Hopkins University–SAIS, *Silk Road Paper*, October 2006.

Cutler, R. 'The Caspian Energy Conundrum', Columbia University, *Journal of International Affairs*, Vol. 56, No. 2, Spring 2003.

Dellecker, A. 'Caspian Pipeline Consortium, Bellwether of Russia's Investment Climate?', *Russie.Nei.Visions,* No. 31, June 2008.

Dellecker, A. 'Kremlin, Inc.: Le système juridique au service de l'industrie énergétique', *Politique étrangère*, No. 4, 2007, December 2007.

Deyermond, R. 'The State of the Union: Military Success, Economic and Political Failure in the Russia–Belarus Union', University of Glasgow, *Europe–Asia Studies,* No. 8, 2004.

Eden, L., Lenway, S. and Schuler, D.A. 'From the Obsolescing Bargain to the Political Bargaining Model', Texas A and M University, *Bush School Working Paper, No. 403*, January 2004.

Garbe, F. 'Energische Integration? Russlands Energiekonflikt mit Belarus', Eurozine, *Osteuropa,* No. 4, 2007.

Gomart, T. 'Politique étrangère russe : l'étrange inconstance', *Politique étrangère*, No. 1, 2006.

Gomart, T. 'Vladimir Poutine ou les avatars de la politique étrangère russe', *Politique étrangère*, No. 3–4, 2003.

Götz, R. 'Gasproms Diversifizierungsstrategie der Exportpipelines und Exportrichtungen', Stiftung Wissenschaft und Politik, *Diskussionspapier*, No. 7, May 2007.

Götz, R. 'Russlands Erdgas und Europas Energiesicherheit', Stiftung Wissenschaft und Politik, *SWP-Studie*, No. 21, August 2007.

Götz, R. 'Energietransit von Russland durch die Ukraine und Belarus. Ein Risiko für die Europäische Energiesicherheit?', Stiftung Wissenschaft und Politik, *SWP-Studie*, No. 38, December 2006.

Guillet, J. 'Gazprom as a Predictable Partner: Another Reading of the Russian–Ukrainian and Russian–Belarusian Energy Crises', Ifri, *Russie.Nei.Visions*, No. 18, March 2007.

Henderson, S. 'Caspian Energy Accords: A Job Half Done', Washington Institute for Near East Policy, *Policy Watch*, December 1999.

Hett, F. 'Belarus unter Druck? Die Belarussisch–Russischen Energiekonflikte und ihre Folgen', Stiftung Wissenschaft und Politik, *Diskussionspapier*, No. 16, November 2007.

Hill, F. 'Seismic Shifts in Eurasia: The Changing Relationship between Turkey and Russia and its Implication for the South Caucasus', The Hellenic Foundation for European and Foreign Policy, *Journal of Southeast European and Black Sea Studies*, Vol. 3, No. 3, September 2003.

Hubert, F. 'Strategic Investment in International Gas-Transport Systems', European Investment Bank, *EIB Papers*, Vol. 12, No. 2, 2007.

Ilkhamov, A. 'Russia Lures Uzbekistan as its Strategic Satellite in Central Asia', Institute of History, University of Basel, *Russian Analytical Digest*, No. 29, October 2007.

Iniutin, I. 'Russian Strategy in Central Asia', Institute for Central Asian and Caucasian Studies, *Central Asia and the Caucasus*, Vol. 2, No. 38, 2006.

International Crisis Group, 'Azerbaijan: Defence Sector Management and Reform', *Europe Briefing,* No. 50, October 2008.

International Crisis Group, 'Central Asia's Energy Risks', *Asia Report*, No. 133, May 2007.

International Crisis Group, 'Central Asia: What Role for the European Union?', *Asia Report*, No. 113, April 2006.

Jafalian,. A. 'L'évolution de la politique américaine dans la région de la mer Caspienne', Centre Thucydide, *Annuaire français des relations internationales*, Vol. 4, 2003.

Kassenova, N. 'Kazakhstan and the South Caucasus Corridor in the Wake of the Georgia–Russia Was', Centre for European Policy Studies, *EU–Central Asia Monitoring*, No. 3, January 2009.

Kassenova, N. 'The New EU Strategy Towards Central Asia: A View of the Region', Centre for European Policy Studies, *Policy Brief*, No. 148, January 2008.

Kenisarin, M. 'Addressing the Problem of Uzbekistan's New Energy Policy Formulation', Institute for Central Asian and Caucasian Studies, *Central Asia and the Caucasus*, Vol. 3, No. 39, 2006.

Kuchins, A.C. , Beavin, A. and Bryndza, A, 'Russia's 2020 Strategic Economic Goals and the Role of International Integration', Paris–Washington DC, Ifri–CSIS, *Europe, Russia and the United States: Finding a New Balance*, July 2008.

Laumulin, M. 'Gazprom as a Transnational Corporation and Central Asia. Part I', Institute for Central Asian and Caucasian Studies, *Central Asia and the Caucasus*, Vol. 5, No. 41, 2006.

Lei, Z. and Qinyu, S. 'Will China Go to War over Oil?', *Far Eastern Economic Review*, Vol. 169, No. 3, April 2006.

Lindner, R. ' "Friendship" Blockaded. The Russia/Belarus Conflict is a Post-Soviet Turning Point', Stiftung Wissenschaft und Politik, *SWP-Comments*, No. 2, January 2007.

Lounev, S. 'Central Asia and the Southern Caucasus: Geopolitical Value for Russia', *Central Asia and the Caucasus*, Vol. 39, No. 3, 2006.

Luong. P.J. and Weinthal, E. 'Prelude to the Resource Curse: Explaining Energy Development Strategies in the Soviet Successor States and Beyond', *Comparative Political Studies*, Vol. 34, No. 4, May 2001.

Lynch, D. (ed.) 'What Russia Sees', EU Institute for Security Studies, *Chaillot Papers*, No. 74, January 2005.

Main, S.J. 'The Bear, the Peacock, the Sturgeon, the Eagle and the Black Black Oil: Contemporary Regional Power Politics in the Caspian Sea', CSRC Defence Academy of the UK, *Caucasus Series*, 05/67, December 2005.

Mankoff, J. 'Eurasian Energy Security', Council on Foreign Relations, *Council Special Report*, No. 43, February 2009.

Mateeva, A. 'EU Stakes in Central Asia', EU Institute for Security Studies, *Chaillot Paper*, No. 91, July 2006.

Milov, V. 'Russia and the West. The Energy Factor', Paris–Washington DC: Ifri–CSIS, *Europe, Russia and the United States: Finding a New Balance*, July 2008.

Milov, V., Coburn, L.L. and I. Danchenko, 'Russian Energy Policy 1992–2005', *Eurasian Geography & Economics*, Vol. 47, No. 3, 2006.

Milov, V. 'The EU–Russia Energy Dialogue: Competition versus Monopolies', Ifri, *Russie.Nei.Visions*, No. 13, September 2006.

Moshes, A. 'Mezhdu Vostokom i Zapadom: Ukraina i Belorussiya na evropeiskom prostranstve', ['Between East and West: Ukraine and Belarus in the European Space'], Moscow Carnegie Center, 2003.

Naumkin, V. 'Uzbekistan's State Building Fatigue', *The Washington Quarterly*, Vol. 29, No. 3, Summer 2006.

Norling, N. 'Gazprom's Monopoly and Nabucco's Potentials: Strategic Decisions for Europe', *Silk Road Paper*, Washington–Stockholm–Nacka: Central Asia–Caucasus Institute/Silk Road Studies Program, November 2007.

Olcott, M.B. '"Druzhba narodov" v mire energetiki', ['"The Friendship of the People" in the Energy Sector'], *Pro & Contra*, Vol. 10, No. 2, Summer 2006.

O'Neill, J. 'Building Better Global Economic BRICs', New York, Goldman Sachs, *Global Economic Paper*, No. 60, November 2001.

Paramonov, V. and Strokov, A. 'Structural Independence of Russia and Central Asia in the Oil and Gas Sectors', Defence Academy of the UK, Conflict Studies Research Centre, *Central Asian Series*, 07/16 E., June 2007.

Pomfret, R. 'Kazakhstan's Economy since Independence: Does the Oil Boom Offer a Second Chance for Sustainable Development?', *Europe–Asia* Vol. 57, No. 6, September 2006.

Riley, A. and Umbach, F. 'Out of Gas. Looming Russian Gas Deficits Demand Readjustment of European Energy Policy', *Internationale Politik (Global Edition)*, No. 1, Spring 2007.

Shaffer, B. 'From Pipedream to Pipeline: A Caspian Success Story', Harvard University, Belfer Center for Science and International Affairs, *Current History*, Vol. 104, No. 684, October 2005.

Smith, M.A. 'Russia and Turkey', Defence Academy of the UK, Conflict Studies Research Centre, *CSRS Paper F80*, September 2002.

Spanjer, A. 'Russian Gas Price Reform and the EU–Russia Gas Relationship: Incentives, Consequences and European Security of Supply', *Energy Policy*, No. 5, 2007.

Spanger, H-J. 'EU–Russland: Was Bleibt von der Strategischen Partnerschaft?', *Internationale Politik und Gesellschaft*, No. 2, 2007.

Stern, J. 'The New Security Environment for European Gas: Worsening Geopolitics and Increasing Global Competition for LNG', Oxford: Oxford Institute for Energy Studies, *NG 15*, October 2006.

Tomberg, I. 'Central Asia and the Caspian: New Stage in the Great Energy Game', *Central Asia and the Caucasus*, Vol. 5, No. 41, 2006.

Tomberg, I. 'The North European Gas Pipeline Project in the Geopolitical Context', *International Affairs*, Vol. 52, No. 2, March 2006.

Torbakov, I. 'The Georgia Crisis and Russia–Turkey relations', *Jamestown Foundation Report*, The Jamestown Foundation, 26 November 2008.

Tsereteli, M. 'The Blue Stream pipeline and the geopolitics of natural gas in Eurasia', *CACI Analyst*, 30 November 2005.

Umbach, F. 'Zielkonflikte der europäischen Energiesicherheit. Dilemmata zwischen Russland und Zentralasien', Berlin, DGAP, *DGAP analyse*, No. 3, November 2007.

Umbach, F. 'Europe's Next Cold War', *Internationale Politik (Global Edition)*, Summer 2006.

Umbach, F. 'The Legs of the Triangle – The EU–China Relations', in W. Jung (ed.) *The New Strategic Triangle: China, Europe and the United States in a Changing International System*, Konrad-Adenauer-Stiftung, KAS-Schriftenreihe No. 76, Beijing: KAS, 2005.

Umbach, F. 'The Wounded Bear and the Rising Dragon – The Sino-Russian Relationship at the Beginning of the 21st Century: A View from Europe', *Asia–Europe Journal,* No. 2, 2004.

Weitz, R. 'The Gabala Gambit and Azerbaijan's Geopolitics', *CACI Analyst*, 13 June 2007.

Wilson, D. and Purushothaman, R. 'Dreaming With BRICs: The Path to 2050', New York, Goldman Sachs, *Global Economics Paper*, No. 99, October 2003.

Zaiko, L. 'Russia and Belarus: Between Wishing and Reality', *Russia in Global Affairs* No. 1, 2006.

Zweig, D. and Jianhai, B. 'China's Global Hunt for Energy', *Foreign Affairs*, Vol. 84, No. 5, September–October 2005.

Policy papers online

Abbasov, I. 'Talysh of Azerbaijan look South and North', Institute for War and Peace Reporting, *Caucasus Reporting Service*, 12 July 2007.

Adegun, K. 'Take or Pay Contracts as Investment Drivers for Gas Development Projects in Developing Countries: Are there Co-pilots?', University of Dundee, Centre for Energy, Petroleum and Mineral Law and Policy, 20 February 2007.

Ali, K. 'Iranian Azerbaijanis Move North', Institute for War and Peace Reporting, *Caucasian Reporting Service*, 13 April 2006.

Baev, P.K. 'Russia Assembles Its Allies and Ponders the Content of its "Privileged Interests"', The Jamestown Foundation, *Eurasia Daily Monitor*, Vol. 5, No. 172, 9 September 2008.

Baev, P.K. 'Putin Protects Islam and Praises Democracy in Grozny', The Jamestown Foundation, *Eurasia Daily Monitor*, Vol. 2, No. 232, 14 December 2005.

Bierman, S. 'Supply May Be the Least of Nabucco's Challenges', *Nefte Compass*, 21 June 2007.

Blagov, S. 'Russia's Gazprom Eyes Governmental Support', The Jamestown Foundation, *Eurasia Daily Monitor*, Vol. 5, No. 233, 8 December 2008.

Blagov, S. 'Russia Moves to Repair ties with Azerbaijan', *Eurasia Insight*, Eurasianet.org, 2 April 2007.

Blagov, S. 'New Deal Will Process Kazakh Gas at Nearby Russian Facility', *Eurasia Daily Monitor,* Vol. 3, No. 185, 6 October 2006.

Blank, S. 'Moscow Brings Pressure to Bear on Baku', *Eurasia Insight*, Eurasianet.org, 4 November 2008.

Blank, S. 'China's Recent Energy Gains in Central Asia: What Do They Portend', Central Asia Caucasus Institute, *CACI-Analyst*, 31 October 2007.

Blank, S. 'The Burgas–Alexandroupolis Pipeline and its Implications', Central Asia Caucasus Institute, *CACI Analyst*, 5 April 2007.

Cornell, S.E. 'So Far, Europe's Approach to Central Asia has been Moralistic and Counter-Productive', *Europe's World*, No. 2, Spring 2007.

Daly, J-C.K. 'Iran Gains from Georgian Confrontation', The Jamestown Foundation, *Eurasia Daily Monitor*, Vol. 5, No. 199, 17 October 2008.

J-C.K. Daly, 'With no Pipelines in Sight, Tankers Roam the Caspian', Central Asia Caucasus Institute, *CACI Analyst*, 16 May 2007.

De Waal, T. 'Heidar Aliyev: A Political Colossus', *IWPR Caucasus Reporting Service*, 18 December 2003.

Farkhad, S. 'Are Kazakhstan's Economic Bonds with Georgia the Price of Stronger Ties with Russia?' The Jamestown Foundation, *Eurasia Daily Monitor*, Vol. 5, No. 190, 3 October 2008.

Farkhad, S. 'Nazarbayev Plays Double Game on South Ossetia', The Jamestown Foundation, *Eurasia Daily Monitor*, No. 167, 2 September 2008.

Felgenhauer, P. 'The Russian–Georgian War Was Preplanned in Moscow', The Jamestown Foundation, *Eurasia Daily Monitor*, Vol. 5, No. 156, 14 August 2008.

Guluzade, K. 'Azerbaijan faces Iran dilemma', Institute for War and Peace Reporting, *Caucasian Reporting Service*, 18 May 2006.

Huang, J. 'Beijing's Perspectives on the Russo–Georgian Conflict: Dilemma and Choices', The Jamestown Foundation, *China Brief*, Vol. 8, No. 17, 3 September 2008.

Ismailzade, F. 'The Georgian–Russian Conflict Through the Eyes of Baku', The Jamestown Foundation, *Eurasia Daily Monitor*, Vol. 5, No. 154, 12 August 2008.

Ismailzade, F. 'Azerbaijan Moves to Shape Image of Islamic World', The Jamestown Foundation, *Eurasia Daily Monitor*, Vol. 4, No. 84, 30 April 2007.

Ismailzade, F. 'Azerbaijan–Russia relations enter turbulent phase', The Jamestown Foundation, *Eurasia Daily Monitor*, Vol. 3, No. 221, 30 November 2006.

Ismayilov, R. 'Azerbaijan, Georgia and Turkey: Building a new transportation triumvirate?', *Eurasia Insight*, 7 February 2007.

Kupchinsky, R. 'The Price of Gas and Russian Democracy', The Jamestown Foundation, *Eurasia Daily Monitor*, Vol. 5, No. 240, 17 December 2008.

Kupchinsky, R. 'The Shadowy Side of Gazprom's Expanding Central European Gas Hub', The Jamestown Foundation, *Eurasia Daily Monitor*, Vol. 5, No. 217, 12 November 2008.

Linn, J.F. 'Central Asia: A New Hub of Global Integration', Brookings Institution, 12 December 2007.

Mamedov, J. 'Azerbaijan flexes military muscles', Institute for War and Peace Reporting, *Caucasus Reporting Service*, 19 July 2007.

Marat, E. 'CSTO Summit Disappoints Moscow, Unites Central Asia', The Jamestown Foundation, *Eurasia Daily Monitor*, Vol. 5, No. 175, 12 September 2008.

Muradova, M. and Bakinsky, K. 'Building bridges for President Aliyev's re-election?', *Eurasia Insight*, Eurasianet.org, 30 May 2007.

Peyrouse, S. 'Sino-Kazakh Relations: A Nascent Strategic Partnership', The Jamestown Foundation, *China Brief*, Vol. 8, No. 21, 7 November 2008.

Roberts, J. 'Azerbaijan looks set to extend plateau production', London, *Platts*, 11 June 2007.

Socor, V. 'Hungary Doubling Efforts for Nabucco Ahead of Budapest Summit', The Jamestown Foundation, *Eurasia Daily Monitor*, Vol. 5, No. 237, 12 December 2008.

Socor, V. 'Azerbaijan Quietly Lending Impetus to Nabucco Project', The Jamestown Foundation, *Eurasia Daily Monitor*, Vol. 5, No. 228, 2 December 2008.

Socor, V. 'Oil Development in Kazakhstan Underscores Significance of Azerbaijani–Georgia Transit Corridor', The Jamestown Foundation, *Eurasia Daily Monitor*, Vol. 5, No. 212, 5 November 2008.

Socor, V. 'Romania Firmly Aboard Nabucco Projects, Hints at Leverage vis-à-vis South Stream', The Jamestown Foundation, *Eurasia Daily Monitor*, Vol. 5, No. 210, 3 November 2008.

Socor, V. 'Gazprom Seducing Romanian Government with South Stream', The Jamestown Foundation, *Eurasia Daily Monitor*, Vol. 5, No. 204, 24 October 2008.

Socor, V. 'OMV Joins with Gazprom to Undercut Nabucco', The Jamestown Foundation, *Eurasia Daily Monitor*, Vol. 5, No. 17, 29 January 2008.

Socor, V. 'Central Asia–Europe Energy Projects: Itemizing What Went Wrong', The Jamestown Foundation, *Eurasia Daily Monitor*, Vol. 4, No. 106, 31 May 2007.

Socor, V. 'Azerbaijan at the forefront Caspian basin energy initiatives', The Jamestown Foundation, *Eurasia Daily Monitor*, Vol. 4, No. 57, 22 March 2007.

Socor, V. 'Agreement Signed on Trans-Balkan Pipeline, Rival to Trans-Caspian Project', The Jamestown Foundation, *Eurasia Daily Monitor*, Vol. 4, No. 53, 16 March 2007.

Socor, V. 'Trans-Caspian Oil Pipeline Planned In Kazakhstan', The Jamestown Foundation, *Eurasia Daily Monitor, Vol. 2*, No. 95, 16 May 2005.

Stern, J. 'The Russian–Ukrainian Gas Crisis of January 2006', Oxford Institute for Energy Studies, 2006.

Torbakov, I. 'Russia and Turkey Forge New Ties on Security, Trade', *Eurasia Insight*, Eurasianet.org, 8 August 2005.

Torbakov, I. 'Eurasian idea Could Bring together Erstwhile Enemies Turkey and Russia', *Eurasia Insight*, Eurasianet.org, 18 March 2002.

Uslu, E. 'Turkey and Iran Sign Accord on Natural Gas Cooperation', The Jamestown Foundation, *Eurasia Daily Monitor*, Vol. 5, No. 224, 21 November 2008.

Valiyev, A. 'Al Qaeda in Azerbaijan: Myths and Realities', *Terrorism Monitor*, 18 May 2006.

Yafimava, K. and Stern, J. 'The 2007 Russia–Belarus Gas Agreement', Oxford Institute for Energy Studies, *Oxford Energy Comments*, January 2007.

Ziyadov, T. 'Putin's Gabala offer gets mixed reaction in Azerbaijan', The Jamestown Foundation, *Eurasia Daily Monitor*, 22 June 2007.

Press articles

Aliyev, A. 'The Inertia of Expensive Oil', *Expert*, 18 December 2008.

Aliyev, A. 'From Bulgaria to Greece', *Expert*, 15 March 2007.

Anderson, G. 'The Kazakhstan Upstream Tax Regime: A Comparative Analysis', *InvestKazakhstan*, March 2004.

Burmistrov, P. and Grebtsov, I. 'Gas dances around the Caspian', *Expert*, 11 September 2008.

Goodman, W. and Kern, P.J. 'Bad Tidings', *The National Interest*, January–February 2008.

Gromov, A. 'There Will Be No Revolution', *Expert*, 24 October 2005.

Guriev, S. Kolotilin, A. and Sonin, K. '*High Oil Prices and the Return of "Resource Nationalism"*', VoxEU.org, 12 April 2008.

Klare, M.T. 'Clearing the Air', *The National Interest*, January–February 2008.

Ross, M.L. 'The Political Economy of the Resource Curse', *World Politics* No. 51, 1999.

Rubanov, I. 'Not by the Right of Force', *Expert*, 21 May 2007.

Sarsenova, D. 'Program to Develop Kazakh Sector of the Caspian Shelf Presented', *The Times of Central Asia*, 16 July 2003.

Steinmeier, F-W. 'Reviving the Silk Road', *Frankfurter Allgemeine Zeitung* (FAZ), 30 June 2007.

Umbach, F. 'Diversifizierung statt Protektorat. Energiepartnerschaft zwischen Russland und der EU', *Die Politische Meinung,* September 2008.

D.G. Victor, 'What Resource Wars?', *The National Interest*, 11 December 2007.

Official Documents

Auty, R. 'Patterns of Rent-Extraction and Deployment in Developing Countries: Implications for Governance, Economic Policy and Performance', Paper Prepared for the Poverty Reduction and Economic Management Unit Seminar, Washington, DC: The World Bank, April 2004.

Bush G.W. and Nazarbayev, N. 'Joint Statement between the United States of America and the Republic of Kazakhstan', *The White House: President George* W. *Bush*, 29 September 2006.

Energy Charter Treaty, 1998.

European Council, '"Baku Initiative" An Energy Dialogue between the EU and the countries of the Black Sea, Caspian Basin and their Neighbours', Brussels, 2004.

European Council, 'The EU and Central Asia: Strategy for a New Partnership', Brussels 2007.

European Council, 'Conclusions of the Council. Emergency Meeting', 12594/08 DE, CONCL 3, Brussels, 31 September 2008.

European Commission, 'An EU Energy Security and Solidarity Action Plan. Second Strategic Energy Review', Communication from the Commission to the European Parliament, the Council, the European Economic and Social Committee of the Regions, Brussels, November 2008.

European Commission, 'An EU Energy Security And Solidarity Action Plan; Communication From The Commission To The European Parliament, The Council, The European Economic And Social Committee And The Committee Of The Regions', Second Strategic Energy Review (SEC (2008) 2794; SEC (2008) 2795).

International Monetary Fund, 'Republic of Belarus: Selected Issues', IMF Country Report No. 05/217, June 2005.

OSCE/ODIHR, 'Election Observation Mission Final Report', 12 November 2003.

Solana, J. 'Toward an EU External Energy Policy', Address at the EU Energy Conference, S324/06, Brussels, 20 November 2006.

World Bank, 'Azerbaijan: Issues and Options Associated with Energy Sector Reform', Report No. 32371-AZ, Washington, DC: The World Bank, 31 March 2005.

Reports and annuals

BP Statistical Review of World Energy, BP, June 2008.

Gazprom Annual Report, Moscow: Gazprom, 2007.

Glagolev, A.I. and Orlov, Y.N. 'Gazotransportnye proekty OAO Gazprom v Srednei Azii', Gazprom, 2004.

IPM Research Center: *Belarusian Monthly Economic Review,* No. 1 (64), January 2008.

IPM Research Center: *Belarusian Monthly Economic Review*, No. 3, March 2007.

IPM Research Center: *Small and Medium Business in Belarus. Quarterly Review No. 1, 2007.*

US Geological Survey, 'World Petroleum Assessment, 2000', Washington, DC: US Geological Survey Digital Data Series 60, 2000.

World Energy Outlook 2008, Paris: International Energy Agency, 2008.

World Energy Outlook 2006, Paris: International Energy Agency, 2006.

Index

Page numbers in *italics* refer to a table